Quantum Mechanics:
An Empiricist View

Quantum Mechanics:
An Empiricist View

Bas C. van Fraassen

CLARENDON PRESS · OXFORD
1991

Oxford University Press, Walton Street, Oxford OX2 6DP
Oxford New York Toronto
Delhi Bombay Calcutta Madras Karachi
Petaling Jaya Singapore Hong Kong Tokyo
Nairobi Dar es Salaam Cape Town
Melbourne Auckland
and associated companies in
Berlin Ibadan

Oxford is a trade mark of Oxford University Press

Published in the United States
by Oxford University Press, New York

British Library Cataloguing in Publication Data
Data available

Library of Congress Cataloging in Publication Data
Van Fraassen, Bas. C., 1941–
Quantum mechanics: an empiricist view/Bas C. van Fraassen.
p. cm.
Includes bibliographical references and index.
1. Quantum theory 2. Science–Philosophy. I. Title.
QC174.12.V34 1991 530.1'2–dc20 90-26302
ISBN 0-19-824861-X
ISBN 0-19-823980-7 (pbk)

Typeset by Keytec Typesetting Ltd, Bridport, Dorset

Printed in Great Britain by
Biddles Ltd, Guildford & King's Lynn

Que, touchant les choses que nos sens n'aperçoivent point,
il suffit d'expliquer comment elles peuvent être ...

René Descartes, *Principes*, iv. 204

PREFACE

QUANTUM theory grew up, from Planck to Heisenberg and Schroedinger, in response to a welter of new experimental phenomena: measurements of the heat radiation spectrum, the photoelectric effect, specific heats of solids, radioactive decay, the hydrogen spectrum, and confusingly much more. Yet this theory, emerging from the mire and blood of empirical research, radically affected the scientific world-picture. If it did describe a world 'behind the phenomena', that world was so esoteric as to be literally unimaginable. The very language it used was broken: an analogical extension of the classical language that it discredits, and redeemed at best by the mathematics that it tries to gloss.

Interpretation of quantum theory became genuinely feasible only after von Neumann's theoretical unification in 1932. Von Neumann himself, in that work, attempted to codify what he took to be the common understanding. Astonishingly, the attempt led him to assert that in measurement something happens which violates Schroedinger's equation, the theory's cornerstone. As he saw very clearly, interpretation enters a circle when its main principle is Born's Rule for measurement outcome probabilities, while at the same time measurements are processes in the domain of the theory itself. Behold the enchanted forest: every road leads into it, and none leads out — or does the hero's sword cleave the wood by magic?

An empiricist bias will be evident throughout this book, but my own interpretation of quantum mechanics does not begin until Chapter 9. The first three chapters provide philosophical background; though they overlap my *Laws and Symmetry*, I have tried to make them interesting in their own right. The next four chapters mainly outline the achievements of foundational research, though with an eye to the philosophical issues to come. The negative part is to show that the phenomena themselves, and not theoretical motives, can suffice to eliminate Common Cause models of the observable world. The positive part is the conclusion that there are adequate descriptions of measurement—in the sense required for Born's Rule—internal to quantum theory. To make the book relatively self-contained, Chapters 6 and 7 introduce all the quantum mechanics needed for the philosophical discussions to come.

From a purely philosophical point of view, the most important clarification reached since 1925 concerns the criteria of adequacy for interpretations of quantum mechanics. It appears at present that more than just one tenable interpretation, already in process of development, can meet those criteria.

I regret that I may have done little justice to the promising interpretations now underway which differ from my own, although I have tried to point to them as often as I could. I regard every interpretation as increasing our understanding, and believe that an awareness of what rival interpretations may be tenable is crucial to clarity. But that attitude already needs defence, for it involves views on what science is, and what philosophy can hope for.

I have also tried to take the philosophical debates somewhat further, into the fascinating cluster of problems that concern quantum-statistical mechanics and identical particles. At every point, but here especially, I was acutely aware of rapid progress in foundational research and of the kaleidoscopically changing philosophical debates. It is true that interpretation focuses on a single theory at one more or less definite historical stage—and yet, what we try to interpret is not static. Every time we understand a little more, we change what we are trying to understand. It is not surprising that scientists often become impatient with philosophy: what is ever achieved if every generation has to face the same questions again, with a new understanding of what is being asked, unable to rest on past answers? But philosophy does not create our predicament. It is only a myth that modern science had arrived at a clear and well-integrated world-picture, or that contemporary science has already effectively given us a new one. At best, we are in process of replacing what never has existed by something that never will. It is only in this *unendliche Aufgabe*, this reaching for what we cannot finally have or hold, that understanding consists.

The pleasures of acknowledgement are always accompanied by a good deal of soul-searching. Debts are subtle, and always so numerous that only a few can be avowed, for philosophy is a thoroughly historical and communal enterprise. For the first part of the book, devoted to general philosophical background, my debts are largely acknowledged already in my previous books. But I must thank above all my teachers Adolf Grünbaum, who led me into the intricacies of determinism and indeterminism, and Wilfrid Sellars, who would not allow me to treat those or any other subjects in isolation. To Henry Margenau I believe I am indebted in two ways, first through what I received from him through Grünbaum, who was his student and my teacher, and then directly as he drew me into his quantum-mechanical questioning during my two years at Yale. In the next year at Indiana University Wesley Salmon took me in, as it were, to instil a preoccupation with causality, probability, and frequency. Salmon was the first to comment on my fledgling ideas about identical particles. It was also around then that I participated in a symposium with Hilary Putnam, who challenged me with a new way to see quantum logic. In the individual chapters I have tried as much as possible to indicate my more specific debts, for example to Enrico Beltrametti and Gianni Cassinelli, whose book became one of my bibles, to my frequent collaborator, R. I. G. Hughes, and to Jeffrey Bub, Nancy Cartwright, Roger Cooke, Maria Luisa Dalla Chiara, Arthur Fine, Clifford Hooker, Simon Kochen, Pekka Lahti, James McGrath, Peter Mittelstaedt, and Brian Skyrms, among others. Alan Hajek and R. I. G. Hughes read large parts of the manuscript and gave many helpful comments. Almost every section of each chapter benefited from the close reading and comments by Sara Foster. The National Science Foundation and Princeton University steadfastly supported my research, while Anne Marie De Meo typed the results and helped me generously through many practical difficulties.

B.C.v.F.

SUMMARY TABLE OF CONTENTS

CONTENTS

1

What Is Science?

As painting, sculpture, and poetry turned abstract after the turn of the century; as Minkowski recast the electrodynamics of moving bodies in four-dimensional geometry; as Hilbert and Russell turned geometry and analysis into pure logic; philosophy of science too turned to greater abstraction. But philosophy of science is still philosophy, and is still about science. Before broaching the philosophy of physics, and the foundations of quantum mechanics, I shall locate those projects in the larger enterprise of philosophical reflection on science as a whole.

1. TWO VIEWS ABOUT SCIENCE

There is quite a difference between the questions 'What is happening?' and 'What is really going on?' Both questions can arise for participants as well as for spectators, and usually no one has more than a fragmentary answer. To the second question the answer must undeniably be more doubtful, because it has to be somewhat speculative in the interpretation it puts on what happens. Yet both questions seem crucially important.

So far I might have been talking about war, a political movement, contemporary art, the Diaspora, the Reformation, or the Renaissance—as well as about science, its current state or its historical development. Scientists, the participants in this large-scale cultural activity, we can consult only about what is happening. Both these participants and the more distant spectators cannot help but attempt some interpretation as well. Indeed, we are all to some extent both participant and spectator, for science has become an activity of our civilization as a whole.

Philosophy of science has focused on theories as the main product of science, more or less in the way philosophy of art has focused on works of art, and philosophy of mathematics on arithmetic, analysis, abstract algebra, set theory, and so forth. Scientific activity is understood as productive, and its success is measured by the success of the theories produced. In the most general terms, of course, the aim of an activity is success, and what that aim exactly is depends on the 'internal' criterion of success. The personal aim of the chess-masters may well be fame and fortune, but the aim in chess—as such, the 'internal' criterion of success in chess—is checkmate. To say what is really going on in science, therefore, we try to determine its aim in this sense.

Both scientists and philosophers have come up with very different answers to this. To some extent, the differences may reflect personal aim. Newton for example may have understood the aim of science as such as uncovering God's design, or arriving at the truth about the most basic laws of nature. To some extent, also, the answers have reflected current beliefs about just what it is possible to achieve in science. Newton believed that there was a unique derivation of the laws of nature from the phenomena. If that is indeed so, it is of course reasonable to say that this is exactly what science aims to do. When Mach answers instead in terms of economy of thought and organization of knowledge, or when Duhem denies that the aim of science includes explanation, that is undoubtedly in part because they believed so much less about what could be achieved in science, or any other way.

Can we find an internal criterion of success that characterizes scientific activity for all ages, and equally for philosophical and unphilosophical participants? The first thing to do is see exactly what the product is whose success is to be assessed by that criterion. When we focus on the scientific theory, as product of science, we turn this into a question about theories. So, first, what sort of thing is a theory? A scientific theory must be the sort of thing that we can *accept* or *reject*, and *believe* or *disbelieve*. Accepting a theory implies the opinion that it is successful; science aims to give us acceptable theories. More generally, a theory is an object for the sorts of attitudes expressed in assertions of knowledge and opinion. A typical object for such attitudes is a proposition, or more generally a body of putative information, about what the world is like.

Given this view about theories, we return to the question of aim.

First answer: realism. At this point we can readily see one very simple possible answer to all our questions, the answer we call *scientific realism*. This philosophy says that a theory is just the sort of thing which is either true or false; and that the criterion of success is truth. As corollaries, we have that acceptance of a theory as successful is, or involves, the belief that it is true; and that the aim of science is to give us (literally) true theories about what the world is like.

That answer would of course have to be qualified in various ways to allow for our epistemic finitude and the tentative nature of reasonable attitudes. Thus realists may add that, although it cannot generally be *known* whether or not the criterion of success has been met, we may reasonably have a high degree of belief about this, and that the scientific attitude precludes dogmatism. Whatever doxastic attitude we adopt, we stand ready to revise in face of further evidence. These are all qualifications of a sort that anyone must acknowledge. They do not detract from the appealing and, as it were, pristine clarity of the scientific-realist position. But that does not mean that it is right. Even a scientific realist must grant that *an analysis of theories*—even one that is quite traditional with respect to what theories are—*does not presuppose realism*. We may grant that theories are the sort of thing which can be true or false, that they say something about what the world is like; it does not follow that we must be scientific-realists.

Second answer: empiricism. There are a number of reasons why I advocate an alternative to scientific realism (see van Fraassen 1980*b*, 1985*a*). One concerns the difference between acceptance and belief; reasons for acceptance include many which, *ceteris paribus*, detract from the likelihood of truth. This point was made very graphically by William James; it is part of the legacy of pragmatism. The reason is that, in constructing and evaluating theories, we follow our desires for information as well as our desire for truth. We want theories with great powers of empirical prediction. For belief itself, however, all but the desire for truth must be 'ulterior motives'. Since therefore there are reasons for acceptance which are not reasons for belief, I conclude that acceptance is not belief. It is an elementary logical point that a more informative theory cannot be more likely to be true; therefore the desire for informative theories creates a tension with the desire to have true beliefs.

Once we have driven the wedge between acceptance and belief, we can reconsider the ways to make sense of science. As one such way, I wish to offer the anti-realist position I advocate, which I call *constructive empiricism*. It says that the aim of science is not truth as such but only *empirical adequacy*, that is, truth with respect to the observable phenomena. Acceptance of a theory involves as belief only that the theory is empirically adequate. But acceptance has a pragmatic dimension; it involves more than belief.

The agreement between scientific realism and constructive empiricism is considerable and includes the literal interpretation of the language of science, the concept of a theory as a body of information (which can be true or false, and may be believed or disbelieved) and a crucial interest in interpretation, i.e. finding out what this theory says the world is like.[1] There is much that the two can explore together. Acceptance has a clear pragmatic aspect: besides belief, it involves commitments of many sorts. Realists may hold that those commitments derive from belief, but they will wish to join empiricists in exploring just what they are. Similarly, empiricists will wish to explore in the same way as realists just what it is that our accepted theories say. Both agree, after all, that theories say something about what the world is like. The *content* of a theory is what it says the world is like; and this is either true or false. The applicability of this notion of truth value remains here, as everywhere, the basis of all logical analysis. When we come to a specific theory, the question: *how could the world possibly be the way this theory says it is?* concerns the content alone. This is the foundational question *par excellence*, and it makes equal sense to realist and empiricist alike.

2. THEORIES AND MODELS

To formulate a view on the aim of science, I gave a partial answer to the question of what a scientific theory is. It is an object of belief and doubt, so it is the sort of thing that could be true or false. That makes sense only for a proposition, or a larger body of putative information. It does not follow that a theory is something essentially linguistic. That we cannot convey information, or say what a theory entails, without using language does not imply that—after all, we cannot say what anything is without using language. We are here at another parting of the ways in philosophy of science. Again I shall advocate one particular view, the *semantic view* of theories. Despite its name, it is the view which de-emphasizes language.

Words are like coordinates. If I present a theory in English, there is a transformation which produces an equivalent description in German. There are also transformations which produce distinct but equivalent English descriptions. This would be easiest to see if I were so conscientious as to present the theory in axiomatic form; for then it could be rewritten so that the body of theorems remains the same, but a different subset of those theorems is designated as the axioms, from which all the rest follow. Translation is thus analogous to coordinate transformation—is there a coordinate-free format as well?

The answer is *yes* (though the banal point that I can describe it only in words obviously remains). To show this, we should look back a little for contrast. Around the turn of the century, foundations of mathematics progressed by increased formalization. Hilbert found many gaps in Euclid's axiomatization of geometry because he rewrote the proofs in a way that did not rely at all on the *meaning* of the terms (point, line, plane, . . .). This presented philosophers with the ideal: a pure theory is written in a language devoid of meaning (a pure *syntax)*. A scientific theory should then be conceived of as consisting of a pure theory (the mathematical formalism, ideally an exact axiomatic system in a pure syntax) *plus* something that imparts meaning and so connects it with our real concerns.

This did help philosophical understanding of science in one crucial case. How could the geometry of the world be non-Euclidean? What sense does this make? Well, as pure theory, both Euclidean and non-Euclidean geometries are intelligible. Suppose now we impart meanings by what Reichenbach called *coordinative definitions*. For example, say that a light ray is the physical correlate of a straight line or geodesic. Then, at that point, it becomes an empirical question whether physical geometry is Euclidean. For it is an empirical question how light rays behave.

For a little while it seemed that the meaning-restoring device would be very simple—a dictionary or a set of 'operational definitions'. By its use, the language of theoretical physics would receive a complete translation into the simple language of the laboratory assistant's observation reports. But the example of physical geometry already shows that the physical correlates do *not* yield a complete translation. Any two points lie on a straight line, but not necessarily on a light ray. If, on the other hand, we translate 'straight line' as 'possible light ray path', the theoretical element is not absent, and criteria of application are indefinite; is this possibility relative to laws and circumstances which could be stated without recourse to geometric language? The opposite view is to consider the mathematics in use an abstraction from the science that uses it, and to leave the reconstruction of language along formalist lines to a different philosophical enterprise.

Despite certain undoubted successes, the linguistic turn in analytic philosophy was eventually a burden to philosophy of science. The first to turn the tide was Patrick Suppes, with his well-known slogan: the correct tool for philosophy of science is mathematics, *not* meta-mathematics. This happened in the 1950s; bewitched by the wonders of logic and the theory of meaning, few wanted to listen. Suppes's idea was simple: *to present a theory, we define the class of its models directly*, without paying any attention to questions of axiomatizability, in any special language, however relevant or simple or logically interesting that might be.[2]

This procedure is common in contemporary mathematics, where Suppes had found his inspiration. In a modern presentation of geometry we find not the axioms of Euclidean geometry, but the definition of the class of Euclidean spaces. Similarly, Suppes and his collaborators sought to reformulate the foundations of Newtonian mechanics, by replacing Newton's axioms with the definition of a Newtonian mechanical system. This gives us, by example, a *format* for scientific theories. In Ronald Giere's recent encapsulation, a theory consists of (*a*) the *theoretical definition*, which defines a certain class of systems, and (*b*) a *theoretical hypothesis*, which asserts that certain (sorts of) real systems are among (or related in some way to) members of that class (see especially Giere 1985, 1988).

This new tradition is called the 'semantic approach' or 'semantic view'. It has been developed, since the mid-1960s, by a number of writers, some scientific realists (like Giere) and some not.[3] In this approach the role of language (and especially syntax or questions of axiomatization) is resolutely de-emphasized. The discussion of models treats them mainly as structures in their own right, and views theory development as primarily model construction. Almost all questions in philosophy of science take on a new form, or are seen in a new light, when asked again within the semantic view. Two books have recently appeared, applying the semantic view in philosophy of biology (Lloyd 1988; Thompson 1989). Most studies within this approach, however, have focused on philosophy of physics.

Families of structures, mathematically described, are something quite abstract. This is true even if we take an example like 'A Newtonian planetary system is a structure consisting of a star and one or more planets and is such that . . .'. The nouns are not abstract, and if a formalist says that we should deduce consequences from this definition only by arguments which rely not at all on the meanings of 'star' and 'planet', that is just his or her predilection. The abstractness consists rather in the fact that the same family of structures can be described in many ways; it is something non-linguistic, but (banal as that is) we can still present it only by giving one specific verbal description. So concepts relating to language must also retain a certain importance. Objectively, the conceptual anchor of informativeness is logical implication: if T implies T' then T is at least as informative as T'. But its informativeness *for us* depends on the formulation we possess, and the extent to which we can see its implications. This brings us from semantics into the area of pragmatics; our pragmatic reasons for accepting one theory rather than another may include that the former is more easily processable *by us*.

How does this affect our conception of the aim of science? The empiricist takes this aim to be to give us empirically adequate theories; the realist says that it is to give us true ones. Now, we identify a theory as a class of models. So is not that aim at once satisfied, in either case, by someone who says: 'I have a nice theory. It has as models exactly those structures which are isomorphic to the real world in the following respects'?[4] I think the problem arises equally for empiricist and realist. But I do not think that we need to revise the statement of the aim of science. We should answer that person: 'Of course we believe that your theory is true (or adequate in the respects you specify). Obviously the real world, properly conceived, *is* one of the models of your theory. But we have no reason to make the other commitments that go into acceptance—such as designing a research programme for it, using it to answer why or how questions, attempting to redescribe phenomena in its terms, and the like. The relevant pragmatic factors are missing because, however informative it is in a strictly objective or semantic sense, it is not informative *for us*.'

It might be countered to this answer that the aim of science is not exactly to produce true or adequate theories, but to produce acceptable ones. That seems wrong to me. Analogously, it is not the aim of mathematics to produce proofs that we can follow; it is merely the case that we cannot know, or have good reason to believe, a given putative theorem unless we can follow the proof. This concerns not the criteria of success, but our ability to see whether the criteria are met. Nevertheless, this is a qualification of the description I gave in *The Scientific Image* of the theoretical virtues. My gloss on 'we want informative theories' was that we want empirical strength, which I characterized as independent of pragmatic factors. The qualification is that, as with other virtues characterizable semantically, whether they are perceptible depends on the formulations of the theory that we actually possess.

3. INTERPRETATION: SCIENCE AS OPEN TEXT

The main project of the latter part of this book will be an interpretation of quantum mechanics. One other sort of interpretation, following the lead given by von Neumann, will also be described in contrast, and still others will be briefly discussed. But what sort of project is this? The semantic approach requires no axiomatization, nor a division into pure syntax *plus* an interpretation in the sense of something that bestows meaning on that syntax. Nor does it propose a dissection of the language of science into a theoretical and pure observational terminology, or even suggest that this is possible. With all those philosophical projects of an earlier generation gone, what can be meant by interpretation of science today?

There is interpretation both *in* science and *of* science. Science does not simply represent the phenomena, but interprets them. That interpretation is incomplete in various ways, even within its own terms, and so itself calls for completion. Here we must distinguish attempted completion by means of *extensions* with new empirical content, and by *interpretations*, which render a fuller account but with no added empirical content. The working scientist is mainly intent on the former, and may be dismissive of the 'empirically superfluous' factors (as Feyerabend called them) in the latter. But the interpretational demands of *What is really going on (according to this theory)?* or even the more modest *How could the world possibly be how this theory says it is?* will not disappear if science is to help us construct and revise our world-pictures.[5]

This too is a view of science, and contrasts with other views. Just as art was once conceived of as a mode of representation pure and simple, so was science:[6]

Now there do exist among us doctrines of solid and acknowledged certainty, and truths of which the discovery has been received with universal applause. These constitute what we commonly call *Science* . . . (Whewell 1840, 1–2)

But in science as in art, the work does not merely represent, and even as soon as Aristotle began to explore this view inherited from Plato, that became apparent. For both the artist and the scientist manage to represent their object *as* thus or so, and this introduces an element of interpretation. In almost parallel passages in the *Poetics* and the *Physics*, Aristotle tells both the dramatist and the physicist to depict events as part of a causal story 'proceeding in accordance with necessity or probability'. That is, the events are to be represented as links in a causal chain.

This Aristotelian view of science need not be accepted to make the general point. Far from deriving the laws from the phenomena, Newton showed us how to conceive of our solar system *as* a system of classical mechanics, and Einstein showed us how to reconceive it *as* a system of relativistic mechanics. The example is not perfect because some phenomena of which Newton had no knowledge did not fit his conception. But the phenomena he did know, we now realize, can be represented as either.

My second point was that such interpretations of the phenomena as science actually gives us are radically incomplete in themselves, and therefore call for interpretation too. The analogous point is certainly more easily made for art or literature; consider for instance the much discussed question: who is the true protagonist of the *Antigone*, Antigone or Creon? In the case of science many open questions are just waiting on new experimental and theoretical developments. They may still come to light with an element of surprise, and answers might parade as really just the old accepted theory, better understood.

Ronald Giere (1988) gives a good example of incompleteness in meaning in his discussion of what Newton's mechanics was. Certainly Newton is outstanding in how much he dictates in the *Scholia* for the interpretation of his work. But when Newtonians tried to model new phenomena, what was to count as an admissible Newtonian force function? Are only central forces allowed, as the paradigm example of gravitational force certainly is? Must phenomena of attraction and repulsion always be due to distinct kinds of force? (What about the magnet?) And later: do systems not described by Lagrangians or Hamiltonians, if any exist, count as Newtonian or not? It was Poincaré who pointed out that the hypothesis of central forces was a distinct eighteenth-century addition to classical mechanics (Poincaré 1905/1952, 101 ff.). The nineteenth century added conservation of energy.[7] These are additions to the theory; they pare the Newtonian family of systems down to a narrower one.

However, Newton did not just leave open questions to be answered by experimentation, observation, and new empirical postulates. Meaning gaps remained as well. This comes out clearly, for instance, in the different treatments of mass in attempted axiomatizations of classical mechanics. The kinematic history of a system places limits on the values that the parameters of mass (and/or force) can take. But kinematic data only sometimes, and not always, suffice to identify those values uniquely. In different axiomatizations the parameter is then allowed to take an arbitrary value, a stipulated don't-care value, or no value at all. This is interpretation; neither theoretical necessity nor empirical phenomena force a choice.[8] It should not be pretended that the answers were already implicit in Newton's *Principia*.

Now we have arrived at the sort of thing currently meant by interpretations of quantum mechanics. Foundational research has laid bare certain structures which can be found in the theory, and produced concomitant reconstructions of that theory. But this work is not merely reconstructive, it is creative as well. Foundational research has two effects: it allows us to see quantum theory as one of a large family of theories, hence placing it in a larger perspective, and it shows us what is open to interpretation and also what must be accepted as rock-bottom common ground for all interpretations.

There is no dearth of topics for interpretation. Quantum theory employs probability—but what is probability? The theory introduced indeterminism at a fundamental level into modern science—but just what is (in)determinism? At the very basis of quantum theory, already we encounter the notion of measurement—but isn't that also a physical interaction, and hence a subject within the theory's own domain? In the probabilities assigned in this indeterministic world we see patterns of distant correlation, so what has happened to action by contact and the limits on causal accessibility that seemed crucial to relativity theory? And on it goes.

If we try to answer such questions, we take for granted that it is legitimate for us to produce our own reading of physics—as a candidate to be assessed, needless to say—and that physics itself does not implicitly dictate entirely how we should understand it. Again, that is one view. It does not denigrate science.

Analogies with art and literature, even if sometimes just mystifying, are of real use here. A *closed* text is one that comes with its meaning dictated by the author.[9] One example would be a novel which does not simply show us the characters but tells us what they feel and think, dictating, as it were, how we are to assign motives to their actions, what the expressed emotions signify. That is characteristic of bad literature; the text shows the heroine with heaving bosom, and says that she realizes (a word that entails the truth of what follows) that she has fallen in love. We are not allowed to remain puzzled for long when she stabs the hero to the heart; she does this because she loves him too much to let him go to perdition in the arms of another woman. The reader is not left so comfortably, in a totally passive role, in *Anna Karenina*. We are shown this life; we have as much, and as little, basis for interpretation as we might in reality.

4. MODELS AND SCIENTIFIC PRACTICE

So far we have concerned ourselves mainly with limit cases, which ignore the complexities of practice, but

> Between the idea/ And the reality
> Between the motion/ And the act
> Falls the Shadow ...

It was right to do so, in my opinion—clear understanding of a subject should begin with its clearest cases—but there must also be a clear way to apply our concepts to practice. Not coincident-ally, the gap may be filled here by explaining how the scientist does exactly what we are doing right now: beginning with the ideal and then attending to its links with the less ideal.

To accept a theory without qualification involves the unquali-fied belief that it is empirically adequate. Empirical adequacy, like truth, admits of no degrees. There is no such thing as just a little bit true, no more than being just a little bit alive. But in both cases the attribute is surrounded by a large cluster of closely related conditions (the heart has stopped but the brain has been without oxygen for only 30 seconds so far . . .). To be empirically adequate is to have some model which can accom-modate all the phenomena. But at any point, science has carried through the model construction only so far, and does not yet know whether certain phenomena can be accommodated. This qualification does not affect the question of empirical adequacy as such, but is certainly a relevant qualification of acceptance. In practice we are *always* in this position.

But in practice, the conscious attempt is not to do something ideal and do it badly: it is to do something feasible and do it well. Our picture of what scientific theories are like must therefore go beyond the ideal case. I shall here present Nancy Cartwright's more nearly phenomenological description of scientific activity, and shall try to show how it can be accepted as a friendly amendment to the semantic approach.[10]

In Giere's scheme the theory defines a certain kind of system —or, in effect, the class of models which represent systems of that kind—plus a hypothesis to the effect that certain real things are of that sort. Nancy Cartwright has disputed schemas of this order of simplicity. In practice, the scientist constructs models that are *overtly* of a sort the real things do *not* fit. The hypothesis which relates real phenomena to the *explicitly worked out* models is much more complex, for the relation is mediated by approximations, idealizations, and fictions.

In classical mechanics, for instance, the texts may state Newton's laws at the beginning and/or present Lagrange's or Hamilton's formalism with mathematical generality. But then they go on to prove theorems about very special mechanical systems: the linear oscillator—the spring, the pendulum—the damped linear oscillator, motion on a frictionless plane, a two-body gravitational system with no imposed external forces, etc. All these are wildly idealized or simplified with respect to the real entities *to which they are nevertheless applied*. Demonstration experiments illustrate the results. 'Illustrate' is the operative word, for it is not literally true that the pendulum on the bench is correctly represented by the model provided. But there is a good approximate fit to the data because the factors not represented—though not irrelevant—have 'negligible' effect relative to the precision and scope of the measurement carried out. The hypothesis is *not* that the real pendulum is a pendulum in the textbook sense, but that it bears this more complicated relation to it.

Does that mean we should reject Giere's schema? We need not reject it, provided we add that the demonstration experiment is also meant to make plausible the tacit claim that there is *some* mechanical model—the textbook pendulum model with a number of extra parameters added into its equations—which does fit the real pendulum. That tacit claim is part of the unqualified claim of truth.[11]

As Cartwright describes it (1983, 132–4), the typical scientific episode has as beginning a two-stage *theory entry* process. The first stage is an *unprepared description* which emerges as starting point. Historically, this is already in midstream. One of her examples is the description of the laser, say after the first working laser was produced, and after the laboratory issued a report on its construction and the effects observed. The report is issued by scientists, of course, but a well-trained laboratory staff without theoretical education should be able to duplicate the reported construction. The next stage is to turn this into a *prepared description* which must be such that the theory, as it has been developed so far, should be able to match an equation:

Shall we treat a CW GaAs laser below threshold as a 'narrow band black body source' rather than . . . 'quieted stabilized oscillator'. . .? Quantum theory does not answer. But once we have decided to describe it as a narrow band black body source, the principles of the theory tell what equations will govern it. (Cartwright 1983, 134)

This prepared description will be explicitly and overtly incorrect, strictly speaking, in the same sense that it is strictly incorrect to describe the pendulum on the bench as a pendulum in the sense of a certain paragraph in a mechanics textbook. But it is useful because the theory, or at least the text, also informs us which sorts of approximation are usable here. The data are accordingly not even *meant* to fit the given equations perfectly. They shouldn't; there would be something wrong if they did! And— here is a crucial point—as the project of better and better theoretical treatment continues, we never get to an essentially different sort of stage:

After several lectures on classical electron oscillators [in an engineering course], Professor Anthony Siegman answered that he was ready to talk about the lasing medium in a real laser. I thought he was going to teach about ruby rods—that ruby is chromium doped sapphire. . . . Instead he began 'Consider a collection of two-level atoms.' (Cartwright 1983, 148)

As a general view which fits this scientific activity better than the picture of a search for truth or even empirical adequacy, Cartwright proposes a view of physics as theatre. This view already fits the discipline of history as conceived by Thucydides, she says; a historical episode is reconstructed not with an eye to veracity in all details, but to veracity in relevant respects—not with, for example, the actual words in a political speech, but with the general sense of what was actually said.

So far it is easy to agree, and Cartwright's account is very illuminating. But now we come to the assessment of what this means for our view of science. In her view, if we never have a science that is empirically adequate, but find models for all phenomena that yield approximations to a required extent of quantities in which we have special interest, then science will still meet the highest and strictest criterion of success. If that is so, then the aim of science is exactly that. With this conclusion I do not agree. I do not believe that complete truth in the description of actual observable phenomena is rejected even as ideal. The final question about an apparent anomaly will never be whether we really are or ever will be interested in certain divergences from our calculation, but whether the divergence is at least in principle compatible with the general theory— interesting or not. (The decision to investigate that final question or not is another matter.) But we can accept everything short of Cartwright's final conclusion as an amendment to the semantic view. Indeed, we can see a parallel, at this level of interpretation of science, to what Cartwright describes as the scientist's procedure in the interpretation of nature. Cartwright's model of two stages of theory entry serves the same function as the textbook pendulum, a very important function indeed: it helps us to understand real episodes in science in a way that the fully general account cannot. And yet it falls under that fully general account.

5. MORE ABOUT EMPIRICISM

The choice between the semantic approach to science and its rivals is entirely independent of the controversies between scientific realism and anti-realism. As Giere (1988) and Hacking (1983) especially have emphasized, scientists in the throes of empirical research express themselves just as would true believers in the reality of the entities postulated. We do not need to take this way of talking as necessarily indicating a point of view embodied in the very enterprise of science—many famous scientists appear not to have done so, and I certainly do not.

But the semantic approach would start with a severe handicap if it embodied a particular stance on such questions.

Although I have tried to present the approach in mutually acceptable form, my exposition here has been from an empiricist point of view. At this point of conclusion I would like to add a little more about the empiricist view: about what are good models, about the strong feelings involved in adherence to a scientific tradition, and about loss of innocence.

To be good, it is not necessary for a model to have all its elements correspond to elements of reality. What is needed instead is that the model should fit the phenomena it was introduced to model. But—this must be emphasized—neither this modest virtue nor a complete and strict mirroring of reality is enough. Even in its strong form, correct representation of facts is by itself too weak a criterion. We also want our theories to be informative; for that virtue we are willing to risk forgoing total empirical adequacy. This too is not enough. We are finite; we always work at one particular stage in history; so we always have an eye on the future. A new model will be the more highly valued if we regard it as suggesting new ideas for how to extend our theories, how to construct yet further models, how to open up new avenues of experimental and theoretical research. This virtue is pragmatic and historically relative—but so are we; we cannot transcend, even in our highest achievements, what is possible for us at our historical stage.

In this reflection we have a glimpse of the strong emotions, loyalty, and overriding commitment a scientific tradition can exact. Newton did not just give the world a nicely working model of mechanical and astronomical phenomena—he gave it, by example, a new way to approach all phenomena, including those still far beyond the reach of his own century. In the terms Kuhn taught us, his example became a paradigm. Scientists forsook all others, and wedded themselves to this new research programme, in hope and in faith. We have but a brief moment in eternity—to choose thus is to wager one's all. Call it true belief, purity of commitment, or cognitive dissonance—surely one cannot make such a choice and preserve perfect intellectual neutrality, theoretical detachment, and an undiminished sense of irony?

But now I have overstated the matter! So do, I think, the

scientific-realists, though perhaps not so blatantly. All I said is true enough, but even deep and true commitment does not require a sacrifice of the intellect. We can still distinguish science from *scientism*, a view in which science, which allows us so admirably to find our way around in the world, is elevated (?) to the status of metaphysics. By metaphysics I mean here a position, reaching far beyond the ken of even possible experience, on what there is, or on what the world is really like. Scientism is also essentially negative; it denies reality to what it does not countenance. Its world is as chock-full as an egg; it has room for nothing else. Commitment to the scientific enterprise does not require this. If anyone adopts such a belief, he or she does it as a leap of faith. To make such a leap does not make us *ipso facto* irrational; but we should be able to live in the light of day, where our decisions are acknowledged and avowed as our own, and not disguised as the compulsion of reason.

PART I
Determinism and Indeterminism in Classical Perspective

WHEN philosophers turned to quantum mechanics, they brought with them conceptual tools that had been crafted in encounters with Newton's physics, statistical mechanics, and relativity theory. They found very soon that the new science called for more; yet what they had went quite far. Indeed, it went farther than was at first appreciated, and the exposition here of matters conceptually prior to quantum mechanics can be drawn on extensively when we come to that theory.

2
Determinism

As science developed from antiquity to the present, it was recurrently faced with the question of whether nature is subject to chance to any extent, or evolves deterministically. For most of that time, the need to include chance was avoided by abstraction; the subject of study was so delimited as to allow its description as a deterministically evolving system. As the theory of probability developed in modern times, even chance came within the fold of predictability, and the degree of abstraction could be lowered. A great deal of the systematic description of deterministic systems still carries over, however, and it is indeed possible now to form a clearer idea of what exactly that ideal of determinism signified.

1. HOW SYMMETRY IS CONNECTED TO DETERMINISM

Two preambles are needed before we try to explicate the idea of a deterministic system. The first concerns symmetry, its associated strong but sometimes deceptive intuitions, and its proper role in the systematic description of mathematical spaces. The second concerns the use of such spaces to study physical systems, by representing their possible states and possible modes of evolution. Here we begin the first preamble.

1.1. *Asymmetries and hidden variables*

When we read that certain conclusions were reached on the basis of considerations of symmetry, we will not always find a pure and a priori demonstration. Instead, there may just be an assumption of actual symmetry in the actual world. To say that a symmetry was broken means then only that the world did not after all go along with the assumption. But at other times it seems that something impossible has happened: where could this newly found asymmetry have come from?

Asymmetries do often seem merely accidental. For example, humans all have their heart on the same side. Does this establish a global asymmetry for the world as a whole? The local asymmetry in one body defines a global asymmetry if its mirror image exists nowhere in nature. Of course, it can't be just the placing of the heart. Imagine we were spheres, rolling around on the earth, with hearts just off centre. Each sphere would have some asymmetry, but it does not even make sense to ask whether all the spheres have their heart on the same side. The significant relationship exhibited by the human species must have to do with some other features as well, like feet, head, navel. ... All this seems so accidental that we speculate at once that the whole of fundamental physics can be written without reference to this 'handedness' of nature. Such a pervasive asymmetry on earth must have been due to some asymmetry in the initial conditions. In technical jargon, *parity* is surely not violated on the most fundamental level.

That speculation is however not a priori certain. If it is not true, how deeply are our intuitions shaken? The recent story of parity violation which changed nuclear physics (theory by Lee and Yang, experiments by Madame Wu) was indeed disturbing. A radioactive nucleus emits an electron when it decays. The nucleus itself has a certain asymmetry describable in terms of spin. If we change our spatial frame of reference by a geometric reflection, then the spin we called 'up' will become 'down'. But whether a given electron is emitted in the direction of the spin cannot have the same answer for both frames of reference. The experiments showed a preponderance of emissions in the direction of spin. This shows that the sort of global asymmetry we can associate with the human race, expressed in terms of hands and heart, can also be found on the level of the nucleus.

We could again speculate that parity will be conserved at some still more elementary level. The disturbing quality of the result was that atomic physics did not appear to allow that option. In the context of this theory, the one we cherish, parity violation is indeed more fundamental. But how shall we analyse the inclination to assume or speculate insistently on such symmetries anyway? The reply is that we are trying to ask 'where the asymmetry came from', what the reason for it was, why things should be this way. To insist that such questions must have an answer is also typical when determinism shows its hold on our imagination, and we shall see below exactly how the two are logically related. In the nineteenth century, there were notable examples of successful postulation of hidden variables to explain the reasons for observed asymmetry. A good example was given by Mach. When a wire is 'electrified', a parallel suspended compass needle will 'choose' a direction in which to turn. So described, it is like a little miracle, for the term 'electrify'—pre-theoretically understood—does not indicate any preferred direction. The air of the miraculous disappears when we postulate a directed electric current. The intuitive reaction which makes us think that there must be such a hidden asymmetry that accounts for the asymmetry in the observed event is also our intuitive reaction to Buridan's ass: if it eats, there must have been some asymmetry after all in its little mind, or in qualities of fodder it can detect and we can't. What is that, if not an insistence on hidden determinism?

Pierre Curie formulated this as an explicit principle, to govern physical theory.[1] I would express his principle as: *an asymmetry can only come from an asymmetry*. Curie expressed it in causal terminology, again suggesting a link with determinism—'When certain causes produce certain effects, the elements of symmetry of the causes must be found in the effects produced'—and asserts that this principle 'prevents us from being misled into the search for unrealizable phenomena' (Curie 1894, 401, 413). But it cannot very well be a priori, since it rules on an empirical question. For there may simply not be any hidden agency; the 'choice' of direction could be a true chance event. Even if it is a persistent or widespread uniformity, it might still be one which does not derive from any hidden mechanism but just *is*.

Several objections could be raised to my suggestion that Curie's principle stands or falls with the principle of determinism. First of all, perhaps the examples show only accidental links of symmetry with determinism, and there is no logical connection. Secondly, perhaps the question of the *possible* existence of such an agency is not an empirical matter: perhaps all possible observable phenomena are at least consistent with a hidden determinism. Both objections can be answered, but not without more preparation. The first I will take up below, in Section 4. The second will be the main topic of Chapter 4, which deals with Bell's Inequalities. There we will see that a photon reaching a filter may 'choose' a polarization in a way which cannot be 'traced back' to any preceding features of the situation, and yet is not random with respect to the rest of nature. If both objections can be answered, we have to conclude that Curie's putative principle (even in my formulation, which did not use causal terms) has no fundamental ontological status. If so, it betokens only a thirst for hidden variables, for hidden structure that will explain, will answer *why?*—and nature may simply reject the question.

1.2. *Symmetries of problems*

> The rhythmical complexity of the songs ... follows its own mysterious laws. In one kind of Moravian dance song, for example, the second half of the measure always lasts a fraction of a second longer than the first half. Now, how can that be notated? The metrical system used by art music is based on symmetry. A whole note divides into two halves, a half into two quarters. A bar breaks down into two, three, or four identical beats. But what can be done with a measure that has two beats of unequal length?
>
> Jaroslav in Milan Kundera, *The Joke*

Although no single mundane symmetry is a priori, and although a given asymmetry may have no explanation at all in terms of 'prior' asymmetries, Curie's principle does have significance. It codifies a methodological tactic: the tactic of seeking solutions which import no asymmetry absent from the problem. That means, first of all, the choice of models with as much internal symmetry as possible. In slogan form, it is the requirement that a solution should 'respect' the symmetries of the problem.[2]

In fact, of course, there is absolutely no guarantee that the problem was posed in a way that exhibits all the factors physically relevant to the phenomenon studied. Sometimes it will be impossible to find a solution at all without introducing new factors—new asymmetries—not found in the problem as posed. Imagine the predicament of Buridan's ass, if it is asked for a rule which takes as input the symmetric arrangement of its mouth and the two piles of food, and has as output a configuration in which its palate is in contact with food. It will get nowhere if it respects the symmetries of the problem. The example is funny exactly because that is what it should try first, and because without further information the tactic obviously will not work. Tactics have their limits; but let us see how this tactic works when it does.

Faced with an intuitively stated problem, we choose a model for the situation. The model is at least somewhat general, with a set P of specifiable variables. The problem now takes on the manageable form: what unique object of type *OUTPUT* is associated with each $INPUT_m$? Here m stands for variables belonging to P. The solution must be a rule, a function, which effects this sought-for unique association. Now it appears that the chosen model has *symmetries*. Intuitively speaking, the situation has both relevant and irrelevant aspects, and certain transformations affect the values of the variables, but leave the problem 'essentially' or 'relevantly' the same. At this point, a tactical dictate comes into play: *Same problem, same solution!* The rule or function offered as solution must 'respect those symmetries'. Denoting that function as f, and one of those transformations as t, we must be able to 'close the diagram' (Fig. 2.1). That is the stringent hidden requirement that makes the diagram 'commute':

$$ft = tf$$

which needs to be met in consequence of the explicitly imposed demand.

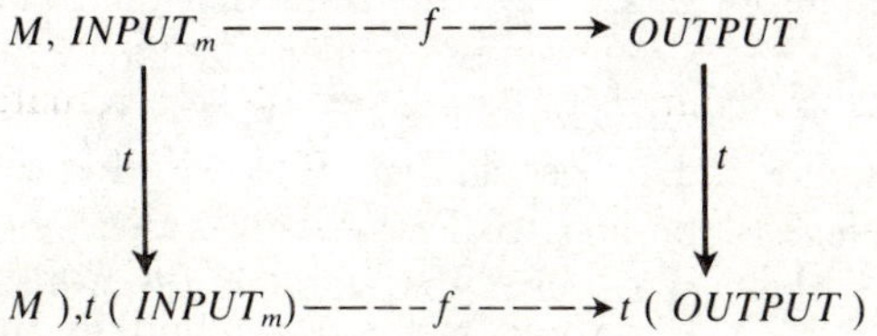

$$M, INPUT_m \dashrightarrow f \dashrightarrow OUTPUT$$

$$t \big\downarrow \qquad\qquad\qquad\qquad t \big\downarrow$$

$$t(M), t(INPUT_m) \dashrightarrow f \dashrightarrow t(OUTPUT)$$

Fig. 2.1

These ideas also allow us to make clear what is meant by the solution *depending* on a parameter. The function f does *not* depend on parameter p if p can be independently varied without affecting the value of f—i.e. if f is an invariant of the relevant transformations which change p without affecting other members of P. Note however that this is relative to the designation of set P as the primitive parameters. That is an aspect of the model: we must not expect to find model-independent notions in this neighbourhood without invoking metaphysics.[3]

One use of this tactic is that, by transforming one problem into another (which we regard as essentially the same), we may come to one that is easier to solve. Perhaps this is indeed crucial to understanding.[4] We have understood a problem 'in its full generality' only when we know exactly what counts as essentially the same problem; that means when we know exactly which transformations do and which do not change the situation in relevant respects. Now, to state a problem in its full generality is to achieve the proper degree of abstraction: to abstract the problem itself from the concrete guise of its appearance. Generality, abstraction, transformation, equivalence of problems— some very old philosophical ideas are here mobilized in new logical form.

2. STATE-SPACE MODELS AND THEIR LAWS

The semantic approach to scientific theories means to refine the analysis of their structure. At the level of analysis of the preceding chapter, we addressed the notion of theory *überhaupt*. To go further, we must consider relativistic and non-relativistic theories separately. With very few exceptions, our discussion of quantum mechanics will remain with its non-relativistic formulation; the difference this makes is that we can keep time as a separate category. A physical system is assumed capable of certain *states*, and to be characterized by physical magnitudes (observables) pertaining to the system, which can take certain *values*. In classical science, the state of the system at a given time can be construed broadly enough so as to determine also the values of the relevant observables at that time. (Whether this is so for quantum mechanics, too, is a matter of interpretation.) This means that we can represent the classical system's history—its evolution in time—simply as a *trajectory* in its space of possible states (*state-space*). Such a trajectory is essentially just a map s: time $\longrightarrow$ state-space (Fig. 2.2). We may classify a certain class of systems together as a *kind* of system if the states they can have form overlapping parts of one single 'large' state-space and/or if the trajectories they can have are in some way similar. Obviously this notion of kind is vague, though some philosophers insist that a given such classification can be natural ('cutting reality at its joints', in Plato's phrase) or merely arbitrary.

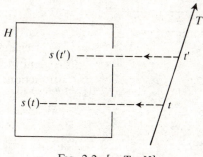

Fig. 2.2 [s: $T{\rightarrow}H$]

A prime example is the classic harmonic oscillator. Imagine an object constrained to move on a given straight line, but subject to a force directed towards a given point and proportional to the distance from that point. With q standing for its position on the line, its motion is described by the equation

(1) $\quad m(\mathrm{d}^2q/\mathrm{d}t^2) = -(1/n)q$

where constant n depends on the force. In terms of its momentum $p = m(\mathrm{d}q/\mathrm{d}t)$,

(2) $\quad \mathrm{d}q/\mathrm{d}t = (1/m)p \qquad \mathrm{d}p/\mathrm{d}t = -(1/n)q$

The possible states are represented by couples in the p–q plane, and the possible trajectories by curves which satisfy equation (2) or, equivalently,

(3) $H = (1/2n)q^2 + (1/2m)p^2 = \text{constant}$

which in effect describes the total energy of the system, and says that it is conserved. Those trajectories are therefore represented by the elliptical curves in Fig. 2.3.

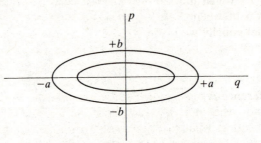

Fig. 2.3   Trajectories in p–q plane

We get a more general representation, applicable to other mechanical systems as well, by expressing p and q in terms of the Hamiltonian H:

(4) $dq/dt = \partial H/\partial p$; $dp/dt = -\partial H/\partial q$

A scientific theory will in general have many such clusters of models, to represent sub-kinds of the kind of system it studies, each with its state-space. So its presentation will describe a class of *state-space types*. (See further *Proofs and illustrations*.)

For such a non-relativistic theory, we can see at once that there must be basic equations of two types, which correspond to the traditional ideas of *laws of coexistence* and *laws of succession*. The former type, of which Boyle's Law $PV = rT$ is a typical example, restricts positions in the state-space. Selection and superselection rules are the quantum-theoretical version of such restrictions. The other type has the classical laws of motion, in Newton's formulation or Hamilton's ((3) above), and

Schroedinger's equation as typical examples: they restrict trajectories in (through) the state-space. Symmetries of the model—of which Galileo's relativity is the classical example *par excellence*—are 'deeper' because they tell us something beforehand about what the laws of coexistence and succession can look like. It is in the twentieth century's quantum theory that symmetry, coexistence, and succession became most elegantly joined and most intricately connected.

The laws in question are 'laws of the model': important features by which models may be described and classified. The distinction between these features and others that characterize the model equally well is in the eye of the theoretician. There is no obvious warrant to think that they correspond to any division in nature.[5] To put it differently, there may not be any objective difference between laws of nature and 'mere' facts in the world; that difference is only 'projected' by us onto the world, characterizing not what it is like, but how we see it. Some revisions which do affect what the laws of the model are have absolutely no effect on its adequacy. Suppose we replace one model by another, whose state-space is just what remains in the first after the laws of coexistence are used to rule out some set of states. Then we have not diminished what can be modelled. Similarly, if we include in the state some functions of time, laws of succession may be replaced by laws of coexistence. We can illustrate this with different ways to write Newton's laws. The most obvious is to say that the second law of motion, $F = ma$, is a law of succession. But another way to think of it is this: the second derivative with respect to time, of position conceived as a function of time, is one parameter represented in the state, just as well as the total force impressed and the mass. The second law of motion, $F = ma$, is then a law of coexistence rather than of succession: it rules out certain states as impossible. This sort of rewriting has been strikingly exhibited in certain philosophical articles about Newtonian mechanics (see Ellis 1965; Earman and Friedman 1973).

Elementary statements. Although the semantical approach focuses on models rather than on axioms, there is still an important place for language in its view of theories. To present a theory is to describe—in natural language, augmented by terms from any relevant branch of mathematics—its family of models. But there are two sorts of statements (*elementary statements*) about a physical system whose truth conditions are clearly represented in models of that system:

State attributions: system X has state s (a state of type S)
                          (at time t)
Value attributions: physical magnitude (observable) Q has
                          value r (value in E) (at time t)

In classical science these are so intimately tied together that the state attributions can encompass the value attributions as a special case. When interpreting quantum theory, we shall not be able to take that for granted (and, moreover, we shall have to think about 'unsharp' value attributions).

It is however characteristic of the semantic approach to abstract from the linguistic form in which such attributions are expressed, and to focus on counterparts in the models. For example:

Elementary state attribution A is *true* about system X exactly if the state of X is (represented by a point) in region S_A of its state-space.

The region S_A is or represents a set of possible states; that set is then called the *proposition* expressed by statement A. In quantum mechanics, when A attributes a pure state, that region is a subspace. The family of propositions in general inherits a certain amount of structure from the geometric character of the state-space. To explore that structure is *ipso facto* to explore the *logic* of elementary statements. General relations of logical interest here include:

A implies B exactly if S_A is a subset of S_B.
C is the *conjunction* of A and B exactly if S_C is the g.l.b. (greatest lower bound) (with respect to the implication relationship) of S_A and S_B in the family of propositions.

Again we must enter the caveat that, unless value attributions are reducible to state attributions, the propositions may not simply be regions of the state-space.

Proofs and illustrations

I will just add some brief remarks here on how we should regard relativistic theories, and on state-space types.

In the case of relativistic theories, early formulations can be described roughly as relativistically invariant descriptions of objects developing in time—say in their proper time, or in the universal time of a special cosmological model. A more general approach, developed especially by Clark Glymour and Michael Friedman, takes space-times themselves as the systems. Presentation of a space–time theory T may then proceed as follows: a *(T) space–time* is a four-dimensional differentiable manifold M, with certain geometrical objects (defined on M) required to satisfy the *field equations* (of T), and a special class of curves (the possible trajectories of a certain class of physical particles) singled out by the *equations of motion* (of T).

Returning to the non-relativistic case, I shall describe briefly how models are classified by state-space type in Elisabeth Lloyd's approach.[6] In Lloyd's account of state-space models, a distinction is drawn between parameters and variables: the former are constants that characterize the model (their values only constrained but not determined by the theory), while the latter vary within the model. In our previous example of the linear oscillator, the total energy of the system may be a parameter of a model, while position, momentum, and kinetic energy are variables. These variables are functions of the state. In a stochastic model, only a probability distribution for the values of the variable may be linked to a state. A *state-space type* is a class of state-spaces (belonging to models of a given theory) produced by changing the values of those parameters that define a given state-space. This division induces a corresponding one of models into model types. In one generic use the word 'model' tends in fact to refer to a model type: 'the Bohr atom' and 'the linear oscillator' are singular terms that denote not a single model but a model-type (compare the generic 'The cow is a herbivore' to 'The cow has been eating the garden vegetables again').

In classical statistical mechanics and also in quantum theory (in a somewhat different way) the variables—physical magnitudes, observables—only have their probabilities determined by the state. Population genetics models are also stochastic, and simple to present. Suppose that in a certain population offspring of brown-eyed parents are more likely to have brown eyes, but no deterministic pattern is discernible. We can nevertheless postulate an underlying determinism in the fashion of Mendel. For example, we *suppose* that each parent supplies one colour gene, of type A or a, to each offspring. We add that an individual that has two A genes has brown eyes, while all the others have blue eyes. How do the frequencies of blue and brown eyes change in the population? Will an equilibrium be reached and maintained?

We have now specified an underlying state-space for individuals (three possible (gene combination) states) and for the population a corresponding three-dimensional state-space; each axis represents the variable which is the frequency of one of the three individual states. The individual states AA, Aa $(= aA)$, and aa are also called *genotypes*; the state of the population is a distribution of these genotypes, with proportions $\#AA$, $\#Aa$, and $\#aa$ (summing to 1). We can express it differently in terms of the gene frequencies:

(1) $$p = \#AA + (0.5)\#Aa$$
$$q = (0.5)\#Aa + \#aa$$

which also sum to 1 then. (Thus p is the proportion of the genes present that are of type A, to which an individual contributes $2/N$ if it is of genotype AA, when N is the total number of genes.) The initial $p : q$ ratio does not generally contain as much information as the genotype distribution.

As dynamical law, let us assume random mating and random contribution of genes by parents, and no correlation of genotype and sex in the parents. That means for example that an offspring of two Aa parents can be of any genotype, but it has probability $1/4$ of being AA, also $1/4$ of being aa, and $1/2$ of being Aa. We quickly calculate that the second generation will have proportions

(2) $p^2 : 2pq : q^2$

for genotypes AA, Aa, and aa. Applying (1) to (2), we see that this is a fixed point of the transformation:

(3) $p' = p^2 + (0.5)2pq = p(p + q) = p$

so the third generation will have exactly the same distribution of genotypes as the second. This is the Hardy–Weinberg law, and the above distribution (2) is the Hardy–Weinberg equilibrium reached from the given initial distribution.

The described model is the simplest, because of the assumption of random mating and mixing of genes. If brown-eyed individuals live much longer, or are more likely to reach mating age, than blue-eyed ones, the model assumption of randomness is independently disconfirmed. That is a stroke against the model whether or not the statistics of the actual population match its probabilities. But the Hardy–Weinberg law can be viewed alternatively as a schema in which other parameters can be inserted. The differential survival to mating age can be reflected in a *fitness* parameter. That parameter will be a function of genotype, and can be set to characterize somewhat more sophisticated models. Depending on that parameter, we will find different equilibria or none at all. That parameter, unlike p and q, does not vary in the model but characterizes a model type.

3. SYMMETRY, TRANSFORMATION, INVARIANCE

The themes of this chapter so far—that opting for symmetry in a model is a good tactic even if not a sure guide, and that the model typically has a state-space at its heart—must now be connected. As already noted, symmetry is the clue to structure. When geometry was being generalized in the last century, a systematic way of describing geometric spaces was needed. By the time Minkowski drew on those new developments to formalize relativity theory, the way was well established: a space is characterized by its symmetries, by the description of its transformations and what they leave invariant. I shall present this subject first through its simplest and most familiar example, then in the abstract.

3.1. *Symmetries of space*

The most instructive and quickly appreciated examples of symmetry are those of (Euclidean) space. Euclid laid down his postulates as evident truths about spatial relations and structure.

It would be hard to maintain that he had a concept of Euclidean space (his sort of space as opposed to what?) or even of space, let alone of *a* space. Today the postulates appear only as clauses in the definition of a Euclidean space, which now can have any dimensionality.[7] We treat all geometric objects as sets of points. A *metric space* is a set (the *points*) with a *distance* function. This is a function *d* which maps every pair of points into a non-negative real number, and such that

$d(x, y) = 0$ if and only if $x = y$

$d(x, y) = d(y, x)$

$d(x, y) + d(y, z) \geq d(x, z)$ (triangle inequality)

Which metric spaces are the Euclidean ones is most easily explained through the representation theorem. The Euclidean space of dimension *k* is isomorphic to the metric space, whose elements are the *k*-tuples of real numbers, with the distance $d(x, y)$ being the norm of their difference; that is, $\sqrt{\Sigma z_i^2}$. All other geometric notions such as angle and congruence can here be defined in terms of distance. Having said all this, though, let us proceed more intuitively.

If I put a cardboard capital letter F on the table, I cannot produce its mirror image just by moving that cardboard letter around on the table-top. I could do it by picking the letter up and turning it upside down. A reflection in the plane can be duplicated by a rotation in three-dimensional space. But a reflection through a mirror (three-dimensional reflection) cannot be produced by rotations, because we have no fourth dimension to go through or into. Reflections and rotations are transformations which leave all Euclidean geometric relations—equivalently, all distances—the same. They are accordingly called symmetries of Euclidean space, for they are all *isometries*, the transformations that leave all distances the same.

The simplest sort of isometry is neither reflection nor rotation, but translation. A translation just moves everything over, a fixed distance, in a fixed direction. But every translation is the product of two reflections. To translate plane figure F from point *A* to point *B*, draw two lines *m*, *n* between *A* and *B*, perpendicular to the line that joins them. Reflect F through line *m*; then reflect the result through line *n*. If you spaced the lines correctly, F will then be situated at *B*. In fact, we can also produce rotations in this way; let *m* and *n* not be parallel but intersect; translate the result of the second reflection back to the point of origin.

Basic theorem: The isometries are exactly the products of finite sequences of reflections.

By a product is here meant simply successive application; so the product of the rotations through 45 and 30 degrees is the rotation through 75 degrees. If the minimum number of reflections needed to produce a given isometry is even, it is called a *rigid motion* or *proper motion*; if odd, it is called an *improper motion*.

To show the important difference, imagine again that the letter F is moved across a plane surface. If the figure traces out a continuous path, then none of these isometries that connect the figures at successive times can be improper motions. Single reflections, for example, can never be used in this process. For if you reflected the figure around some line, at a definite time *t*, then at least one of the tips of the figure would 'make a jump' at that time. It would not trace a continuous path at all. Therefore the proper motions are properly so-called—for they represent real motion of real objects. The proper motions of the plane represent real motion on flat surfaces, and the proper motions of space represent real motions in three dimensions.

A product of proper motions is always again a proper motion. But a product of improper motions need not be improper. So the proper motions by themselves form a natural family, while the improper motions do not. What is meant here by a natural family is exactly the topic of the next subsection.

Besides symmetries of the space as a whole, we can also study symmetries of specific figures. A symmetry of a figure is an isometry which leaves this figure invariant, i.e. acts like the identity on this figure. The letter R has no symmetries except the identity itself. Of course, O has a large family: all rotations around the centre, and all reflections in lines that pass through its centre, and all products of series of these rotations and reflections. For each figure there is therefore a specific family of symmetries, and it is again a natural family, in that the product of any two members is again a symmetry of that figure.[8]

3.2. *Groups and invariants*

Having seen several examples of natural families of symmetries, we turn now to the study of such families. This is the theory of groups. I shall try to show how group structure can be looked at in three different ways, each instructive in its own way, but equivalent. Let us begin with the philosophical idea of abstraction. For example, if we first discuss individual animals, and then abstract ('the mouse is long-tailed', 'the shark is one of the oldest extant species', 'the human is an omnivore'), we have switched attention from the individuals to types. We have thereby divided the individuals into classes and produced a partition:

A *partition* of a set K is a class X of subsets of K which is:
exhaustive: the union of X is K
disjoint: no two members of X overlap

The members of the partition are called cells. Notice that this abstraction proceeded by classifying the individuals as the same in some respect. That sameness is an equivalence relation; and the cells are just the classes of animals grouped together by this relation:

R is an *equivalence relation* on K exactly if R is:
reflexive: x bears R to x
symmetric: R is its own converse—if x bears R to y then y bears R to x
transitive: if x bears R to y, and y does to z, then x also bears R to z
for all members x, y, z of K.

Here is the first result: equivalence relation and partition are effectively the same concept. Corresponding to partition X, we have the equivalence relation, call it *X-equivalence*, of belonging to the same cell of X. Conversely, any relation R on K associates with each member x of K the class $R(x) = \{y \in K : x$ bears R to $y\}$. If R is an equivalence relation, these are called (*R-*)*equivalence classes*, and they form a partition of K. This correspondence between partitions and equivalence relations is one-to-one.

We introduce now the idea of a *transformation* of the set—a one-to-one mapping of K onto itself—and consider just those transformations which 'respect' the cells. That means, if t is such a transformation, then $t(x)$ is in the same cell as x itself. It may come as no surprise that this is an example of a set of transformations which corresponds exactly to the equivalence relation:

> If T is the family of all transformations t of K such that $t(x)$ is in the same cell of X as x is, then two elements y and z of K are X-equivalent if and only if $z = t(y)$ for some member t of T.

The interesting question which leads directly to the concept of group is now, What must a family of transformations of K be like, in order to be related in this way to a partition (or equivalence relation)? We define:

> Class T of transformations of domain K is a semi-group (a *transformation semi-group on K*) exactly if tt' belongs to T if t and t' belong to T.
>
> A *monoid* is a semi-group which contains the identity transformation $I(x) = x$ of its domain.
>
> A *group* is a monoid which contains for each member t also an inverse
> $t^{-1} \colon t(x) = y$ if and only if $t^{-1}(y) = x$.

We have already seen that there is an exact correspondence between partitions and equivalence relations. We add now a similar correspondence of the two to groups of transformations of the same domain.

> If T is a family of transformations of set K, then x and y in K are called *T-equivalent* if x can be transformed into y, i.e. if $t(x) = y$ for some member t of T.

> *Theorem:* (a) If T is a group of transformations of K, then T-equivalence is an equivalence relation on K.
>
> (b) If S is an equivalence relation on set K, then there is a group G of transformations of K such that S is the relation of G-equivalence.

(See *Proofs and illustrations*.) The geometrical examples we had before will illustrate this very well. In Euclidean geometry, the most basic equivalence relationship is *congruence*; the corresponding group is the family of isometries of the space, also called its *symmetry group*. The family of proper motions is also a group, a subgroup of the isometries. In the case of a specific given figure F, the family of symmetries of F is also a group; the partition is then a very simple one, for the figures are just divided into F and all the others. In the case of a class of figures, we must distinguish the symmetries of the class (which transform all and only the members of that class into members of that class) and the transformations which are symmetries of all its members. But in both cases we find a group.

We call a subset S of K *invariant* under g (under T) exactly if $g(x)$ is in S for all x in S (for all g in T). Similarly, a relation R on K is *invariant* under g (under T) exactly if $g(x)Rg(y)$ whenever xRy (for all g in T).

Invariance is related to our previous notions as follows. Suppose that S is invariant under T. Then $T[x] = \{g(x) : g \in T\}$ is part of S for each x in S. Now if T is a group, then $T[x]$ is exactly the set of objects which are T-equivalent to x—it is a cell of the partition formed by the T-equivalence classes. Hence S is equivalently a set closed under the operations of T, a set closed under T-equivalence, and a union of cells of the corresponding partition of T-equivalence classes.

Proofs and illustrations

To prove the theorem, note first that, since a group is a semi-group, the relation of T-equivalence is transitive. Because a group is a monoid, it is also reflexive. And because the group is closed under inverses, the relation is symmetric. For part (b) let us take as G simply the class of all one-to-one functions t of K onto itself such that $t(x)$ bears S to x. Recalling that S partitions K into the family of equivalence classes $S(x)$, we see that each member t of G is just pasted together from functions t_x which map $S(x)$ one-to-one onto itself. Conversely, any such pasting together produces a member of G. So if y and z belong to $S(x)$, then there is a member t of G such that $t(y) = z$, and otherwise there is not. That G is a group is easy to prove too, so S is just G-equivalence.

One word of caution, however. The correspondence we found

between partitions and equivalence relations is one-to-one, but this is not so for the correspondence with groups. Two distinct groups can generate the same equivalence relation. As an example, let $K = \{1, 2, 3, 4\}$, and let t be defined by

$$t(1) = 2, \qquad t(2) = 3, \qquad t(3) = 1, \qquad t(4) = 4$$

We call the result of applying t successively N times in a row the Nth power of t. This set of powers is a transformation group on K, for the inverse of t is just its second power, and the identity operation is the third power of t. (So every power of t is the same as either t or tt or ttt.) The corresponding equivalence relation holds between the numbers 1, 2, and 3, but relates 4 to itself alone. Yet it is easy to see that this group is not the group which the proof of the theorem constructs for this job, for the powers of t do not include the transformation which maps 1 and 3 into each other, but which maps 2 as well as 4 into itself. So here we have two groups with the same corresponding equivalence relation.

4. SYMMETRIES OF TIME: CLASSICAL (IN)DETERMINISM

Determinism and indeterminism are related to symmetry in two ways. One has to do with the very meaning of determinism, the other with its empirical implications.

We cannot even ask the question whether the world, or a given (kind of) system, or a theory is deterministic unless we have a precise meaning for the term first. I shall not occupy myself with the question of what kind of theory should be called deterministic; following the semantic approach, we must address ourselves primarily to systems. Suppose that a given system has a state-space H. A *trajectory* of such a system is a map s: *Time* $\longrightarrow H$. What exactly must be true of these trajectories to make the system a deterministic one? To approach this question, we must think about the relation of the possible trajectories of the system to its state-space. In the case of Newtonian science, for example, space is Euclidean, which means that the symmetries of Euclidean space induce also symmetries of the state-space as a whole. This means in turn that, if we describe a possible history of a Newtonian system, and then change this description so that the result depicts an exactly similar history but with all geometric configurations transformed by an isometry, then that is also a possible history of this kind of system.

1. If S is a symmetry of the state-space, and s a possible trajectory, then $s' : s'(x) = S(s(x))$ is also a possible trajectory.

This is a general principle which says something about the set of possible trajectories as a whole, as opposed to a law of succession which constrains the trajectories individually. But laws of succession must be of a form which allows this principle to hold, and that fact constrains theory. The symmetries of the state-space do not 'reach across time', however, except in such implications. We must add diachronic symmetry separately.

Bertrand Russell (1953) suggested that determinism requires that time itself not appear as a factor in the evolution of the state. That means that there must be a function f such that, for all times t and positive numbers b, $s(t + b) = f(s(t), b)$. This ascribes a certain *periodicity* to the actual history, for it means the same as:

For any times t, t', if $s(t) = s(t')$ then $s(t + b) = s(t' + b)$.

Periodicity is a certain kind of symmetry in time. It is not mirror-image ('bilateral') symmetry, but symmetry under *translation $t \longrightarrow t + x$* of time: the symmetry of identical repetition.

This is still not enough if we require it only of the actual history of the system we are looking at.[9] The system was capable of different possible states, including ones that it never actually has. The idea of determinism certainly requires that the evolution also would not have depended on the absolute time if the system had followed one of its other possible trajectories.[10] We shall require then of a deterministic system:[11]

2. There is a function f such that, for all times t, all positive numbers b, and all possible trajectories s', $s'(t + b) = f(s'(t), b)$.

Another way to put this, which shows that it is not vacuous, is this:

2a. For all times t and t', for all $b > 0$, and possible trajectories u and v, if $u(t) = v(t')$, then $u(t + b) = v(t' + b)$.

Time translation thus induces a symmetry which imparts periodicity to each individual trajectory but also relates trajectories.

Define now the operations U_b on the possible states of our system, by means of the equation

4. $U_b v(t) = v(t + b)$

This defines U_b uniquely, if the above condition of determinism is satisfied. The family of operators $\{U_b : b \geq 0\}$ is a semi-group with identity—a monoid—and is also called the *dynamic semi-group* of the system. Obviously U_0 is the identity, and the product of two operators is given by

5. $U_b U_c = U_{b+c}$

We can use this to rephrase the condition we had arrived at:

6. The system S has a dynamic semi-group, namely a family $\{U_b : b \geq 0\}$, such that 4 and 5 hold.

And this we can equivalently take as the definition of determinism.

There is also a stronger notion in the literature, *bi-determinism*, which means that the past must also have been the same if the present is. This is defined equivalently by the conditions

7. For *any* real number b, any times t, t', any possible trajectories u and v, if $u(t) = v(t')$ then $u(t + b) = v(t' + b)$.
8. The system has a *dynamic group*, namely a family $\{U_b : b$ a real number$\}$, such that 4 and 5 hold and $U_b^{-1} = U_{-b}$.

The operators U_b are in all cases called *evolution operators*. One way of looking at mechanics is to say that the laws of motion (or Schroedinger's equation, in quantum mechanics) describe the dynamic group.

Let us now step back, and put this discussion in larger perspective. Since only the possible states and possible trajectories played a real role, the above conditions pertain properly to *kinds* of systems. Rather than say that a given system is deterministic, we should say that it belongs to a deterministic kind. But kinds may be sub-kinds of other kinds, so it may also belong to a kind which is not deterministic. In the semantic view a theory typically, in presenting a class of models, describes a kind of system, with an implied classification of sub-kinds. Whether a given real system—such as the world, or you, or I—is deterministic is therefore definitely not a univocal question. This also suggests various options for the interpretation of theories.

First, it does not matter how a theory presents its class of models. So it might just say, for example, that the models all have a certain state-space type, and the possible trajectories satisfy certain equations *and* are deterministic. If the equations are not everywhere defined, for example if there is a discontinuity, the stipulation of determinism may not be redundant.[12] One might object that this will make the theory radically incomplete, if there is more than one maximal set of possible trajectories which satisfy (singly) the equations where defined, and (jointly) condition 6 of determinism. But, first of all, incompleteness is the rule rather than the exception when it comes to theories; secondly, it would mean that we appear to have certain options of interpretation. Let the theory be T; clearly, T-systems are not a deterministic kind in this case, but they divide e.g. into T_A-systems and T_B-systems which are deterministic sub-kinds. Option 1 is to give as theoretical hypothesis that any two real systems belong to the same sub-kind if they belong to either, but we do not know which. This is essentially an ordinary sort of incompleteness. Option 2 is to say that there are real systems belonging to each. Then we face the further question of whether they are distinguished by certain physical magnitudes, not treated in theory T. To say that they are—option $2(a)$—is to attribute to T again an ordinary sort of incompleteness. To say that they are not—Option $2(b)$—appears to be an unequivocal assertion of indeterminism in the real world.

Two qualifications must be entered here, for in one way that assertion looks stronger than it is, and in another way it looks weaker. From an empiricist point of view, at least, the most we can really say about a real system X in this regard is:

System kind K is (is not) deterministic, and is such that one system of this kind has a history which coincides with the history of real system X, in the time interval d, with respect to the values of the relevant physical magnitudes.

This means that it makes sense to say of a real system not that it is or is not deterministic, but only that it can be thought of as (or as not) deterministic. Typically, the same system can be thought of as a (temporal part of a) system of many different kinds.

More interesting, and equally valid for the empiricist, is the question of whether the phenomena are not only such that we can think of X as a system of an interesting indeterministic kind, but such that we *cannot* think of it as deterministic. Is that even possible? Despite the great value the nineteenth century apparently attached to the ideal of determinism, and its consequences for the scientific world-view, this question was not truly broached. Transposing Kant's theory of our constitution of the empirical world, neo-Kantians had the conviction that any phenomena whatever could in principle be fitted to a deterministic account, and they regarded theory construction as not finished until such an account is produced. But even Hans Reichenbach, in his influential attacks on this conviction, went no further than insisting on the possibility of choosing Option 2(b) above. In the case I described there, we have *ipso facto* a possible deterministic completion of T, by postulating that the subscripts A and B correspond to so far undescribed physical quantities. Reichenbach pointed out that it is purely contingent whether or not there would be observable quantities to play that role, and insisted that it was reasonable to postulate sometimes that there are no such quantities at all in reality.[13] But there is a worse case, not like this schematic example. Let us say that what Reichenbach envisaged here is *classical indeterminism*, characterized by the fact that it is at least *possible* to conceive of the case as incompletely described determinism. It was only after the advent of quantum theory that we came to see the possibility of *non-classical* indeterminism in the phenomena, which is not even conceptualizable as compatible with an underlying determinism.

Proofs and illustrations[14]

In classical mechanics, systems are typically described by a set of differential equations

(1a) $\mathrm{d}x_i/\mathrm{d}t = X_i(x_1, \ldots, x_n, t)$

or in matrix form

(1*b*) $dx/dt = X(x; t)$ with $x = \langle x_1, \ldots, x_n \rangle$

The single higher-order differential equation,

(2) $d^n x/d^n t = f(x, dx/dt, \ldots, d^{n-1} x/dt^{n-1}, t)$

is also equivalent to a system of form (1). The general solution of (1) consists in the fact that under suitable conditions (e.g. if all the partial derivatives of X_i in x_j exist and are continuous) it is equivalent to

(3) $x(t + b) = x(t) + \displaystyle\int_t^{t+b} X(x(t); t)\,dt$

Owing to the linearity of integration, this describes evolution in accordance with a dynamic group $x(t + b) = U_b x(t)$, provided the integrated part is independent of t. That is not always the case, illustrating essentially the fact that classical indeterminism is formally indistinguishable from the case of a system which is one part of (and affected by another part of) a larger deterministic system.

5. CONSERVATION LAWS AND COVARIANCE

Descartes's most basic physical principle was a conservation law: that the same amount of motion is always preserved in the universe. From this principle, via his concept of quantity of motion, he purported to derive the law of inertia and the laws of collision. Though he was not successful in this, the form of his attempt set a great precedent for modern science.

As noted above, an *observable* or physical magnitude or measurable quantity—terms often used interchangeably—has possible *values* and bears certain relations to both states and trajectories. In classical conception the values are real numbers, or numerical constructs, such as vectors with real-number components, and are functionally related to the state. In contemporary physics and its interpretations we come across quantities construed rather differently. One departure is that a quantity capable of taking precise real-number values may be said to have sometimes no precise value, or an 'unsharp value' represented by an interval. In another departure, the possible values of a quantity are related to the state indeterministically, through a probability distribution. Although I focus here on classical ideas, I do still want to explain what quantities are in a way that is classically correct, yet suited to generalization. Suppose m is an observable and E an interval of real numbers.[15] With this supposition we can in all these cases divide the states into two classes. In the simple classical case, the division picks out those states in which m has a specific value which lies in E. In the other cases, it may pick out those states in which m has a sharp or unsharp value represented by some subset of E, or else those for which the corresponding probability distribution gives 1 to E. (The latter could be read as: in these states, m must certainly have value in E, though it may have a sharper value than this indicates.) Formally, we would say that the observable m is represented by a function which maps the intervals of real numbers to sets of states. Proceeding somewhat more intuitively, let us introduce the notation $[m, E]$ and its equivalent readings:

1(a) State x belongs to $[m, E]$,

1(b) $[m, E]$ is the set of states in which m must have value in E,

1(c) m must have value in E if the system to which it pertains has a state in $[m, E]$.

In the terminology of Section 2, these are equivalent elementary statements of the sort I called *state-attributions*, and $[m, E]$ is the *proposition* which they express. We can say a little more now about their logical interconnections, e.g.

2. $[m, E \cap F] = [m, E] \cap [m, F]$.

The simple classical case is distinguished by the fact that, if m must have value in a certain interval, then it must have one of its possible sharp values which lies in that interval. Then we have equivalently:

3a. $[m, E \cup F] = [m, E] \cup [m, F]$ for all intervals E and F.

3b. There is a function m' which maps the state-space into the real numbers and such that x is in $[m, E]$ if and only if $m'(x)$ is in E.

If observable *m* is thus, we can call it a classical state-attribute. In the simple classical case, all observables are classical state-attributes.

A conservation law is a principle to the effect that some quantity *m* is conserved (retains its value, is constant) in time, under certain conditions. We can immediately restate this as: if a possible trajectory (of the relevant sort) enters [*m*, *E*]—for *E* chosen appropriately—it will not leave it again. Equivalently, those state-attributions are invariants of the dynamic (semi-) group:

4. For all *t*: if *x* is in [*m*, *E*], then $U_t(x)$ is in [*m*, *E*].

So that is what a conservation law is: a description, in terms of an observable, of invariants of the dynamic (semi-)group. If 4 holds for all intervals *E*, then the observable *m* is a constant of the motion; the observable represents a conserved quantity. Note that, in the case of a group, the if–then arrow can be replaced by an equivalence: a trajectory which has a point in [*m*, *E*] has all its points, both past and future, in that region.

Formally related to this is the somewhat similar notion of covariance. If states *x* and *y* are related by a symmetry, one expects the values of the observables of systems in those states also to bear some relation to each other. An observable may be called an 'absolute' quantity if it is related to the symmetries of the state-space in the above way. (The corresponding state attributions are invariants of the symmetry group.) In the case of classical mechanics, the symmetries are the Galilean transformations, and acceleration is one of the invariants. Velocity, on the other hand, does not remain invariant under these transformations. There is an intermediate case, in which a quantity varies systematically in a certain way with the transformations. This is the case for velocity with a Euclidean reflection through a plane (multiplication by −1 of the relevant velocity component), but not with a Euclidean translation (the velocity is unaffected). We shall not need to exploit this notion of covariance for quantities, but must note that it is also applied in a certain way to equations that relate quantities. When we write for example that $PV = rT$, we mean that the values of these quantities are always related so as to provide a solution for the corresponding numerical equation. In that case, let us say that the quantities (taken in the right order) *satisfy* the equation. The general concept of covariance for such an equation is then:

An equation EQ is covariant with respect to group G exactly if the following is the case: for all transformations g in G, if $\langle x_1, \ldots, x_n \rangle$ satisfies EQ, so does $\langle g(x_1), \ldots, g(x_n) \rangle$.

When G is the group of evolution operators, this can of course be rewritten as a conservation law: instead of writing $PV = rT$, for example, we can state that the quantity $PV - rT = 0$, in which case covariance of the equation means that a trajectory which enters $[PV - rT, \{0\}]$ does not leave it again. Hence covariance of an equation with respect to a group of symmetries of the state-space is formally analogous to a conservation law.

Proofs and illustrations

Conservation of energy; stability. In 1669 Huyghens proposed that in perfectly elastic collisions not only the momentum is conserved, but also another quantity (soon thereafter called the *vis viva*), namely the sum of the products of mass with velocity squared. Over the next century this idea was extended to other mechanical situations; proportional to the *vis viva* is the *kinetic energy*, in addition to which another factor is to be recognized, the *potential energy*. The form of the extended conservation laws which appeared in Lagrange's *Analytical Mechanics* (1788) may be summed up as follows:

In a *conservative* system the state depends on a certain real vector x, its derivative dx/dt, and two real continuous functions, the *kinetic energy* $KE(x; dx/dt)$, which is non-negative with $KE = 0$ iff $dx/dt = 0$, and the *potential energy* $V(x)$, determined only up to a constant, such that $KE + V$ is constant.

Lagrange's theorem (proved only later by Dirichlet) says that, if in position $X(0)$ the potential energy V is at a minimum, that position is in a *stable equilibrium*. This means that for any $\varepsilon > 0$ there is a $\delta > 0$ such that, if $|x| + |dx/dt|$ is smaller than δ at $t = 0$, it remains smaller than ε for all $t' > 0$. Thus the trajectory is such as to stay in a significant sense 'near' its critical point at $t = 0$.

There is a well-known slogan especially associated with the results of Emmy Noether: 'For every symmetry a conservation law'. In classical mechanics as in quantum mechanics, the cash value of this slogan consists in the theorems which say that, if the Hamiltonian has a certain symmetry, then a certain quantity is conserved. This is really a more beautiful subject in quantum than in classical mechanics, but the main classical results are also striking. In the Hamiltonian formalism, the system with Hamiltonian H is *conservative* exactly if $dH/dt = 0$, an equation which is covariant with respect to Galilean transformations. For a conservative Hamiltonian system, with Hamiltonian function H:

(a)   if H is invariant under space translation, then the total momentum is conserved.

(b)   If H is invariant under time translation, then the total energy is conserved.

(c)   If H is invariant under spatial rotation, the total angular momentum is conserved.

These three facts, (a)–(c), are the principal classical conservation laws. To instantiate the slogan, they should be phrased in the form: if spatial translation (time translation, rotation) is a symmetry of H, then total momentum (total energy, total angular momentum) is conserved (see e.g. Aharoni 1972, 295).

3

Indeterminism and Probability

No doubt the reader will be astonished to find reflections
on the calculus of probabilities in such a volume as this.
What has that calculus to do with physical science?

Henri Poincaré, *Science and Hypothesis* (1905)

PURE chance and indeterminism were discussed by Aristotle,
Lucretius, and Aquinas, but were relegated to illusion and
ignorance in modern times. Yet when they reappeared in
contemporary physics, they found the way prepared—in the
interval, through that very connection with ignorance, they had
become conceptually tractable.

1. PURE INDETERMINISM AND THE MODALITIES

Let us say that the present and past are settled fact, a system X
has evolved up to time $t = now$ along its state-space trajectory
$u(t)$, and many possible futures stretch out before it. Is this
world-picture unintelligible?[1] Not at all; as soon as we have a
precise conception of determinism, we have one *ipso facto* of its
opposite. As we found in the preceding chapter, we can imagine
our ignorance of what the future will bring to be irremediable.
Nature may lack those hidden factors needed to extend the true
history of the phenomena into a deterministic story.

There may still be limits on that future. All possible trajectories which agree with $u(t)$ up to $t = now$ may lead the system
eventually into region S of the state-space. This might be just
the set of states in which observable m has value in E: the ones
which, in our previous symbolism, make the proposition $[m, E]$
true. In that case we say, equivalently:

All possible futures of X are such that $[m, E]$ will be true.
It is necessary that $[m, E]$ will be true.

This reveals several logical facts about necessity and possibility. The first is that the words 'necessary' and 'possible' must behave logically like 'all' and 'some'. For example, just as 'Not all Greeks are virtuous' is the same as 'Some Greeks are not virtuous', so 'It is not necessary for the population to grow exponentially' is the same as 'It is possible that the population will not grow exponentially'. In contemporary logic, necessity is symbolized by a square and possibility by a diamond. Hence we express this point, and several related ones, as follows:

$$\text{Not (All } X \text{ are } Y) \Leftrightarrow (\text{Some } X \text{ are not-}Y)$$
$$\text{not-}(\Box A) \Leftrightarrow \Diamond(\text{not-}A)$$
$$(\text{All } X \text{ are } Y) \longrightarrow X_1 \text{ is } Y$$
$$\Box A \longrightarrow A$$
$$X_1 \text{ is } Y \longrightarrow \text{Some } X \text{ is } Y$$
$$A \longrightarrow \Diamond A$$

This formal analogy serves as a good guide to reasoning with the *modalities*, as possibility, necessity, actuality, and contingency are called.

The second point our discussion reveals is that there is something relative about this sort of necessity and possibility. After all, it is in virtue of having its past represented by the trajectory function u, ranging from the far past to $t = now$, that the statement of necessity for $[m, E]$ in its future is true for system X. It might not have been true if the system had a different past and/or present. What we called the possible futures for X were the trajectories bearing a certain relation to its actual trajectory. Let us introduce some relative modalities as illustration:

Trajectory v is possible$_1$ relative to trajectory u at time t:
exactly if $u(t') = v(t')$ for all $t' \le t$.

A system with actual trajectory u:
necessarily$_1$ has feature F at time t exactly if any and every system with a trajectory v which is possible$_1$ relative to u at t also has feature F;
necessarily$_2$ has a certain state-attribute exactly if any and every system with a trajectory v which is possible$_1$ relative to u has that state-attribute for all $t' \ge t$.

The duals 'possibly$_1$' and 'possibly$_2$' are of course defined dually. Other varieties of possibility and necessity can also be introduced, for the modalities are neither univocal nor absolute.

But let us continue for a moment with these. In the characterization of determinism, a crucial role was played by the time translation symmetry. We can postulate this symmetry also for indeterministic systems in a weaker form.

Postulate: Let g be a time translation. Then:
  (i) u is a possible trajectory exactly if ug is;
  (ii) v is possible$_1$ relative to u at t if and only if vg is possible$_1$ relative to ug at $g(t)$.

Given this postulate, we can introduce an indeterministic counterpart to the dynamic (semi-)group.

For any subset S of the state-space, and $b \geq 0$, define:
$$T_b(S) = \{x : \text{for some possible trajectory } u, \text{ time } t, \text{ and state } y \text{ in } S, u(t) = y \text{ and } u(t + b) = x\};$$
$$T_b^{-1}(S) = \{y : \text{for some possible trajectory } u, \text{ time } t, \text{ and state } x \text{ in } S, u(t) = y \text{ and } u(t + b) = x\}.$$

A given set S may be 'fragmentary', in that it is only a proper part, and not the whole, of $T_b T_b^{-1}(S)$, so that these 'vague' evolution operators do not form a group. But they do form a semi-group with identity, since, for b and c both non-negative, $T_b T_c = T_{b+c}$. (See *Proofs and illustrations*.)

So formulated, we see that what is necessary$_2$ about the future as a whole depends exactly on what is invariant under the dynamic semi-group $\{T_b : b \geq 0\}$. This is a rather interesting, if modest, formal structure. Yet there is no doubt that a state-space model equipped with no more than an indeterministic set of possible trajectories would give us a very weak theory. It is time to look at probability, the grading of the possible.

Proof and illustrations

Let us designate as the *t-cone* of trajectory u the set of possible trajectories which agree with u up to and including t (i.e. the set of those defined as possible$_1$ relative to u at t). Next, let the *b-surface for u at t* be the set of states $v(t + b)$: v in the *t*-cone of u. We can think of this as a time slice or instantaneous snapshot of the t-cone, taken at the later time $t + b$.

1. The b-surface for u at t equals the b-surface for ug at $g(t)$.

For by the postulate of time translation symmetry, v is in the t-cone of u, exactly if vg is in the t-cone of ug and $vg(t + b) = v(g(t + b)) = v(g(t) + b)$, given that g is a time translation.

2. For any k, $T_b(S)$ = the union of all the b-surfaces at $t = k$ of trajectories u such that $u(k)$ is in S.

That $T_b(S)$ includes the described set is obvious by definition. Now suppose that x is in $T_b(S)$ because $u(t') = y$ is in S and $u(t' + b) = x$. Then we can set $g = +t' - k$, and so $ug(k) = u(t') = y$ in S and $ug(k + b) = u(t' + b) = x$. Therefore $T_b(S)$ is also included in the described set.

3. $T_b T_c(S) = T_{b+c}(S)$ for $b, c \geq 0$.

By the preceding, $T_c(S)$ is the union of all the c-surfaces at $t = 0$ of trajectories u such that $u(0)$ is in S. Also, $T_b(T_c(S))$ is the union of all the b-surfaces at c of trajectories v such that $v(c)$ is in $T_c(S)$. That means: the set of states $v(b + c)$ of trajectories v such that $v(0)$ is in S, i.e. in $T_{b+c}(S)$.

At time t, what is necessary$_1$ for a system with actual trajectory u can be summed up as: the system's trajectory lies in the t-cone of u. Let $C_t(u)$ be the union of all b-surfaces of u at t, for all $b \geq 0$. That is just the set of all states after t in the trajectories possible$_1$ relative to u at t. So $C_t(u)$ is invariant under the semi-group $\{T_b : b \geq 0\}$. Also, $C_t(u)$ must be included in $[m, E]$ for that state-attribute to be necessary$_2$ at t for a system with actual trajectory u. Relevant to necessity$_2$, we have the following:

4. $C_t(u) = \cup \{T_b(\{u(t)\}) : b \geq 0\}$

5. $C_t(u)$ is invariant under $\{T_b : b \geq 0\}$

6. S is invariant under $\{T_b : b \geq 0\}$ iff $C_t(u) \subseteq S$ for all u such that $u(t)$ is in S

7. $[m, E]$ is necessary$_2$ at t in a system with actual trajectory u exactly if $C_t(u) \subseteq [m, E]$

as was to be shown for its connection with invariance.

2. PROBABILITY AS MEASURE OF THE POSSIBLE

Reflecting on our possible future, we do gauge some things as more likely to happen than others. In modelling the evolution of a physical system, we would like similarly to grade some continuations of its trajectory beyond a given time t as more or less probable. What exactly that means is not at all transparent, and gives rise to many philosophical problems. For the most part I shall leave these aside here, and concentrate on the structure of probability while relying on our intuitive under-standing.[2]

To have a well-defined probability, one needs to specify to what objects the probability is assigned. This may be a family of *events* that may or may not occur, of *propositions* that may or may not be true, or of (purported) *facts* that may or may not be the case. The family must have at least the simple sort of structure that allows representation by means of sets. From measure theory, probability inherited the concentration on fields (Boolean algebras of sets) for which I shall show the motivation below.

A *field* of sets on a set K is a class F of subsets of K such that:

K and Λ are in F
If A, B are in F, so are $A \cap B$, $A \cup B$, $A - B$

A field F on K is a *Borel field* or *sigma field* on K if in addition F contains the union of any countable class $A_1, \ldots, A_n, \ldots$, of its members.

It follows automatically that, if $A_1, \ldots, A_n, \ldots$ are in a Borel field, so is their intersection.

Intuitively, measure is a generalization of the familiar notions of length, area, volume, etc. Precisely, a *measure* is a non-negative, countably additive function defined on a sigma field, which need not have an upper bound; indeed, some sets may have infinite measure.

Why this insistence on fields; why not have the measure defined on every subset of K? To understand this we must take a brief look at the history of this subject. Measure theory began in the 1890s with some rather tentative and sceptical attempts to use Cantor's set theory in analysis, and only finite additivity was originally noted as a defining requirement for measure. Countable additivity was made part of the definition in Emile Borel's monograph (1898). The measure which Borel defined on the unit interval $[0, 1]$—which we now call *Lebesgue measure*—is not defined for all sets of real numbers in that interval. The definition runs as follows:

> The measure of an *interval* of length s has measure s; a countable union of disjoint sets with measures $s_1, \ldots, s_n, \ldots$ has measure Σs_i; and if $E \subseteq E'$ have measures s_1 and s_2, then $E' - E$ has measure $s_2 - s_1$.

The sets encompassed by these clauses Borel called *measurable*, and the measurable subsets of $[0, 1]$ we now call the *Borel sets* on that interval. It is clear that they form a Borel field, so here we see the origin of our terminology. The Borel subsets of the whole set of real numbers, of the plane, of three-space, and of the n-dimensional space R^n, are defined in the same way. The role of the intervals is there played by the generalized rectangles $\{(x_1, \ldots, x_n): \ a_1 \leq x_1 \leq b_1, \ldots, a_n \leq x_n \leq b_n\}$. The class of Borel sets in that space is the smallest class that contains these and is closed under countable union and set-difference.

Henri Lebesgue posed the explicit problem of defining a measure on all the subsets of a space. There are trivial measures of that sort, for example those which assign 1 to any subset of $[0, 1]$ that includes a given number and 0 to the others. The question is whether the requirement to be everywhere defined would eliminate important or interesting functions. Lebesgue (1962, 236) introduced his Measure Problem as follows: can we have a measure defined on all subsets of a space which satisfies the following conditions:

(i)   There is a set whose measure is not zero.

(ii) Congruent sets have equal measure.

(iii) The measure of the sum (union) of a finite or denumerable infinity of disjoint sets is the sum of the measures of these sets.

Congruence is the relation between sets which can be transformed into each other by symmetries of the space. In the Euclidean case, one of the symmetry transformations is translation, and so we see at once that the measure Lebesgue calls for is already essentially uniquely determined on all the Borel sets:

Lebesgue measure is (up to a multiplicative constant) the unique measure defined on the Borel sets in the n-dimensional space R^n, $n = 1, 2, 3, \ldots$, which is translation-invariant.

For in an n-dimensional space, all the Borel sets can be set-theoretically approximated by choosing the Borel field generated by the family of the generalized rectangles congruent to a single very small generalized cube C^n. The approximation gets progressively better as we take C^n smaller and smaller. But it is obvious that, if the generalized unit cube has measure 1, it can be divided into m^n disjoint generalized cubes which must (by additivity) all receive measure $1/m^n$. Thus, the measure of all little generalized cubes is uniquely determined, and hence, by continuity, is the measure of all the Borel sets.

Lebesgue's Measure Problem was solved in a rather curious way. Around 1905 there was a heated debate about the Axiom of Choice in set theory. Borel was one of its staunchest and most vocal opponents. Lebesgue was more moderate but rejected it as well. In his own proofs, however, Lebesgue seemed to have tacitly relied on the Axiom of Choice. It turned out that, if the Axiom of Choice is true, Lebesgue's 'Measure Problem' has no solution.[3] Therefore the requirement to have measures defined everywhere is unacceptable.

Turning now to probability, this is, since Kolmogorov, standardly defined as a measure with maximum 1. For future reference, it will be useful to have a summary of the defining characteristics in more perspicuous form:

P is a *probability measure* on set K, defined on sigma field F—and $\langle K, F, P \rangle$ is a *probability space*—exactly if

(1) $P(A) \in [0, 1]$ for each member A of F
(2) $P(\Lambda) = 0$, $P(K) = 1$
(3) $P(A \cap B) + P(A \cup B) = P(A) + P(B)$
      for any members A, B of F
(4) $P(A) = \lim_{n \to \infty} P(A_i)$ if

A is the union of the series $A_1 \subseteq \ldots A_n \subseteq \ldots$ of members of F.

Here (3) and (4) have some more familiar equivalents:

(3*a*) $P(A \cup B) = P(A) + P(B)$ if $A \cap B = \Lambda$
(4*a*) $P(\cup A_i) = \Sigma P(A_i)$ if $A_i \cap A_j = \Lambda$ for all $i \neq j$
(4*b*) $P(\cap A_i) = \lim_{n \to \infty} P(A_i)$ if $A_1 \ldots \supseteq \ldots \supseteq A_n \supseteq \ldots$

The property described by (3*a*) is *finite additivity* and that described by (4*a*) is *sigma additivity* or *countable additivity*. It is clear from formulations (4) and (4*b*) that the countable additivity is just finite additivity plus a continuity requirement.

There are also derivative notions, of which the most important is *conditional probability*. What is the probability that a certain atom will decay in the next five minutes, *given that* it is either radium or radon? There is an immediate temptation to read 'given that' as 'if'—perfectly all right when speaking informally, but thoroughly treacherous otherwise. For if we do so, we shall assume the conditional probability of A given B to be the ordinary probability of some third proposition, i.e. *if A then B*. It is almost impossible to construe that assumption so as to make it tenable or even sensible (see Harper *et al.* 1981; Hajek 1989). Discussions of quantum mechanics have sometimes foundered on those shores. Formally, the most perspicuous characterization is:

If $P(A) = 0$, then $P(\ |A)$ is undefined; otherwise, $P(\ |A)$ is the unique probability measure P' such that $P'(A) = 1$ and $P'(B) : P'(C) = P(B) : P(C)$ for all subsets B, C of A in the domain of P.

As immediate corollary, $P(X|A) = P(X \cap A)/P(A)$ when defined. There are 'deeper' characterizations of conditional probability, but this will suffice for us (see van Fraassen 1989, ch. 13). A second derivative notion is the *expectation value* of a quantity: if m has possible values x_1, x_2, ..., with probability $P(m = x_1)$, $P(m = x_2)$, ..., then the expectation value of m equals $\Sigma x_i P(m = x_i)$.

3. SYMMETRY AND A PRIORI PROBABILITY[4]

In the eighteenth and nineteenth centuries, probability was conceived as objective—in the sense that its assignment was unique—and yet not empirical, for the unique assignment was correct on a priori grounds. In the slogan form associated with Laplace, the probability of an event is the ratio of cases favourable to the event to the total number of equipossible cases. The question is obviously how to make this precise. Indeed, while the attempts to do so were in the main abandoned by the turn of our century, there have been noteworthy new attempts thereafter. That the conception kept a hold on the imagination even after its official abandonment is nowhere more evident than in the reactions to Bose–Einstein and Fermi–Dirac statistics in quantum mechanics. It will therefore be worthwhile surveying both the downfall of the classical conception and the reach of its success in circumscribed contexts.

Harking back to the earliest disputes at the time of de Méré, Poincaré (1905) asked: what is the probability that at least one of two thrown dice turns up *six*? The obvious answer is that there are $6 \times 6 = 36$ equipossible cases, of which 11 are favourable. But if we just ask what are the possible (unordered) pairs of numbers that may come up, we count only 21, of which six are favourable. Which division really divides by equipossibility? Since the example is finitary, we may be tempted to answer: it *must* be the most exhaustive relevant division, into 36. The danger now lurks in 'relevant', but in any case, the problem examples are not all finitary. Von Kries, and later Bertrand, asked: what is the probability that a meteorite falling on earth hits Eurasia? If I am given no prior information as to areas, should I count for each continent the number of countries, or should I just count the number of continents?

In this case we may insist that the probability assignment must go by area. But does that mean then that the a priori determination of probability is either (*a*) impossible to carry out in the absence of certain types of relevant information, or (*b*) relative to the statement of the problem? Two further difficulties emerge immediately. Area means Lebesgue measure; but on what basis is that privileged among the infinite diversity of measures of the regions on which it is defined? And secondly, what happens when two problems are not merely equivalent but logically equivalent? The answer—driven home by Bertrand's family of paradoxes published around 1890—was truly disturbing. Ask for instance what is the probability that a given square of area ≤ 4 has an area ≤ 1. Judged by area, using Lebesgue measure for the division into equipossibility, the answer is $\frac{1}{4}$. But judged by edge length, the question is for the probability of length ≤ 1 given length ≤ 2, which by the same type of division equals $\frac{1}{2}$.

There are explanations of why these foundational problems should not have prevented fruitful applications of the classical idea of probability in the preceding history of physics. But they are explanations in terms of the contingently good common sense of the physicists (so classifiable, of course, only in retrospect). Nevertheless, we can also outline the sorts of explicit and tacit *assumptions* under which a certain form of reasoning does lead to correct prior assignments, with an air of resting on no more than considerations of 'indifference' or 'sufficient reason'. The clue is once more symmetry.

We have to list Henri Poincaré, E. T. Jaynes, and Rudolph Carnap among recent writers who believed that the Principle of Indifference could be refined and sophisticated, and thus saved from paradox.[5] Their general idea applies to all apparent ambiguities in the Principle of Indifference: a careful consideration of the exact symmetries of the problem will remove the inconsistency, provided we focus on the symmetry transformations themselves, rather than on the objects transformed.

As first example, let us take the following problem. I have drawn a square S, and inside it a square s of area $\frac{1}{100}$ of S, and have tossed a (negligibly small) coin into the larger square. What is the probability that the coin has landed in the smaller square? It is certainly true that I can define other measures on the relevant space (say, surface of the earth) besides area. Whatever measure m I use, my answer will be the ratio $m(s):m(S)$. But if I am to give a rule which answers this type of problem, it will take as input two squares with diagonal corners at certain coordinates. The specific coordinates are not relevant to the problem: if we give other coordinates related to the former by an isometry, the problem has not been essentially changed. The rule we give must respect the symmetries of the problem. But that means that the measure it uses is translation-invariant, and so (as we saw above) it is Lebesgue measure, i.e. area. Hence the answer for this problem is $\frac{1}{100}$, as of course we should have expected. Note however that this depended on my requiring a *rule*, so that the 'Same problem, same solution!' symmetry requirement could be applied. It depended also on my tacitly using a model in which the squares were drawn in a Euclidean plane, or a sufficiently similar geometric structure to have the required translation symmetry.[6]

But now consider the problem: I have found a table of numerical entries (e.g. amounts of chlorine found in water supplies of various cities) and look at a randomly chosen entry. What is the probability that the first digit after the decimal point is 5? Note that the significant entry is really the original minus its integer part, hence a decimal representing a number in [0, 1]. We don't know to what extent the number was rounded off, so the question is the same as: what is the probability that this decimal is in [0.5, 0.6)? The answer should surely be *in some sense* no different if we had asked the question about 3, but we must not at once assume that the sense is that found in the preceding problem. The sameness must again be that of 'Same problem, same solution!' In this case, imagine we take the table in question, and convert it to a different unit of measurement. This means multiplying by a constant k; suppose it is $\frac{6}{5}$. Then the interval [0.5, 0.6) turns into [0.6, 0.72). Surely the probability that the original number chosen is in the former interval equals the probability that its conversion is in the latter? The rule we should give as solution should assign corresponding probabilities for corresponding questions about the two tables (see Rosenkranz, 1977, 63–8; 1981, 4.2–2 ff. and sect. 4.1*). We conclude that we need a scale-invariant measure M:

$$M(a, b) = M(ka, kb)$$

for any positive number k (invariance under *dilations*).

There is indeed such a measure, and it is unique in the same sense. That is the *log uniform* distribution:

$$M(a, b) = \ln b - \ln a$$

where ln is the natural logarithm. This function is defined on the positive real numbers; is increasing, one-to-one, and non-negative; has value 0 for argument 1; and has the nice properties:

$$\ln(xy) = \ln x + \ln y,$$

so

$$M(ka, kb) = \ln k + \ln b - \ln k - \ln a = M(a, b)$$

$$\ln(x^n) = n \ln x, \qquad \text{so } M(b^m, b^{k+m}) = M(1, b^k)$$

but should be used only for positive quantities, because it moves zero to minus infinity. The first of these equations shows already that M is dilation-invariant. The second shows us what is now regarded as equiprobable:

The intervals (b^n, b^{n+k}) all receive the same value $k \ln b$, so within the appropriate range, the following are series of equiprobable cases:
$(0.1, 1), (1,10), (10,100), \ldots, (10^n, 10^{n+1}), \ldots$
$(0.2, 1), (1, 5), (5, 25), \ldots, (5^n, 5^{n+1}), \ldots$.

and so forth. It is reported that the physicist Frank Benford found empirical support for the log uniform distribution when inspecting the first significant digit in numerical table entries.[7]

This is striking. But we can see again that the answer found was arrived at because we asked for a *rule* that would apply 'in the same way' to a conceptual family of problems which we singled out as 'relevantly' the same. Whether or not this way of solving a given empirical problem succeeds is therefore empirically contingent; nature may or may not have adopted a solution with exactly this sort of generality. For that generality consists in the criteria for what the relevant variables are, and hence what the symmetries of the problem are which that rule must respect. To imagine that it would not be—that empirical predictions could be made a priori, by 'pure thought' analysis—is feasible only on the assumption of some metaphysical scheme such as Leibniz's, in which the symmetries of the problems which God selects for attention determine the structure of reality.

In our century Harold Jeffreys introduced the search for invariant priors into the foundations of statistics; there has been much subsequent work along these lines by others.[8] However impressive the results, we must see them as conditioned in the above fashion.[9] The most general version of the approach in terms of symmetries outlined above is due to Jaynes.[10] This draws on mathematical theorems to guarantee that under certain conditions there exists indeed only one possible probability assignment to a group.

The general pattern of the approach is as follows. First, one selects the correct group of transformations on our set K which should leave the probability measure invariant. Call the group G. Then one finds the correct probability measure p on this group. Next, define

$$(1) \qquad P(A) = p(\{g \text{ in } G: g(x_0) \in A\})$$

where x_0 is a chosen reference point in the set K on which we want our probability defined. If everything has gone well, P is the probability measure 'demanded' by the group. What is required at the very least is that (a) p is a privileged measure on the group; (b) P is invariant under the action of the group; and (c) P is independent of the choice of x_0. Mathematics allows these desiderata to be satisfied; if the group G has some 'nice' properties, and if we require p to be a left Haar measure (which means that $p(S) = p(\{gg' : g' \in S\})$ for any measurable subset S of G and any member g of G) then these desired consequences follow, and p, P are essentially unique.[11] The required conditions may indeed hold in sufficiently nice geometric models. The important points are two. First, there is no valid a priori reasoning from principles of sufficient reason or indifference which dictate our models of the phenomena. But second, once we settle on some way of modelling them, even if still only in a very general way, the symmetries of our model may go a long way to dictate the assignment of probabilities.

4. PERMUTATION SYMMETRY: DE FINETTI'S REPRESENTATION THEOREM[12]

Appeals to sufficient reason or indifference typically attempt to motivate the use of a uniform distribution. On a single toss of a die, with no information about its constitution, such an appeal may introduce the assignment of equal probability to all six faces. But there is another use: to motivate treating different events or factors as mutually independent. Faced with the fact that so few men are both handsome and intelligent, one might be tempted to conclude that either factor inhibits the other. But instead, one could reason that these factors have probabilities p and q, both less than $\frac{1}{2}$, and in the absence of any information about connections between them, one would assign their conjunction the probability pq, which is considerably less than either and certainly smaller than $\frac{1}{4}$. This includes a tacit appeal to indifference to motivate a probability assignment which renders these factors (statistically or stochastically) independent. Even if such an appeal carries no logical weight, the model may be a good one.

Imagine next that, in settling on such a model, we do not have determinate probabilities for the individual event. As an example, take the tossing of a die which may be either fair or biased in several ways. We have in effect a mixture of several hypotheses, according to each of which the tosses are mutually independent. As our probability assignment, we can accordingly choose a *mixture* (*convex combination*) of the corresponding several probability measures. What is the character of this mixture? It still has, like its components, permutation invariance. This is the subject of a deep theorem due to De Finetti, with many interesting applications. We begin with a definition and a lemma.

P is *symmetric* on set S of events if and only if, for all members $E_1, \ldots, E_n$ ($n = 1, 2, 3 \ldots$), the probability of the sequence $[E_1, \ldots, E_n]$ is invariant under permutation of indexes; i.e., $P([E_1, \ldots, E_n]) = P([E_{t1}, \ldots, E_{tn}])$ for any permutation t of $\{1, \ldots, n\}$.

Lemma: If P and P' are symmetric on S, so is their mixture $cP + (1 - c)P'$ with $0 \le c \le 1$.

To illustrate: whether the die is fair or biased, the sequence 123 has the same probability of coming up as 312, namely, the product of the three individual probabilities. This feature is evidently preserved under mixing.

But correlations are not preserved under mixing. (This fact is sometimes called *Simpson's paradox*, and shows up in many interesting places.) For example, whether the die tossed is fair or biased, there is zero correlation between successive tosses. If the die is fair, the probability of *ace* is always $\frac{1}{6}$, if it is a certain biased die it is always $\frac{1}{3}$, and so forth. But on the mixture, successive tosses are not independent. Intuitively: if you see *ace* come up on the first toss, you have some evidence that the die is biased, and that figures in your calculation of probability for the next toss. Thus, if those two are the only possibilities, and $c = 0.5$, then the overall probability of *ace* on the first toss is $\frac{3}{12}$. Similarly, the probability of two *aces* in a row is the average of the squares of $\frac{1}{6}$ and $\frac{1}{3}$, that is, $\frac{5}{72}$. The conditional probability of *ace* on the second toss, given *ace* on the first, is therefore $(\frac{5}{72}) : (\frac{3}{12}) = \frac{5}{18}$, and not the same as the absolute probability $\frac{3}{12}$.

In this example, we constructed a symmetric probability function out of what might be called *chance functions*, which treat the different tosses as independent. De Finetti's representation theorem says (with the qualifications given below) that every symmetric probability function is a mixture of chance functions. This is certainly a representation of the more complex in terms of the simpler. For if I tell you that P is symmetric on S, then to know exactly which function P is on S, you need to know the numbers

$$P(k \text{ occurrences among } n \text{ members of } S) = p_{kn}$$

But given that P is a chance function on S, then to determine P uniquely, you need only the probability p of occurrence of each member of S.

To make this precise, it must be stated in terms of random variables on product spaces. A random variable represents a classical observable when the probability space is a state-space. Suppose F is a Borel field on a set K, and m a measure with domain F. Then a function $f : K \longrightarrow R$ is a *random variable* (also called a *measurable function*) provided the inverse images $f^{-1}(E)$ are in F for all the Borel sets E. That means intuitively that the proposition that variable f takes a value in E is assigned a measure (a probability if the measure is a probability function). Given many such pairs $\langle K_i, F_i \rangle$, we construct the product space $\langle K, F \rangle$ by taking as elements for K the infinite sequences $s : s(i)$ is in K_i for $i = 1, 2, 3, \ldots$ To construct a new field, we take to begin with the sets:

$$A_1 \times \ldots \times A_n = \{s \text{ in } K : s(i) \text{ is in } A_i \text{ for } i = 1, \ldots, n\}$$

where each A_i belongs to corresponding field F_i. These sets are a family closed under countable intersection; F is the smallest Borel field of sets which contains them all. Given the sequence $m : m(i)$ of measures defined on the field F_i, the function

$$m(A_1 \times \ldots \times A_n) = m(1)(A_1) \ldots m(n)(A_n)$$

is uniquely extendible to a measure on F, called a *product measure*. These product measures correspond to what I called chance functions. If for instance the event A_i is the event of *ace* on the ith toss of the chosen die, then m treats that as statistically independent of the event A_j.

There are evidently other measures on F. With the notation only a little adjusted, the above definition of symmetric measures applies here. The random variables on $\langle K, F \rangle$ too are constructible (by limit operations) from those on the component spaces. There is for example for f defined on K the variable f^n which takes on s the value which f takes on $s(n)$, and so forth. By concentrating on these, we can introduce an equivalent concept of permutation symmetry definable for arbitrary probability spaces, ignoring the complexities of product construction. That is the concept of *exchangeability*. Given K, F, and probability measure P defined on F, we define:

$[f < r] = \{x \text{ in } K : f(x) < r\}$, the event that f takes a value $< r$; random variables $f(i)$, $i = 1, \ldots, m$ are *exchangeable* exactly if $P(\cap\{[f(i) < r(i)] : i = 1, \ldots, m\}) = P((\cap\{[f(ti) < r(i)] : i = 1, \ldots, m\})$ for any permutation t and any numbers $r(i)$.

A countable sequence of random variables is called exchangeable if the condition holds for all its finite subsequences. Not every finite sequence of exchangeable random variables is extensible to a countable such sequence.[13]

In this context we can present De Finetti's theorem as follows. Given a sequence of exchangeable random variables $f(i)$, we can represent their individual distributions $P([f(i) < r])$ as all produced by integration on a set of distributions deriving from probability measures for which these variables are independent (that is, measures p such that $p([f(i) < r]$ $\cap [f(j) < q]) = p([f(i) < r])p([f(j) < q]))$.

5. ERGODIC THEORY: UNDERLYING DETERMINISM

In many cases, the evolution of probabilities will be induced by evolution in their domain, and that is the subject to which we now turn.[14]

A physical system has a variety of possible states, which together constitute its state-space. Even if the evolution of its states is deterministic—as in the case of an isolated classical mechanical system—we may have to use probabilities, because the initial states may be uncertain. 'Initial' refers here to an arbitrarily picked time *zero*. So the problem we deal with—the abstract form of the problem of statistical mechanics—is characterized by three factors:

a state-space H;
a probability measure m on H;
a dynamic semi-group $\{U_t : t \geq 0\}$ of evolution operators for H.

This dynamic semi-group we encountered in the discussion of determinism in the preceding chapter. Here I shall not consider only the case of bi-determinism (dynamic group) but allow that a system could reach a given end-state from more than one possible initial state.

The meaning of the probability $m(A)$ is that, at an arbitrarily chosen instant, the probability that the state is in part A of the state-space equals $m(A)$. Of course this is not consistent unless that is also the probability that five minutes ago it was in a state which could evolve into one in A by a process lasting five minutes. This gives us the basic condition relating measure and evolution operators. If T is any operator acting on H, define $T^{-1}(A) = \{x : T(x) \in A\}$. In other words, $T^{-1}(A)$ is the set of points that the elements of A could come from by transformation T. Therefore the condition in question can be written this way:

Basic Condition: For each member T of the dynamic semi-group, and each set A in the domain of m, $m(A) = m(T^{-1}(A))$.

Notice this requires automatically that if the probability m is defined for A, then $T^{-1}(A)$ also belongs to its domain of definition.[15]

For purposes of illustration, let us define a simple special case. Let the transition times be minutes or seconds or other discrete units; the dynamic semi-group is $\{U_t : t = 0, 1, 2, \ldots\}$. Now U_0 is just the identity, and $U_2 = U_{1+1} = U_1 U_1$. In general, if we define

$$T = U_1$$
$$T^0 = U_0$$
$$T^{n+1} = TT^n$$
$$T^{-n} = (T^n)^{-1}$$

we have $U_n = T^n$ and $\{U_t : t = 0, 1, 2, \ldots\} = \{T^n : n = 0, 1, 2, \ldots\}$ or $\{T^n\}$ for short. Call this a *discrete* case. Any real case can be 'approximated' by discrete cases, by taking the unit successively smaller: minute, second, millisecond, etc.

Henri Poincaré approached this subject in terms of the question of eternal recurrence. Drawn to popular attention by such writers as Heine and Nietzsche, the question of whether the world returns eternally to its same states over and over again was of some interest (see van Fraassen 1985b, ch. III, sect. 1). Nietzsche himself sketched a rudimentary proof in *The Will to Power* that in a simple enough universe (as modelled by a set of dice) it is so. Our actual universe isn't that simple, but in the nineteenth century it was still assumed to be deterministic. And Poincaré was able to give an almost completely positive answer to the question for a world conceived as a dynamic system of the sort we have now described.

Define x in H to be a *recurrence point* of set A exactly if $U_t(x)$ is in A for some $t > 0$ and an *eternal recurrence* point of A if $U_t(x)$ is in A for infinitely many intervals t.

(*Poincaré*) If $m(A)$ is positive, then almost every point in A is an eternal recurrence point of A.

'Almost every' has the usual probabilistic meaning: the probability measure of the set of points that are not recurrence points of A, equals zero. I shall omit the proof, which may be found in the references.[16]

Of course, this result is compatible with such a return

occurring hardly ever at all. That is the problem with the infinite long run. For example, if you look through the numbers 1, 2, 3, ... it will happen infinitely many times that you will come across an even number. In fact, in a straightforward sense, you find an even one exactly half the time. You will also come across powers of 10 infinitely many times. But after an initial flurry, it happens less and less often, and indeed, the fraction of powers of 10, among the inspected numbers, gets smaller and smaller as you go along (1/10, 2/100, 3/1000, ..., which converge to *zero*). In the same sense, therefore, it happens hardly ever at all.

The next problem of ergodic theory is therefore to quantify Poincaré's result, and prove exactly what fraction of its time a system will spend back in set A, once it has passed through there. This is the problem solved by Birkhoff's ergodic theorem. In statistical mechanics, Boltzmann had in effect speculated that, for every possible dynamic system, the answer would be determined in the same way: the amount of time spent in set A is proportional to the measure of A (*ergodic hypothesis*). This is certainly not true in the abstract, and the ergodic theorems which were eventually proved in the twentieth century require special assumptions.

I will first explain and state Birkhoff's general theorem pertaining to the ergodic hypothesis. It makes no special assumptions. Then I will define the subclass of ergodic systems and show how the ergodic hypothesis is true of them. Because of the nice relationship between dynamic systems and the approximating discrete cases they contain, I shall now focus entirely on such a discrete case.

Let us first define the *relative frequency* with which a system with initial state x returns to a state in A in the series of times $t = 0, 1, 2, \ldots$

$$\bar{A}(x) = \lim_{n \to \infty} (1/n) \sum_{k=0}^{n-1} I_A(Tk_x)$$

Here I_A is the characteristic function of A; that is, $I_A(x) = 1$ if x is in A and $= 0$ otherwise. As illustration, let us take as trajectory the series 1, 2, 3, ... and A = set of even integers, B = set of positive integral powers of 10. Then $\bar{A}(1)$ is the limit of the series 0/1, 1/2, 1/3, 2/4, 2/5, 3/6, ..., and this limit is 1/2.

The number $\bar{B}(1)$ is the limit of $0/1$, ..., $0/9$, $1/10$, ..., $1/100$, $2/100$, ..., $2/999$, $3/1000$, ..., and this limit is *zero*. Limits do not always exist, of course, but only if the sequence converges.

> (*Birkhoff*) For a dynamic system with probability measure m
> and dynamic semi-group $\{T^n : n = 0,\ 1,\ 2,\ \ldots\}$
> and measurable set A,
> (1) for almost all x, $\bar{A}(x)$ exists;
> (2) $\bar{A}(T^nx) = \bar{A}(x)$ if that exists;
> (3) the expectation value of the function $\bar{A}(\cdot)$
> equals $m(A)$.

The set of points $\{T^k x\}$ is called the (forward) *trajectory* of point x. Call a set A *invariant* exactly if it contains the whole trajectory of each point in it. In other words, if a state lies in set A, evolution in time from this state leads only to more states in A.

Secondly, define the dynamic system to be *ergodic* exactly if all its invariant sets have measure 0 or 1. This means that, if there is any positive probability at all that the system's state could be outside, as well as inside, set A, then it must be possible to have some state in A eventually evolve into one outside A. We can now state three results for ergodic systems:

> *Theorem*: If $\langle H, m, \{T^n\}\rangle$ is ergodic and $m(A) \geq 0$, then:
> (1) almost every trajectory enters A;
> (2) the forward trajectory of A 'fills' the space,
> i.e. has measure 1;
> (3) for almost all x, $\bar{A}(x) = m(A)$.

The important part is of course the third, which says that the ergodic hypothesis is true for ergodic systems.

6. A CLASSICAL VERSION OF SCHROEDINGER'S EQUATION

How do probabilities evolve if we do not assume an underlying determinism? As a general problem, that is not well posed, but we can easily think of examples in which we take there to be a lawful evolution of probability without first imagining an underlying mechanism. Examples might be: what is the probability that a given radium atom has decayed by time t, the probability that a customer in a bank has been served in the first n hours of waiting, the probability at time t that Vesuvius will erupt in the next five minutes?

We can approach this by taking the states of the system to be themselves probability measures, and describing change, as before, as a transformation of state. I will discuss a finite case. The set of simplest possibilities will be a disjoint finite set $E_1, \ldots, E_n$. The probabilistic states of the system will be the probability measures on the field generated by these. This field is itself finite, consisting of all the unions of simple events (with $K = E_1 \cup \ldots \cup E_n$). Hence such a probability measure P is completely specified by giving $p_1 = P(E_1), \ldots, p_n = P(E_n)$. So P is adequately represented by the vector $\langle p_1, \ldots, p_n \rangle$.

It will make our discussion still simpler if we switch to odds, i.e. probability ratios. Since we know that $P(E_1) + \ldots + P(E_n) = 1$, we would also have specified P completely if we were just given certain ratios, such as p_1/p_2, p_1/p_3, ..., p_1/p_m. To give an example, suppose the probabilities that the television or radio, or neither, is on (when at most one can play at any given time) initially (at $t = 0$) equal 0.5, 0.3, 0.2. That information is conveyed equally well by saying that the odds are $5:3:2$. So this probability function can be represented by the *probability vector* $\langle 0.5, 0.3, 0.2 \rangle$ or by the *odds vector* $\langle 5, 3, 2 \rangle$. And of course, to say that the odds are $50:30:20$ is also the same. One form of evolution that is easy to imagine is that the odds are just raised to the power of the time elapsed:

Time	Odds			Probabilities		
1	5	3	2	0.5	0.3	0.2
2	25	9	4	0.66	0.24	0.1
3	125	27	8	0.78	0.17	0.05

This evolution of probabilities is itself not deterministic, because the initial probabilities at $t = 0$ recur at $t = 1$, but not at $t = 2$. In fact, there is a strange discontinuity at time zero. So though imaginable, this is not plausible. To find something more lawlike, we need to pay attention to the symmetries of the state-space we have chosen here.

To introduce some notation, say that $\bar{x} = \langle x_1, \ldots, x_n \rangle$ is an (*n*-ary) *odds vector* iff $x_1, \ldots, x_n \geq 0$. An odds vector is a *probability vector* iff its components sum to 1.

(1)   Odds vectors $\bar{x}$ and $\bar{y}$ are *equivalent* iff there is a positive real number k such that $\bar{x} = k\bar{y}$ (i.e. $x_i = ky_i$ for $i = 1, \ldots, n$).

We have now embedded the set of probabilistic states of the system—our basic concern—in a large set, the space of odds vectors. Obviously, we have not added at all to our descriptive capacity—an odds vector is just another way to represent probabilities. We have added ease. The reason is that the requirement that the numbers add up to 1, for probabilities, always means that you write one number more than you need: $x_n = 1 - x_1 - \ldots - x_{n-1}$ after all. Yet that number conveys lots of information; for example, that the last possibility E_n is 17 times more probable than the first E_1. By writing down the odds, there is no redundancy, and the arbitrariness of setting total probability equal to 1 has been removed.

What kind of structure does this space of odds vectors have? We have already seen the relation of *equivalence*, which we could have called total congruence. There are also partial congruences; for example,

the odds $E_1 : E_2$ are the same for both probability measures P_1 and P_2;

and, more generally, there are partial comparisons, like

the odds $E_1 : E_2$ according to P_1 are 17 times the odds $E_1 : E_2$ according to P_2.

Let us codify this as follows:

(2)   The equation $(x_i/x_j) = k(y_i/y_j)$ is an *odds comparison* of vector $\bar{x}$ with vector $\bar{y}$.

The equation is not defined, and the odds comparison does not exist if either denominator equals *zero*.

We know all there is to be known about the relations between two odds vectors $\bar{x}$ and $\bar{y}$ if we know the set of well-defined numbers $(x_i/x_j) : (y_i/y_j)$. And it is easy to check that equivalence can be explicated as follows:

(3)   Odds vectors $\bar{x}$ and $\bar{y}$ are equivalent exactly if all their

odds comparisons with any third odds vector $\bar{z}$ are the same.

The symmetries of the space of odds vectors are the transformations which leave the structure so defined entirely intact.

(4) A *symmetry* (of the space of odds vectors) is a one-to-one transformation U such that the odds comparisons of $U\bar{x}$ and $U\bar{y}$ are the same as those of $\bar{x}$ and $\bar{y}$, for all odds vectors $\bar{x}$ and $\bar{y}$.

Of course, the definition applies equally to probability vectors.

Theorem 1: A transformation U of the set of odds vectors is a symmetry if and only if it is positive and linear; that is, iff there are positive numbers $u_1, \ldots, u_n, u_n$ such that, for all $\bar{x}$, $U\bar{x}$ is equivalent to $\langle u_1 x_1, \ldots, u_n x_n \rangle$.

For proof see below.

Without loss of generality, I shall now take as symmetries the transformations U such that there are positive numbers $u_1, \ldots, u_n$, such that $U\bar{x}$ is not only equivalent to, but *equal* to, $\langle u_1 x_1, \ldots, u_n x_n \rangle$ for all $\bar{x}$.

With respect to geometry we discussed uniform motions, which were changes over time that preserved all symmetries. We already saw there and in the discussion of determinism that the motion must be described by at least a semi-group of 'evolution operators'. If $x(t)$ is the probabilistic state of t, then it yields at $t + d$ the state $x(t + d) = U_d x(t)$ and $U_{d+e} = U_d U_e$, with $U_0 =$ identity, as we saw at that time. So let us repeat the definition for our new context:

(5) A *uniform motion* of the odds vectors is a one-parameter semi-group $\{U_z : z \le 0\}$ such that each U_z is a symmetry.

Theorem 2: If $\{U_z : z \le 0\}$ is a uniform motion, then there are positive numbers $k(1), \ldots, k(n)$ such that, for each time interval z, $U_z \bar{x} = \langle e^{k(1)z} x_1, \ldots, e^{k(n)z} x_n \rangle$.

In matrix formalism, $U_z \bar{x} = e^{Kz} \bar{x}$; again, see below for proof. Readers familiar with various parts of science will recognize a familiar form here—that of Lambert's law of light absorption, the law of radioactive decay, the calculation of continuously compounded interest, and, most interesting of all, Schroedinger's equation in quantum mechanics.

In the example we had before, let us set $k(i) = i$ for $i = 1, 2, 3$. Then the evolution by uniform motion takes the form (with time $t = 0$ now not at all a singularity):

Time	Odds			Probabilities		
$t + z$	$5e^z$	$3e^{2z}$	$2e^{3z}$			
$t + 0$	5	3	2	0.5	0.3	0.2
$t + 0.5$	8.2	8.1	8.9	0.33	0.32	0.35
$t + 1$	13.6	22.1	40.2	0.18	0.29	0.53
$t + 2$	37	164	807	0.04	0.16	0.80

Proofs and illustrations

Theorem 1. If U is positive and linear, then it is a symmetry. To prove the other half, let $\bar{1}$ be the vector all of whose components equal 1, and let $U\bar{1} = \langle u_1, \ldots, u_n \rangle$. Then if U is a symmetry and $\bar{x}$ any vector, and we write $\bar{x}' = U\bar{x}$, we have

$$(x_i'/x_j') = k(u_i/u_j) \text{ iff } (x_i/x_j) = k$$

and also

$$(x_i'/x_j') = (x_i/x_j)(u_i/u_j) = (u_i x_i/u_j x_j).$$

Thus $U\bar{x}$ must be equivalent to $\langle u_1 x_1, \ldots, u_n x_n \rangle$.

We can sum up the idea of the proof this way: the odds comparison of $U\bar{x}$ with $\bar{x}$ is the same for all $\bar{x}$. And equivalently, we can say: the symmetries are the one-to-one linear transformations.

Theorem 2. First recall the preceding theorem, and notice that, for each index z, U is positive and linear, so we can write

$$U_z(x) = \langle u_1(z)x_1, \ldots, u_n(z)x_n \rangle$$

We now focus on the functions $u_i(z)$. Because we deal with a semi-group, we have

$$u_i(z + w) = u_i(z)u_i(w)$$

Switching to the logarithmic formulation $f_i(z) = \ln u_i(z)$, we have an additive function, so by a lemma from calculus, $f_i(z) = k_i z + b$. But since U_0 is the identity, we have $f_i(0) = 0$; thus $f_i(z) = k_i z$. Turning back to u_i, we have $\ln u_i(z) = k_i z$, so $u_i(z) = e^{k(i)z}$. This ends the proof.

7. HOLISM: INDETERMINISM IN COMPOUND SYSTEMS

The most striking feature of quantum theory is perhaps its *holism*: when a system is complex, the states of the parts do not determine what the state of the whole will be. This is not unrelated to its probabilistic character. I shall first explain the holistic character already found in classical probability theory, and then discuss abstract modelling of interaction. The latter discussion applies indirectly to measurement and will help to clear up an apparent circularity concerning the foundations of quantum theory.

Suppose that we have two systems X and Y, each with a space of possible states H_X and H_Y. On each of these state-spaces we assume specified a field of subsets, F_X and F_Y, on which probabilities are to be defined. For example, if A is in F_X, then one significant proposition about X is: *X has a state in A*. The members of H_X are states of X; let us call probability functions defined on F_X the *statistical states* of X—and similarly for Y.

We can now regard X and Y as together forming one system $X + Y$, and indeed, this will be especially apt if the two are somehow related to each other or interacting. We are here in a purely classical context; there it is natural to make up a *product space* as in the discussion of De Finetti's theorem above:

$$H_{XY} = (H_X \times H_Y) = \{\langle x, y \rangle : x \in H_X \text{ and } y \in H_Y\} \text{ with field}$$
$$F_{XY} \text{ generated by } \{A \times B : A \in F_X \text{ and } B \in F_Y\}$$

What are the statistical states of $(X + Y)$? The fact is that for each choice of statistical states p_X and p_Y for X and Y, there are *many* corresponding statistical states for $X + Y$. Each of these will incorporate some form of dependence or independence between the two component systems. The one in which X and Y appear as totally independent is the product measure:

$$p_{X+Y} (A \times B) = p_X(A)p_Y(B)$$

But suppose that X and Y are so coupled that X can be in any of its possible states, but when it is in state x then Y is in state y_X. This is the case of total or strict dependence:

$$p_{X+Y}(A \times B) = p_X(\{x \in A : y_x \in B\})$$

There are many cases between, of less strict correlation.

On the other hand, the state of the whole does determine unique states for the parts, called the *marginals* (marginal probabilities or probability distributions). Suppose p is a statistical state for the compound system $X + Y$; let us write then:

$$\#p(A) = p(A \times H_Y)$$
$$p\#(B) = p(H_X \times B)$$

and the holism consists in the fact that p determines, but is not itself determined by, these two reductions $\#p$ and $p\#$.

When we are dealing with three or more components, there is a marginal probability for every finite subset thereof. For example, $(X + Y + Z)$ has state-space $H_X \times H_Y \times H_Z$, while $(X + Z)$ has state-space $H_X \times H_Z$. A statistical state p_{XYZ} for the whole has marginals p_X, p_{XY}, p_{XZ} and so forth. These are *consistent* in the sense that $p_X = \#p_{XZ}$ and $p_Z = (p_{XZ})\#$ and so on. Conversely, if we have a collection of statistical states for subsystems of a given compound system, which are consistent in this sense, then there exists a statistical state for the whole of which they are all marginals (see e.g. Loève 1955, pt. I, ch. 1, sects. 4.1–4.3).

Turning now to time and interaction, consider the evolution of a compound system $X + Y$ in the space $H_X \times H_Y$. If that evolution is deterministic, we have a dynamic (semi-)group

$$U_t(\langle x, y \rangle) = \langle x_t, y_t \rangle$$

but this need not mean that each component is a deterministic system in itself. The reason is obviously that the connection with Y may for example constrain the evolution of X. We have then this pattern:

$$U_t(\langle x, y \rangle) = \langle U^X_{t,y}(x), U^Y_{t,x}(y) \rangle$$

Looking at the component system X by itself, for instance, this shows us the indeterministic transition law:

$$x \xrightarrow{\ t\ } \{U^{X}_{t,y}(x) : y \in H_Y\}$$

where H_Y now serves only for bookkeeping, as an index set for the states in which X may land time t after initial state x. There is really no information conveyed by that indexing.

The total system may also be indeterministic, in which case we only have a map

$$\langle x, y \rangle \xrightarrow{\ t\ } \{\langle x^t_i, y^t_i \rangle : i \in I\}$$

the image being the set of states in which the compound system may end time t after that initial state. This does induce a similar indeterministic transition for the component X:

$$x \xrightarrow{\ t\ } \{x^t_i : i \in I\}$$

We may however do better at this point with probabilities. Suppose compound system $X + Y$ does not evolve deterministically in its proper state-space, but does do so in its statistical state-space (in the fashion of either of the two preceding sections). This is the space

$$S_{X+Y} = \text{the set of probability measures defined on } F_{XY}$$

Suppose the evolution has the form

$$p \xrightarrow{\ t\ } U_t(p)$$

Given statistical state p of the whole, components X and Y have marginal states $\#p$ and $p\#$ respectively. Taken by themselves, these do not evolve deterministically:

$$p_X \xrightarrow{\ t\ } \{\#U_t(p) : p_X = \#p, p \in S_{X+Y}\}$$

I have kept this discussion in the classical context of the preceding section—but that was not necessary. Let us consider an abstract compound system with its components:

(a) state-space H_{XY}, H_X, H_Y
(b) fields of events F_{XY}, F_X, F_Y
(c) a map $h_X : H_{XY} \longrightarrow H_X$ such that $h_X^{-1}(A) \in F_{XY}$ for each A in F_X
(d) a map $h_Y : H_{XY} \longrightarrow H_Y$ such that $h_Y^{-1}(B) \in F_{XY}$ for each B in F_Y
(e) $h_X^{-1}(H_X) = h_Y^{-1}(H_Y) = H_{XY}$

Intuitively, $h_X^{-1}(\{x\})$ are the states which the compound system could have, given that X is in state x. Now we define marginals in an abstract sense.

Given statistical state p on $\langle H_{XY}, F_{XY} \rangle$, define

$$\#p(A) = p(h_X^{-1}(A))$$

$$p\#(B) = p(h_Y^{-1}(B))$$

It is easily verified that $h_X^{-1}(A)$ and $h_X^{-1}(A')$ are disjoint if A and A' are, and that $h_X^{-1}(A \cap A') = h_X^{-1}(A) \cap h_X^{-1}(A')$, and also that $\#p$ and $p\#$ are indeed probability functions. The evolution of the compound system now induces evolutions for the state of the components in just the same way as we have discussed above, except that the indexing may have a less neat looking form—but that was not informative anyway. Yet we are not restricted now to such a classically produced construction for the compound state-space. The generalization is entirely natural, and it will be well to remember how naturally this holism—such a puzzling feature of quantum mechanical states—emerges from classical probability theory.

PART II
How the Phenomena Demand Quantum Theory

To a traditional mind, quantum theory is perplexing—and we all start with traditional minds. Need our physics be so peculiar? What do the phenomena demand? The early quantum theory startlingly depicted matter as not continuous, and energy as transmitted only in quanta. But more curious features would appear: correlations between separate, distant events, looking almost like telepathy in nature. Although the quantum theory clearly developed in response to experimental results, the suspicion could not help but arise that physics could have succeeded in a more traditional format. Some historical studies have lately reinforced such suspicions: cultural pressures perhaps inclined physicists to greater sympathy than their predecessors had with indeterminism and holism.

After a half-century of 'no hidden variable' theorems, proofs by John Bell clearly showed that some of the most disturbing features reside in the phenomena themselves. They are not contributed by theory, for any theory which fits the phenomena must accommodate them. It is important to take this up before we look at quantum *theory*, so as not to disguise or hide the solid empirical basis. Of course, the empirical data still underdetermine theory; they do not establish the truth of the theory. But they rule out a whole range of possible theories, of the sort which certain traditional intuitions about causality would require. The next step, to develop one particular way of modelling those phenomena, is not uniquely determined. To prepare for quantum theory itself, we will also inquire into the general form which this next step can take.

4

The Empirical Basis of Quantum Theory

To begin, I will try to show that the appearance of paradox cannot be dismissed as an artefact of quantum theory, nor as merely a matter of indeterminism. Before we look at quantum theory, I shall describe such strange phenomena, and show that they cannot possibly be fitted into traditional ('causal') models.[1] I shall also discuss the extent to which actual experiments indicate or establish that such phenomena indeed exist. The two philosophically significant points are (*a*) that the world *can* harbour such phenomena; and (*b*) that, if it does, then physical theory *must* provide us with models which are not deterministic, or 'causal' in a certain wider sense, and which also cannot be embedded in ones that are. In that case, in other words, the phenomena themselves *demand* the main peculiarities of the new theory.

1. THREAT OF INDETERMINISM

Radioactive decay and the photoelectric effect were studied and recognized at the beginning of our century. The models proposed were ones in which matter is particulate and energy transmitted only in quanta. Let us begin by looking at the threat immediately posed by them for determinism. How conservative could a theory remain, while confronting that threat?

The most familiar law of radioactive decay is just this: the half-life of radium is 1600 years.[2] Roughly: a sample of radium, left alone for 1600 years, will have partially decayed into radon, with exactly half the amount of radium remaining. Now you can see at once that this so-called law must, strictly speaking, be false. For what if the sample consists of an odd number of radium atoms?

The same point emerges if we look at the fully fledged phenomenological law of radioactive decay. Let the amount of material be $N(t)$ at time t, with $N(0)$ the initial amount at time $t = 0$. Then

(1) $$N(t) = N(0)e^{-At}$$

where A is the decay constant characterizing the substance. But this fraction e^{-At} ranges over all the positive numbers from 1 to 0, many of them irrational. Now, whether the initial amount was an odd or an even number of atoms, an irrational fraction of them is not a number of atoms at all. Think especially of the case of *one* radium atom: what does this fraction signify about when it will decay?

The answer given—and how could there be another?—is that e^{-At} is the probability that a given, single radium atom will remain stable for time interval t. Then formula (1) defines the expectation value of the quantity of remaining radium, in the sense defined in probability theory. Now determinism is threatened because this law has taken on a statistical form. The threat, however, looks like one that might in principle be countered. Could it not be that each individual radium atom has some hidden feature which determines its exact time of decay? Our ignorance then introduces the probability—it is the probability of remaining stable, *given only* that it is a radium atom in which the hidden parameter has one of its many possible values.

Nothing counts against that speculation at this point. Consider next the case of polarized light and polarization filters (such as the lenses of some sunglasses). A vertically polarized light-wave encounters such a filter astride its path. The filter may be oriented in various directions. If it is vertical, it lets all this light through; if horizontal, none; and if oriented at intermediate angle θ it lets through proportion $\cos^2 \theta$. (This produces the odd phenomenon that two filters placed together, at right angles to each other, let no light through at all—but do let light through if we place a third filter between them at an angle!) In the new models, light comes in photons, and we have the same problem as before. The function $\cos^2 \theta$ will often enough turn whole numbers into fractions, even irrational ones. The result cannot be a number of photons. Thus we answer again that, for an individual photon, the probability of passing through equals $\cos^2 \theta$.

What would the idea of a hidden determinism look like here? Each photon comes equipped with a certain hidden parameter, which determines for each direction whether or not it will pass through a filter with that orientation. But then, it cannot be that the filter merely lets the photon through. For if we place a vertical filter behind the one with orientation θ, it diminishes the beam again to a proportion of $\cos^2 \theta$. So many of the photons had lost the characteristic of being certain to pass through a vertical filter. Therefore this hidden parameter would be interacting with the filter, and changed by it. Still, we cannot yet rule out the possibility of such a deterministic model. So far, indeterminism is only a threat—the probabilities *might* be merely superficial.

2. CAUSALITY IN AN INDETERMINISTIC WORLD

How much could be saved of the modern world-picture if determinism were given up? Physicists certainly had no logical guarantee that these probabilities for atomic decay and photon absorption had to be ultimate and irreducible. Yet they were receptive to the idea of indeterminism in the world, and were daily less impressed with the traditional conviction that a theory is not complete until it is deterministic. In the preceding Part we saw that the philosopher Hans Reichenbach tried to analyse what conceptual departures were possible. Later he went beyond the argument I noted above, that the question of indeterminism is an empirical one, and attempted to show how causality still makes sense in a universe of chance.

Reichenbach had worked extensively on the foundations of relativity theory, and had shown how it could be conceived as describing aspects of the causal structure of the world. Causality was therefore, in his view, the key to the understanding of this novel and revolutionary theory. We must add at once, however, that the aspects of causality which entered this analysis of physics were very minimal—certainly not adding up to what any philosopher would regard as the entire concept of causality. The structure of space–time, according to Reichenbach, derives from the relational structure constituted by connections between events through transmission of energy and matter (signals and transport of bodies). These are at best a very special example of what had traditionally been understood as causation.

Faced with the still more radical divergences from the classical world-picture in atomic physics, Reichenbach turned again to the notion of causality. Classically, causality and determinism had been inseparable notions. But in Lucretius' world of atoms, all connections between events were through transport of matter—travelling atoms—and yet this world was not deterministic: the atoms were subject to a slight, unpredictable swerve. From Reichenbach's minimalist perspective, we have here a world with a definite causal order, although an indeterministic one. Could the quantum-mechanical world not be understood in some such way?

But when is a theory to be regarded as complete, in this case? Completeness cannot require a deterministic account if an exhaustive survey of all *real* factors does not provide the wherewithal for such an account. Individual events may be spontaneous, may happen for no reason at all—that is what it means for the world to be indeterministic. But then we need a new criterion to judge theories—what must they provide, if not that? Reichenbach's answer was: *a causal explanation* not for individual events but *for every correlation*.[3] An indeterministic universe may still display pervasive and large-scale regularities, in the form of statistical correlations, and these must all be accounted for in a manner which is not trivial or vacuous. His great paradigm in the theory of relativity had been the principle of action by contact: all influences are transmitted in such a way as to fit into a recognizable order of continuous processes. Something approaching this ideal must also be found in quantum mechanics, he thought; our task is only to analyse it.

Clearly, some correlations can be explained, as it were, logically. A coin comes up tails if and only if it does not come up heads. But that is because tails and heads are just two distinct values of a single parameter; once we see that, we do not regard it as a coincidence crying out for explanation. This is a special case, and presumably a limiting case; only an extreme rationalist could imagine that all correlations could eventually be reduced to such logical identities. The correlation between smoking and lung cancer (rate of cigarette consumption at *t and* presence or absence of lung cancer) is not so perfect. But in this case too the two factors, call them A and B, are correlated in the statistical sense that $P(A|B) \neq P(A)$. This is a symmetric relationship in A and B if neither has probability zero; it can be written as $P(A \text{ and } B) \neq P(A)P(B)$. But we explain it by tracing it back to a *common cause*: in this case, the smoking history of the patient.

You can imagine this easily enough. The doctor says that his unknown next patient has, say, a 10 per cent probability of having lung cancer. As the patient comes in he is lighting one cigarette with the butt of another. The doctor raises the probability to 15 per cent. But if the doctor had known the patient's smoking history, the first announced probability would not have been changed upon seeing the patient's present behaviour. This is a familiar story. The patient's smoking history contributed to his present smoking habits and also to his present state of health. More: there is no independent contribution to his *presently* having cancer or not by his *present* smoking—the latter can only be an indicator of the history which did make a contribution. Thus we have for this correlation between A and B a *common cause* in factor C exactly if:

(1) C precedes A and B
(2) $P(A|C) > P(A|not\ C)$ and $P(B|C) > P(B|not\ C)$
(3) $P(A \text{ and } B|C) = P(A|C)P(B|C)$

Condition (3) is the crucial condition of *causality*: preceding factor C not only raises the probabilities of A and B, but also *screens off A from B*. Note that if we were to omit (1), then (2) and (3) could easily be satisfied by setting $C = (A\&B)$. But if we omitted (1), the temporal precedence, we couldn't very well speak of a *causal* explanation at all. However, it could be a 'logical' explanation of the sort indicated above.

There is no attempt here to explicate the notion of causality as such. That may be very intricate; its intricacies may also have nothing to do with physics. But physical science does often satisfy us with an account of an observed correlation in the way described, and this has generally been given as a pervasive example of causal explanation. If that could always be done, we might well be persuaded to say that we still live in a true causal order. If on the other hand even these conditions must be violated by *any* account of the phenomena, we should surely admit a radical breach in the modern world-picture.

By a *causal model* of the correlation of A and B, we shall understand a model in which that correlation is traceable to a common cause in the above sense.[4] As we shall see below, we need to look very carefully at the constraint indicated by (1), which precludes for example a statistical dependency of C on the actions of the experimenter at the time of occurrence of A and B.

Reichenbach's view of science, as recently elaborated, amended, and defended by Wesley Salmon (1984), could be called *causal realism*. For this view holds that science is to provide us with models of phenomena, subject to two conditions:

(a)   The putative physical factors represented in these models must all correspond to elements of reality.
(b)   These models must be causal models for all the represented correlations in the phenomena.

Perhaps the demand is actually stronger: perhaps the models must be causal models as well for all correlations seen inside them, even among factors which do not represent observable characteristics. Whether this was intended will not affect our discussion. We can also qualify the demand a little: the above conditions are to be met by a theory which is ideal, and so the models of a young or developing theory are required only to be embeddable, in principle, in models which satisfy (a) and (b). The completeness criterion has been weakened from determinism to causality in a sense that could be found in an indeterministic world as well. If quantum theory meets Reichenbach's demands, then it does still describe a causal order, although not a deterministic order.

Modest though the requirements are in comparison with the traditions against which Reichenbach rebelled, the view is not tenable. In the quantum-mechanical world, not only determinism but also causality is lost, even in this minimal sense.

Proofs and illustrations

Before turning to a rigorous and general argument to this effect, I can illustrate the crucial bite in Reichenbach's demands with a simple if fanciful example. Imagine two chameleons in separate cages. These chameleons, a special breed, can be only red or green, and each changes colour in apparently random fashion. But we notice a perfect correlation: each is red when and only when the other is green. Now postulate a common cause; this means a factor C which can be absent or present (a moment beforehand) and such that the information it provides about whether or not either chameleon is red, say, cannot be augmented by information about the other chameleon.

But now we can reason as follows. We already know that the probability of the first chameleon being red if the second is red equals zero. What then of P(first red$|C$ and second red)? Well, either it is not well defined, or it equals 0. In the first case the antecedent has probability 0, and in the second case we can use our assumption that common cause C 'screens off' the first fact from the second. Hence

either $P(C$ and second red$) = 0$, i.e. P(second red$|C) = 0$;
or P(first red$|C) = 0$.

Since green gets probability 1 when red has probability 0, we see that C is in all cases a perfect predicator of what will happen. The presence of C guarantees either the situation (first chameleon red and second green) or the situation (first chameleon green and second red). Thus the correlation has a common cause to explain it only if the two-chameleon system is deterministic.

This insight, that the principle of the common cause entails determinism for the special case of perfect correlation, provides the central lemma in the deduction of Bell's Inequalities, to which we now turn.

3. DEDUCTION OF BELL'S INEQUALITIES

It is not true that every possible phenomenon admits of a causal model. The first proof of the fact, without assuming that the model must be of the sort described by quantum mechanics or any other theory, was given by John Bell (1964, 1966). He showed that any such model must satisfy certain conditions— *Bell's Inequalities*—and that their satisfaction is not a logical but purely a contingent matter. I shall give the deduction in an easy form, illustrated by a schematic example—not in full generality, but sufficiently general for our purposes.[5]

3.1. *Surface description of a phenomenon*

There are two generals, Alfredo and Armand, who wish to strike a common enemy simultaneously, unexpectedly, and very far apart. To guarantee spy-proof surprise, they ask a physicist to construct a device that will give them a simultaneous signal, whose exact time of occurrence is not predictable. (This is a science-fiction story—their physics is like ours but their technology is much advanced. It happened in a galaxy long ago and far, far away. ...) The physicist gives each a receiver with three settings, and constructs a source which produces pairs of particles at a known rate, travelling towards those receivers. In each receiver is a barrier; if a received particle passes the barrier, a red light goes on, and otherwise a green one is on. The probability of this depends on the setting chosen. But when the two generals choose the same setting, one member of the pair of particles passes if and only if the other does not. Alfredo and Armand agree to choose a common setting and agree to turn on their receivers for 1 minute every other morning at eight o'clock (starting on a certain day), and then Alfredo will strike the first time his light is green while Armand will strike as soon as his light is red.

The story makes clear that no theory is presupposed in the description of what happens to the generals; and that is where the important correlation is found. Before looking at possible theories that might explain this curious correlation (which in itself is perfectly possible so far, even from a classical point of view), I shall make this description precise, and general.

3.2. *The experimental situation*

Two experiments will be made, one on each of a pair of particles produced by a common source, referred to as the left (L) and the right (R). Each experiment can be of three sorts, or be said to have one of three *settings*. The proposition that the *first* kind of experiment is done on the *left* particle will be symbolized $L1$; and so forth.

Each experiment has two possible outcomes, 0 ('green') or 1 ('red'). The proposition that the *second* kind of experiment is done on the *right* particle and has outcome 0 will be symbolized $R20$; and so forth. Note that $L1$ is equivalent to the disjunction of $L11$ and $L10$. To allow general descriptions, I shall use indices i, j, k to range over $\{1, 2, 3\}$ and a, b over $\{0, 1\}$. In addition, let $\bar{x} = 1 - x$, so that $\bar{a}$ is the opposite outcome of a.

A situation in which the two experiments are going to be done on a single particle-pair can be described in terms of a small field of propositions, generated by the logical partition

$$PR_{surface} = \{ Lia \ \& \ Rjb : i, j = 1, 2, 3 \text{ and } a, b = 0, 1 \}$$

which has 36 distinct members. These propositions include for example:

Setting 1 used on the left = $L1$ = the disjunction of the propositions

$$L1a \ \& \ Rjb \text{ for } a, b \text{ in } \{0, 1\} \text{ and } j \text{ in } (1, 2, 3)$$

and so forth. So we have here the resources for any sort of factual description at the surface level.

3.3. *Surface probabilities*

Probabilities for these propositions come from two sources. First, we may have some information about how the two settings will be chosen (possibly, to ensure randomness, by tossing dice). This gives us probabilities for the proposition in a coarser partition:

$$PR_{choice} = \{ Li \ \& \ Rj : i, j = 1, 2, 3 \}$$

Secondly, we may have a hypothesis or theory which gives information about how likely the outcomes are for different sorts of experiments. Because there may be correlation, the information optimally takes the form of a function

$$P(Lia \ \& \ Rjb \,|\, Li \ \& \ Rj) = p$$

giving the probability of the (a, b) outcome for the (Li, Rj) experimental setup. Let us call this function P a *surface state*. Note that it is not a probability function on our field; but it can be extended to one by combining it with a probability assignment to PR_{choice} (which may be called a *choice weighting*). Such a probability function on the whole field may be called a *total state*. Hypotheses concerning the surface state are directly testable: we simply choose the settings, start the source working, and do the relevant frequency counts—see how often the lights flash—to follow our story.

3.4. *Perfect correlation*

The special case I wish to examine satisfies two postulates for the surface states:

I.	Perfect Correlation	$P(Lia \ \& \ Ria \mid Li \ \& \ Ri) = 0$
II.	Surface Locality	$P(Lia \mid Li \ \& \ Rj) = P(Lia \mid Li)$
		$P(Rjb \mid Li \ \& \ Rj) = P(Rjb \mid Rj)$

It should be emphasized again that these probability assertions are directly testable by observed frequencies.

The Perfect Correlation Principle can be stated conveniently as: Parallel experiments have opposite outcomes. Similarly, the Surface Locality Principle says that the *outcomes* at either apparatus are statistically independent of the *settings* on the other. This assertion, verifiable at the surface level of detectable phenomena, rules out that one general can signal the other by twirling the settings. Both these conditions, symmetric in L and R, are simple conditions on the surface states. If both principles hold, we obviously have

$$P(Lia \mid Li) = P(Lia \ \& \ Ri\bar{a} \mid Li \ \& \ Ri)$$

but the reader is asked to resist counterfactual (and dubious) inferences such as that, if the $L1$ experiment has outcome 1, then, *if* the R1 experiment *had been* done, it would have had the outcome 0!

3.5. *Common causes as hidden variables*

When principles I and II hold, there is a clear correlation between outcomes in the two experiments. What would a causal theory of this phenomenon be like? It would either postulate or exhibit a factor, associated with the particle source, that acts as common cause for the two separate outcomes in the examined probabilistic sense. I shall refer to this as 'the hidden factor'—

not because I assume that we cannot have experimental or observational access to it, but because it does not appear in the surface description (i.e. in the statement of the problem).

Symbolizing as Aq the proposition that this hidden factor has value q, the space of possibilities now has the still finer partition

$$PR_{total} = \{ Lia \ \& \ Rjb \ \& \ Aq : i,j = 1,2,3;$$
$$a,b = 0,1; \text{ and } q \ \varepsilon \ I \}$$

where I is the set of possible values of that factor. A total state must be a probability function defined on the (sigma) field generated by this partition. (Let us not worry about how to restate this in case I is uncountable; as will shortly turn out, that precaution is not needed.) A causal model of the phenomena in terms of this hidden factor must satisfy the following conditions:

III. Causality[6]

$$P(Lia \ \& \ Rjb | Li \ \& \ Rj \ \& \ Aq)$$
$$= P(Lia | Li \ \& \ Rj \ \& \ Aq)P(Rjb | Li \ \& \ Rj \ \& \ Aq)$$

IV. Hidden Locality[7]

$$P(Lia | Li \ \& \ Rj \ \& \ Aq) = P(Lia | Li \ \& \ Aq)$$
$$P(Rjb | Li \ \& \ Rj \ \& \ Aq) = P(Rjb | Rj \ \& \ Aq)$$

V. Hidden Autonomy

$$P(Aq | Li \ \& \ Rj) = P(Aq)$$

This requirement V on the causal models is usually specified only informally, as ruling out certain kinds of cosmic conspiracy or pre-established harmony, which would make the hidden factor statistically dependent on the apparatus setting to be chosen then or later. (Notations in which q appears as a subscript to the probability function cannot express this condition.) We can of course imagine that V is false, for example if hidden forces are determining the experimenter's choices, or his random number generator if he is taking precautions. I shall leave it to the reader to investigate what, if anything, could be established if condition V is dropped, and what level of scepticism would require that.

I shall break up the ensuing argument into three sub-arguments, in which these postulates are separately exploited. But I shall say a few words here to defend the idea that a proper causal theory must satisfy all three.

Causality is just the probabilistic part of the Common Cause Principle stated before. The other two, Hidden Locality and Hidden Autonomy, are meant to spin out implications of the idea that it is the common cause alone, and not special arrangements, or relationships between the two separate experimental setups, that account for the correlation. If we had only III to reckon with, it could be satisfied simply by defining

$$Aq = (Li_t a_t \ \& \ Rj_t b_t)$$

where the actual settings chosen are (Li_t, Rj_t) and the actual outcomes are (a_t, b_t). But the common cause is meant to be located at the particle source, in the absolute past of the two events that have space-like separation. Hence its character should be totally settled by the facts, before choice of setting or actual outcome. The choices of the experimental settings, and of the particular type of source used, can all be made by means of any chance mechanisms or experimenters' whims you care to specify.

Principle II governs the surface phenomena, and is part of what is to be explained. It will play no role in the deduction below. In fact, although II does not follow from Hidden Locality alone—the familiar point of Simpson's paradox—it does follow from IV and V combined.[8]

Of course, I am not saying that nature must be such as to obey these postulates—quite the opposite. These postulates describe causal models, in the 'common cause' sense of 'causes', and the question before us is whether all correlation phenomena can be embedded in such models.

3.6. *Causality alone: a deduction of partial determinism*[9]

Principles I and III alone already imply that when parallel settings are chosen the process is deterministic; the common cause determines the outcomes of the experiments with certainty. For abbreviating '$Li \ \& \ Rj \ \& \ Aq$' to '$Bijq$', we derive from those two principles:

$$0 = P(Lia \,\&\, Ria | Li \,\&\, Ri)$$
$$= P(Lia \,\&\, Ria | Li \,\&\, Ri \,\&\, Aq)$$
$$= P(Lia | Biiq) P(Ria | Biiq)$$

But since the product is 0, one of the two multiplicands must be 0; the other will be 1. For example, if $P(Li1 | Biiq) = 0$, then $P(Li0 | Biiq) = 1$. But setting $a = 0$ in the above deduction, we conclude that if $P(Li0 | Biiq) \neq 0$ then $P(Ri0 | Biiq) = 0$, and hence $P(Ri1 | Biiq) = 1$. So we see that, conditional on $Biiq$, all experimental outcomes have probability 0 or 1.

I doubt very much that Reichenbach can have perceived this consequence of his principle, because he had explicitly designed it so as not to require determinism for causal explanation. Had Einstein read Reichenbach and perceived this consequence in time, he could have added a little codicil to the Einstein–Podolsky–Rosen paradox: according to the Common Cause Principle, conditional certainties of the sort found in that paradox can exist only if they are the result of a hidden deterministic mechanism; so quantum mechanics is incomplete. See how much we have got—and we have hardly begun!

3.7. *Hidden locality: a deduction of complete determinism*

We have just deduced that, conditional on the antecedent $(Li \,\&\, Ri \,\&\, Aq)$, all probabilities for outcomes are 0 or 1. But Hidden Locality (our condition IV) says that this antecedent contains irrelevant information as far as the outcome at either side, separately, is concerned. Hence we deduce:

$$P(Lia | Li \,\&\, Aq) = P(Lia | Li \,\&\, Ri \,\&\, Aq) = 0 \text{ or } 1$$

This follows therefore from I, III, and IV together. It says that, given the value of the hidden variable that acts as common cause, the outcome of any performable experiment on either side is determined with certainty.

3.8. *Hidden autonomy: the testable consequences*

It was a tenet of modern philosophy, owed to Kant, that the mere assertion of causality, or even determinism, has no empirical consequences. Any phenomenon at all can be embedded in a causal story; only specific causal hypotheses have testable consequences. Nothing we have seen so far refutes that tenet, for all the consequences drawn have been about the hidden variables and not about the surface phenomena themselves. But we come now to the peculiar twist that Bell discerned.

Wigner (1970) observed that, given the preceding, there are only eight relevant classes of values for the hidden variable. (And, accordingly, no generality in the causal theory will be lost if we say that the variable has only eight possible values.) For these values can be classified by their answers to the questions:

(a) Suppose Li. Is it the case that $Li1$?

(b) Suppose Rj. Is it the case that $Rj1$?

Given Aq, as we have now seen, each of these questions receives a definite yes or no answer (with probability 1). And indeed, the answers to the second type of question are determined by those to the first, since each outcome must be 0 or 1, and opposite to the outcome at the other side for the same setting. (In this reasoning, we rely on I, III, IV.)

Since there are three questions of form (a), each with two possible answers, these answers divide the hidden variable values into $2^3 = 8$ types. Let us say that q is of type (a_1, a_2, a_3) when this value q predicts outcomes a_1, a_2, a_3 for arrangements $L1$, $L2$, $L3$ respectively. And let us abbreviate the assertion that the actual value is of this type to $Ca_1a_2a_3$. Precisely:

$Ca_1a_2a_3$ = the disjunction of all the propositions Aq such that

$$P(L1a_1|L1 \& Aq) = P(L2a_2|L2 \& Aq)$$
$$= P(L3a_3|L3 \& Aq) = 1.$$

This is an ordinary finite disjunction of form (Aq_1 **or** ... **or** Aq_m) if the set of values of the hidden variables is finite (and we know now that we can assume that without loss of generality). Thus we have not introduced new propositions; we are still working within the field generated by PR_{total}.

Suppose now that we have chosen settings $L1$ and $R2$; what is the probability that we shall get outcomes $L11$ and $R21$? Well, let us put it a different way. Supposing Aq, what must the value q be like if we are to get outcomes $L11$ and $R21$? It must clearly be of type $(1, 0, b)$ for some value b or other. In other words, this outcome will happen only if ($C101$ **or** $C100$) is the case. But this proposition has a probability of its own—and that probability is our answer.

The argument I have just given tacitly presupposes Principle V of Hidden Autonomy, for it assumes that the choice of $L1$ and $R2$ as settings does not affect the probabilities for value q. Let us state the argument precisely. To begin, define a new notation:

$$p(1; 2) = P(L11 \ \& \ R21 | L1 \ \& \ R2)$$

We note the theoretical equivalence:

$$p(1; 2) = \Sigma\{P(Aq)P(L11 \ \& \ R21 | L1 \ \& \ R2 \ \& \ Aq) : q \ \varepsilon \ I\}$$

We notice now that in the summation the conditional probability equals 0 except in cases where q is of type $(1, 0, 1)$ or of type $(1, 0, 0)$. Hence we have

$$p(1; 2) = \Sigma\{P(Aq) : q \text{ is of type } (1, 0, 1) \text{ or } (1, 0, 0)\}$$

$$= P(C101 \ or \ C100)$$

$$= P(C101) + P(C100)$$

In just the same way, we deduce

$$p(2; 3) = P(C110) + P(C010)$$

$$p(1; 3) = P(C110) + P(C100)$$

Adding up the first two equations, we get the sum of four probabilities, two of which appear again in the equation for $p(1;3)$. Hence

$$p(1;2) + p(2;3) \geq p(1;3)$$

Mutatis mutandis for the cyclic permutations of 1,2,3. These are Bell's famous Inequalities in the form Wigner gave them.

It hardly needs pointing out that the numbers $p(i;j)$ are surface probabilities by their definition (in which the hidden variable does not occur). So these inequalities are testable directly by means of observable frequencies. Thus, our quite metaphysical-looking principles have led us to an empirical prediction!

4. THE EXPERIMENTS

We have now seen that there are *possible* phenomena for which it is logically *impossible* to have a causal model. Hidden determinism was ruled out along the way, for the argument proceeded in two main steps: (1) a causal model for such a phenomenon must in fact be a deterministic model; and (2) any such deterministic model must satisfy Bell's Inequalities. But violation of these inequalities is certainly conceivable. The next question is of course: are any such possible phenomena actual? That is not so important to the philosopher, but certainly is of some interest in its own right.

Experiment cannot establish so much. Of course, we are dealing here with the probabilities of observable events—lights flashing, bells ringing, numbers appearing on computer print-outs. But there are two obstacles to saying that we have seen violations of Bell's Inequalities—one relatively trivial, the other insuperable. The first is that in actuality we have only the relative frequencies of the events, displaying a good or bad— perhaps even excellent—fit to the probabilities. This leeway between the actual frequencies and the probabilities is always there, in every experimental investigation; that is why I call the point relatively trivial. The second obstacle is that, for the phenomena to be telling, it must be asserted that the correlated events were simultaneous, or at least temporally sufficiently close together, to prevent the existence of signal-connections between them. The required time intervals are so short that the check must rely on instruments *whose theory itself belongs to atomic physics*. Properly appreciated, the point becomes very general: the best we could hope for is a demonstration that the given theoretical model of the experimental setup (which we already have in physics) cannot be extended to a causal model of the correlations in the experimental outcomes.

The results of the new experiments must be very carefully described. It would be no news at all now to say that the *theory* of quantum mechanics cannot provide a causal model for certain correlations which it admits as possible. This is for us already implied by the Einstein–Podolsky–Rosen paradox of 1935. It is at best a little news to say that we have actual experiments, whose quantum-mechanical description includes correlations of the sort in question. But that is still news about this particular, historically given, theory. Logically speaking, we go further yet when we say: (*a*) the best theoretical model we have of the measurement apparatus entails that the frequency counts in question are of events with space-like separation (i.e. as good as simultaneous), and (*b*) without any reliance on theory, we can show that these frequency counts display an unacceptably—indeed, incredibly—bad fit to the probabilities of any causal model whatsoever. *This* is the correct description of what has been found. And it is, even in principle, the best we could have from experimental inquiry.

To provide more food for our starved imagination, I shall briefly describe some of the actual experiments.[10] Their design was of course inspired by quantum-theoretical predictions of correlation. When an excited atom cascades towards its ground state, it emits photons, and their polarizations are correlated.

The 0–1–0 case. A calcium atom, excited from the 4^1S_0 ground state to the 6^1P_1 state by absorption of a photon of 2275Å light, cascades first to a $(j = 0)$ state, then, with the emission of a photon of 5513Å, to a $(j = 1)$ state, and finally to the $(j = 0)$ ground state with the emission of a photon of 4227Å.

The 1–1–0 case. A mercury atom, excited by electron bombardment to one of the $(j = 1)$ states, cascades via a $(j = 1)$ state to the $(j = 0)$ triplet ground state, emitting in this course one photon of 5677Å and one of 4048Å.

The mean decay times are very short in both cases (between 10^{-8} and 10^{-9} seconds). The correlations to be inspected are in the polarizations of the described photons (other emitted photons being eliminated by colour filters).

Photon detectors are placed in the $+z$ and $-z$ direction from the emitting atoms, with colour filters so that the first detector is sensitive only to the first photon and the second detector sensitive only to the other photon. We can measure the coincidence counting rate $R(0)$ of the two detectors—a coincidence consisting in detection by both detectors within an adjustable time interval.

Let us now place perpendicularly to the $+z$-axis a polarization filter $F(1,\theta)$, with its polarization direction oriented at angle θ to the x-direction. The coincidence rate changes from $R(0)$ to $R(1,\theta)$. If *instead* we place such a filter $F(2,\theta)$ perpendicular to the $-z$-axis, the coincidence rate changes instead to $R(2,\theta)$. In both our cases, according to atomic theory, the light is unpolarized, so the filters pass 50 per cent of the emitted light. Thus we have, whatever angles θ,γ we choose,

$$R(1,\theta) = R(2,\gamma) = (\tfrac{1}{2})\, R(0)$$

We now consider the coincidence count $R(\theta,\gamma)$ when we put both filters, $F(1,\theta)$ and $F(2,\gamma)$, in place. If there were no correlation at all, the probabilities of detection are just multiplied, and we would have

No correlation: $R(\theta,\gamma) = (\tfrac{1}{4})\, R(0)$

because the factor by which $R(0)$ is decreased is just the probability of photons reaching the detectors.

But if quantum mechanics is correct, then the experimental outcome must be otherwise. Indeed, the rate $R(\theta,\gamma)$ depends systematically on the angle $(\theta - \gamma)$ between the two filters. The results are

$$\textit{0-1-0 case: } R(\theta,\gamma) = \frac{\cos^2(\theta - \gamma)}{2}R(0)$$

$$\textit{1-1-0 case: } R(\theta,\gamma) = \frac{\sin^2(\theta - \gamma)}{2}R(0)$$

Therefore, the coincidence count drops to zero in the 0–1–0 case if the filters are placed at right angles towards each other, also in the 1–1–0 case if the filters are parallel. These are examples of perfect correlation, of the sort more schematically considered in our 'two-generals' thought-experiment.

The experiments by Clauser and Holt were along these lines, and showed the expected good fit between actual frequency counts and prediction. The experimental tests by Aspect and his collaborators are more recent and very impressive. First proposed by Aspect in 1975 and 1976, the tests were carried out and the results were reported in 1981 and 1982. The basic scheme of the Aspect optical version of the Einstein–Podolsky–Rosen (EPR) experiment is given in Fig. 4.1. The pair of photons v_1 and v_2 is analysed by linear polarizers I and II (with orientations $\overrightarrow{a}$ and $\overrightarrow{b}$). There is a coincidence if both photons pass the polarizers, and this coincidence is monitored and its relative

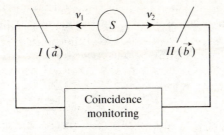

F<small>IG</small>. 4.1 Aspect experiment schema

frequency measured. The coincidence rate $N(a, b)$ depends on the orientations, and Bell's Inequalities can be deduced for these rates, on the assumption that they fit the probabilities in some causal model or other. The deduction does not depend on the details, but yields an invariant relationship for all causal models, as we saw in the preceding section.

However, with such a static setup one could challenge the locality assumptions that the events at polarizer *II* are independent of the *orientation* chosen for polarizer *I*, and that the state in which the photons are emitted at source *S* is also independent of these orientations. (We called these the assumptions of hidden locality and hidden autonomy in our deduction.) This challenge can be given experimental form, to some extent, by introducing mechanisms to vary the orientations of *I* and *II* after the particles leave the source, and late enough so that no signal with a speed of at most that of light could communicate the new setting from one polarizer to the other before the event. With all times somewhat indeterminate, this challenge requires sufficient margins to ensure definite space-like separation of all hypothetical events—including the ones imagined for the hidden parameter. There can then still be a further challenge raising the possibility of a pre-established harmony, set far enough back in time, to co-ordinate the mechanisms which vary the orientation. Verification therefore can not be logically watertight, but can still have great cogency.

In Aspect's experiment, the distances between switch and source are of the order of 6 meters, so that the time interval for signals to traverse them is appreciable (c.40 ns). Switching occurs by means of an acousto-optical interaction with an ultrasonic standing wave in water, about every 10 ns. The lifetime of the intermediate level of the cascade used (type 0–1–0 in calcium) is about 5 ns. The form of Bell's Inequality used for calculation was that of Clauser, Horne, Shimony, and Holt, which has the form

$$-1 \leq S \leq 0$$

where S is a function of the coincidence rates for the settings which occur during the experimental run. The reported experimental results in 1982 yielded

$$S = 0.101 \pm 0.020$$

a violation of the condition $S \leq 0$ by 5 standard deviations. The quantum-mechanical prediction for the same situation was $S = 0.112$, comfortably within the experimental range.

We must be very careful again to distinguish observational finding from theoretical interpretation. A good deal of atomic physics is assumed to construct the basic theoretical model of this experimental situation. This model does not necessarily require extant quantum theory for its intelligibility. We can think of it as theoretically accessible around, say, 1918, after Einstein's theory of photons and Bohr's theory of the atom were assimilated by physics, but before the theoretical unification of Heisenberg or Schroedinger in 1925/6. The question is then: how can this model be developed further so as to accommodate the found coincidence rates? One putative answer is: as does the new quantum theory. And it is found that a model of this sort does exist which fits the experimental phenomena. Another putative answer is: as a causal model incorporating some common cause for the coincidence rate for parallel polarizers. It is then found that for *no* value of the hidden parameter, which represents that common cause, is there such an extended model which fits the phenomena.

In other words, we do not have a presuppositionless result here, but we do have a clear verdict on how the old quantum theory could be amended and developed if the new quantum theory were to be empirically adequate. The verdict is: *not* such as to turn it into a causal theory.

5. GENERAL DESCRIPTION OF CAUSAL MODELS

A surface model of given phenomena may or may not satisfy Bell's Inequalities. In this section I shall present a description (essentially due to Arthur Fine) of the range of theoretical models which can fit those surface models that do satisfy the inequalities. The surface state is *always* a fragment of some classical probability function, but that function may not emerge from any 'simple' model—that is how the matter was put intuitively above. The question is now: what does 'simple' mean here?

The general idea is this: to construct a 'simple' model, we assume that the parameters measured always have some value (in the range of possible outcomes), and this value is faithfully revealed when any measurement is made.[11] Such a model is exactly of the sort in which correlations in measurement outcomes can be traced to a common cause, namely the condition of the object which is independent of but faithfully revealed by measurement. To this relation between a classical conception of measurement and causality I shall return in the next chapter. Suppose for example that our surface model instantiates the measurements in a black box schema, with Ri the proposition that the same measurement is made as described by proposition Li, on the same object, but at a different time and with the outcomes $0, 1$ relabelled as $1, 0$. The assumption underlying the 'simple' model now implies that Lia, for example, is true if and only if the relevant parameter A is measured (Li) and A has value a ($A = a$). Moreover, the probability of A having this value is not affected by whether any measurement is performed:

$$P(A = a | Li) = P(A = a).$$

Thus the situation in such a 'simple' model is as follows. There is a partition Q whose members are the sets $q = (A_1 = a \ \& \ A_2 = b \ \& \ A_3 = c)$, and the following hold:

$$Lia = [Li \ \& \ (A_i = a)] \qquad Ria = [Ri \ \& \ (A_i = \bar{a})]$$

$$P(q | Lj) = P(q) = P(q | Rj) \text{ for all } q \in Q$$

for all i, j, a, b, c. It is easily deduced now that

$$P(Lia \ \& \ Rjb | Li \ \& \ Rj) = P(A_i = a \ \& \ A_j = \bar{b}).$$

We have therefore immediately arrived at the very last stage of the Bell's Inequality proof, in effect, and Bell's Inequalities hold. Indeed, we can cite Arthur Fine's results (1982b, 1986) to prove that such a 'simple' model exists if and only if Bell's Inequalities hold.

Fine proved the empirical equivalence of four sets of conditions, each of which defines a class of models. (See below for the meaning of 'empirical equivalence'.) One of these is what I have called the class of simple models; Fine's results provide a definitive picture of that model class.[12]

For ease of exposition, we limit our set of observable quantities to four: A and A' are measured by different settings of one apparatus (the 'left') while B and B' are measured by different settings of another apparatus (the 'right'). Assume also that each of these four admits only two values, say 0 and 1. The surface state gives us the probabilities which we shall denote as $P(A)$, $P(AB)$, $P(A - B)$, . . . , to be read as:

$P(A)$ = the probability that a measurement of A, if made, yields value 1

$P(AB)$ = the probability that measurement of A and B both yield value 1

$-S$ = the observable which has value $1 - x$ if S has value x.

The observables A, A', $-A$, $-A'$ I shall call the *left* observables, and B, B', $-B$, $-B'$ the *right* observables. Note that the surface state will include the joint probability $P(SS') > 0$ only if one of S and S' is a left, and the other a right, observable, though a theoretical model could of course give joint probabilities for other combinations.

A model for this sort of situation is any model which specifies at least such a surface state. Now we describe various conditions which such a model could satisfy. It will be seen that class I, the first to be described, is the class of 'simple' models intuitively discussed above.

I. A *deterministic hidden variable (h.v.) model* has a set Q of *hidden states* ('complete' or 'hidden' state specifications, in Fine's terminology), a probability density function m on Q, and a *response function* h: if S is any of the above observables, and v is in Q, then $h(S, v)$ is 0 or 1. The *surface state* in this model is the function

$$P(S) = \int h(S,v)m(v)dv$$

$$P(ST) = \int h(S,v)h(T,v)m(v)dv$$

where S and T are any two observables, one right and one left.

II. A *joint probability model* is any model in which the surface state P is part of a 'hidden' probability function P^* which assigns probabilities to all combinations $SS'TT'$, with one of S and T a left and the other a right observable, such that

$$P^*(AA'B) = P^*(AA'BB') + P^*(AA'B - B')$$

$$P^*(AB) \quad = P^*(AA'B) + P^*(A - A'B)$$

and so forth (the surface state represents the 'marginals' of the 'hidden' probability function).

III. A *Bell/Clauser–Horne (CH) model* is any model in which the surface state P satisfies the inequality

$$-1 \le$$
$$P(AB) + P(AB') + P(A'B') - P(A'B) - P(A) - P(B')$$
$$\le 0$$

and also the other seven inequalities formed by permuting A with A', then B with B', and then A with A' and B with B' together.[13]

IV. A *stochastic h.v. model* has a set Q of *hidden states*, a probability density function m on Q, and a *stochastic response function* r: if S is an observable and v is in Q, then

$$r(S, v) \text{ is in } [0, 1]$$

additivity: $r(S, v) + r(-S, v) = 1$

factorizability: $r(ST, v) = r(S, v)r(T, v)$

whenever one of S and T is a left and the other a right observable. The surface state in this model is the function

$$P(S) = \int r(S, v)m(v)dv.$$

$$P(ST) = \int r(S, v)r(T, v)m(v)dv$$

where S and T are two observables, one right and one left.

A few comments: that the probability function on the hidden states can be written in the integral form with density function m is a very small restriction only in this context, and would not have affected our discussion in the preceding sections. The more general form given here to the Bell Inequalities in III was introduced by Clauser and Shimony specifically to suit the discussion of less than totally perfect correlations, arising perhaps from some stochastic element in the response.

In the stochastic h.v. models, the response function does not assign a specific value to each observable in each hidden state, but only a probability distribution on its possible values. This widening of the discussion thus lets in an appreciable element of chance. But Fine's results establish that, if the phenomena admit such a factorizable stochastic model, then they also admit another model in which the element of chance is entirely absent—a deterministic model.

It is important to note clearly just what is established and what is not. *The four model classes are not identical.* On the contrary, the joint probability models, which include both the deterministic and the stochastic h.v. models, constitute logically a significant widening of the class of simple models (i.e. of class I). However, Fine's results establish the equivalence of the four conditions, in the sense that each fits the same family of surface models (*empirical equivalence*). Because all the members of that family fit the Bell Inequalities, these model classes are not rich enough in structure to be adequate to phenomena involving peculiarly quantum behaviour.

6. DOES LOCALITY REALLY PLAY A ROLE?

It has been argued that violation or satisfaction of Bell's Inequalities has nothing to do with causality (or locality, or separability, or other physically significant concepts) because these inequalities can be derived as theorems in the classical probability calculus. This point is then supported by noting that, in the derivation of such results as those of Arthur Fine, discussed in the preceding section, no premises are included concerning time, causes, localization, and so on.[14] If that is so, our whole discussion in this entire chapter has been misguided and misleading.

The argument rests on an equivocation. There is a result which can be proved in classical probability theory, and the formulas in that result are indeed Bell's Inequalities. But that result describes a general feature which is (trivially) satisfied by all the experiments which we cite as violations of Bell's Inequalities. There is another result of classical probability theory, which was proved in Sections 3.6–3.8 above, in which those same formulas occur. It concerns only the special case of causal models; it describes a feature not found in those experiments, and that is their significance.

Let us spell this out precisely. At the beginning of the preceding section I described a 'simple' model one could offer of the two-generals situation. It involved the assumptions that the experimental outcomes were results of measurements which faithfully revealed the values of certain parameters A_i and B_i. These parameters were represented by random variables on a classical probability space. Other such variables represented whether or not a certain kind of measurement takes place. The postulate of hidden autonomy said then simply that the values of the first two kinds of variables are statistically independent of whether any measurement is carried out or not. The hidden variable now became just the specification of values for the first two sets of variables, and the Bell Inequalities can then be deduced by the same argument as in Section 3.8. Having told this whole story, however, we can think about it again in a more abstract way. Given this independence and faithfulness of measurement, let us just ignore the whole discussion of measurement, which now makes no essential difference. We realize then that the proof in Section 3.8 proves in effect a little theorem in classical probability theory, namely:

Result 1(a)   If A_i and B_j are 0–1 valued random variables on a classical probability space and P is a classical probability function such that $P(A_i = B_i) = 0$, and we define

$$p_{ij} = P(A_i = 1 \ \& \ B_j = 1)$$

then

$$p_{mk} + p_{kn} > p_{mn}$$

for k, m, n distinct.

The corresponding result using the alternative formulation of Bell's Inequalities preferred by Clauser and Horne is as follows.

Result 1(b)   If there is a set of events $\{A_i : i \in I\}$ such that $p_i = P(A_i)$ and $p_{ij} = P(A_i \cap A_j)$ for i,j in I and P is a classical probability function defined on the field generated by that set of events, then

$$-1 \le p_{mn} + p_{mq} + p_{kq} - p_{kn} - p_m - p_q \le 0$$

for k, m, n, q in I.

This follows at once from Fine's results reported in the preceding section.

But if we rewrite 1(a) with all the probabilities conditional on corresponding antecedents C_{ij}, it is no longer a theorem. In that rewriting, the elements of the inequalities become probabilities conditionalized on different antecedents, and then the argument no longer goes through. The reason is simply that, if the set $\{C_{ij}\}$ is a partition, the probability calculus places no restriction on the numbers p_{ij}, except that they be in [0,1]. Each cell of the partition can be thought of as a separate probability space. The one extra condition we do have, namely $p_{ii} = 0$, will obviously not force restrictions on the case $i \ne j$. It will entail that the inequality is satisfied when two or all of m, k, and n are identical, but that is all.

The argument I reported therefore hinges on the confusion of results 1(a) and 1(b)—which are theorems about unconditional probabilities—with results not provable in any probability theory concerning conditional probabilities. Sections 3.1–3.7 used the assumptions of the Common Cause model to reduce that special case of the conditional 'result' to the unconditional one.

We can illustrate these abstract points with another brief look at Fine's results. The first thing we must note is that locality is really built into Fine's definition of stochastic h.v. models, tacitly. For it is clear in his interpretation that the probabilities are tacitly conditional on measurement. Hence in that definition

$P(S)$ is elliptical for $P(S|S$ is measured), and $P(ST)$ for $P(ST|S$ and T are both measured). Similarly for the hidden response function: $r(S,v)$ is elliptical for $r(S|v$ and S is measured). Therefore what is there called 'factorizability' means:

$r(ST|v$; S and T are both measured)

$$= r(S|v; S \text{ is measured})r(T|v; T \text{ is measured})$$

which is literal factorizability pure and simple *plus* the contraction into the right-hand side of the product

$r(S|v$; S and T are measured)$r(T|v$; S and T are measured).

Similarly for the definition of joint probability models: there the marginality condition for the theoretical probability function P^* really means

$P^*(A|A$ is measured) $= P^*(AB|A$ and B are measured)

$$+ P^*(A\bar{B}|A \text{ and } \bar{B} \text{ are measured})$$

Given how B and $\bar{B}$ are related, their conditions of measurement are the same; and conditional on that measurement, one or other must take the value 1. Therefore the condition implies $P^*(A|A$ is measured) $= P^*(A|A$ and B are both measured) — the very premiss which I called Hidden Locality. So we see that the locality conditions are only apparently absent from Fine's derivations.[15]

Of course, this whole discussion now throws doubt on the intuitive possibility of regarding the experimental results as outcomes of measurements of classically conceived observables which faithfully reveal their pre-existing values. This is exactly what we shall take up first in the next chapter.

5

New Probability Models and their Logic

No causal model can fit the phenomena that violate Bell's Inequalities. What sorts of models will fit? That is again an empirical question, and the exact answer cannot be a priori. Quantum theory will give us one answer, by offering models which are not causal, and do allow such violations. But the theory is not empirically empty, so these models too will have their limits.

To understand this situation, to see how this question could be answered at all, we need to broaden our concept of statistical model. This wider concept must allow for phenomena that fit into no Common Cause pattern. It must also point the way to some interesting model constructions that are worth exploring, and prepare us for the study of quantum mechanics proper. To this end I shall introduce the general notion of *geometric probability models*, of which the quantum-theoretical models will be a natural development, abstractly speaking. This will also introduce us to some basic elements of *quantum logic* (so-called) and its role in interpretation. This role is important, in my view, but there will be no suggestion of any revolution in logic generally.

1. WHEN ARE VALUES INDETERMINATE?

Much undeserved mystery has surrounded Bell's results. The proof that causality is lost, that conceivable and apparently actual phenomena rule out determinism and even the Common Cause pattern, is astonishing. But the argument is perfectly intelligible to the classical mind. The logic and the ideas about probability involved in the argument are also all of a perfectly familiar sort. Yet it has been proposed that a proper appreciation requires us to turn to non-classical logic and/or non-classical probability theory.

We can divide such arguments nicely into those which involve specific considerations of modality, and those which do not. In this section we shall look at the latter, leaving possibility and counterfactuals to a later section. That means that, for now, we shall be looking at a sort of argument that predates the Bell literature, at least in essence.

Consider a strange situation in which we measure three physical parameters, A, B, and C. The data are that, for each, measurement yields values 0 and 1 with equal frequency. We are able to measure them two at a time, and always find opposite values. Thus:

1. for X, $Y = A$, B, C we extrapolate probabilities
   (*a*) $P(X = 1) = 0.5$
   (*b*) $P(X = 1$ or $X = 0) = 1$
   (*c*) $P(X = 1$ & $Y = 1) = 0$ if $X \neq Y$

What kind of model can represent this situation?

Here is the argument that none can, if we do not violate classical probability theory and/or logic. Consider the statements:

2. No two of A, B, C have the same value.
3. Each of A, B, C has value 0 or 1.

It is clear that 2 and 3 are inconsistent with each other. Yet 1(*b*) entails that statement 3 has probability 1 and 1(*c*) entails that statement 2 has probability 1. By ordinary probability theory, their conjunction must then have probability 1 also. But it is also a principle of that theory that a self-contradiction receives probability 0!

Yet this entire argument is spurious. For the data at the beginning are about measurement outcomes, while the extrapolated probabilities were absolute and unconditional. As we did conscientiously in the preceding chapter, so we should here state explicitly the condition of measurement. Let $M(X)$ be the proposition that parameter X is measured. Then 1 must be replaced by, or be regarded as a sloppy ellipsis for, the following:

4. for X, $Y = A$, B, C we extrapolate the probabilities
   (*a*) $P(X = 1 | M(X)) = 0.5$

(b) $P(X = 1 \text{ or } X = 0 | M(X)) = 1$

(c) $P(X = 1 \text{ and } Y = 1 | M(X) \& M(Y)) = 0 \text{ if } X \neq Y$

Now this certainly has models allowed by classical probability theory and logic. We must note simply that we have here three parameters which cannot be jointly measured. That is, we can *deduce*

(d) $P[M(A) \& M(B) \& M(C)] = 0$

and that is our solution.

How good a solution is it? Conditions 1(a)–(c) were impossible to satisfy by any classical probability function. But 4(a)–4(c) are satisfiable by many—they do not specify any one completely. Nor do they do so if we just add the probabilities of $M(A)$, $M(A) \& M(B)$, etc. What value does C have, while A and B are measured?

Should we try to answer that question here at all? The only thing we really can do at this early stage is to eliminate some wrong answers. Obviously, a measurement is some way of getting information about the object measured. But classical intuitions (if such beings exist) may suggest two postulates:

Value Definiteness: Each physical parameter always has some value, one of the values which may be found by measurement.

Veracity in Measurement: Measurement of a parameter faithfully reveals the value it really has.

These two postulates can be consistently added to our above story, but then they imply some sort of conspiracy about when measurements are made: when A and C both have value 1, we are lucky or clever enough not to measure the two of them! That is unacceptable. Perhaps we should put it this way. The conjunction of these postulates would be an attempt to say that the world is basically the same, whether things are being measured or not. But given the above story, the two postulates are both true only if things are not basically the same in the two cases, because the measured situations are all of a special sort. So the attempt fails: some difference between measured and unmeasured world will have to be admitted.

That first postulate of Value Definiteness was called *Classical*

Principle C by Paul Feyerabend (1958). It is not logic, but this principle that must be rejected, he argued. From Bohr to Feynman, physicists have expressed similar opinions: an observable (measurable parameter) might not have a specific value outside the context of measurement. That is part of the orthodox or Copenhagen interpretation. This rejection may well be palatable from other points of view. Perhaps it is not even all that radical. In ordinary common sense, we attribute colours to liquids; but when the liquid is vaporized, the question of what colour it is *then* has no answer. However, the second postulate—Veracity in Measurement—has also been much looked upon as a candidate for rejection or revision. To keep the first postulate and reject the second—the explanation through disturbance by measurement—would not be a happy option. It would imply some sort of conspiracy again: if A and C do sometimes both have value 1, how does the uncontrollable disturbance in measurement carefully and systematically hide that fact?

To reject classical principle C is so far the only palatable option. It does itself imply a weakening of Veracity in Measurement. For if, at a certain time, parameter A has no value, and is measured, then this measurement *yields* a value as outcome, but clearly does not *reveal* a value. When we specify what counts as a measurement of A, we describe a physical arrangement which must have one of two outcomes (indicator values), in this case 0 or 1. It would indeed not be a measurement if this outcome gave us no information at all about the system which is subjected to measurement. But what sort of information it does yield, and how much, we shall have to consider very carefully. Imagine for instance a pollster who receives the answers *yes*, *no*, *don't know* from his informants, but notes always *yes* or *no* for each. Depending on two factors—our interests and his procedure—this may constitute an acceptable measurement. (Do we care about the percentage of undecided informants? When he hears *don't know*, does he automatically write *yes* or *no* or does he flip a coin?) As long as we know what he is doing, we can cull information from his results. We had better not say more just now. Exactly how measurement is to be conceived in the new physics is a central issue of interpretation, and we should not try to settle it here a priori.

But the rejection of Classical Principle C is surely not classical? No, it is not classical, in *some* sense; but neither does it contradict classical logic. That a given parameter should sometimes have no value at all may look like a violation of the logical principle of Excluded Middle. But it is not. Logic cannot compel us to regard the family $\{A$ has value $x : x \in R\}$ for example as a logically exhaustive division of possibilities. What if we *defined* the parameter A so as to guarantee this form of Excluded Middle by definition? Well, we could do that; but then we would have no guarantee that the defined construct must correspond to reality.[1]

Proofs and illustrations

The example of the strange situation which I gave at the beginning of this section is a mere scientifiction and so could not really prove anything. But the lines of argument displayed in it, in simple form, are just those in the real arguments in the literature, i.e. that quantum theory leads to a violation of classical probability and/or logic.

The plot thickens when we look to where logic and/or probability may be going wrong. We can expand statement 3 to

3 (a) $(A = 1$ or $A = 0)$ and $(B = 1$ or $B = 0)$ and $(C = 1$ or $C = 0)$

(b) $(A = 1$ and $B = 1$ and $C = 1)$ or $(A = 1$ and $B = 1$ and $C = 0)$ or ... or $(A = 0$ and $B = 0$ and $C = 0)$

It is clear that each conjunct in 3(a) receives probability 1 by 1(b), and that each disjunct in 3(b) receives probability 0 by 1(c). Therefore 3(a) has probability 1 and 3(b) has probability 0. But finally—here is the paradox—3(a) and 3(b) are logically derivable from each other. All this clear reasoning nevertheless relies on some theorems of classical logic and probability theory, of which the most important is:

5. X and $(Y$ or $Z)$ is logically equivalent to $(X$ and $Y)$ or $(X$ and $Z)$

That is needed to show that 3(a) and 3(b) are logically equivalent. But statement 4 is the logical law of *Distribution*, which has been questioned by various authors who considered abandoning classical logic for the sake of the quantum-mechanical world. (See further Section 6 below.)

Let us briefly look at the simplest of the familiar, somewhat more realistic, examples, the Two-Slit experiment. Here an electron source sends out a stream of particles towards a screen. There is a barrier, with two slits which can be open or closed. This is a crude position measurement: if one slit is closed and an electron hits the screen, then it must have passed through the other slit. If we place detectors at the slits, we find that any electron found on the screen side was detected at one or other slit. There are now three important probabilities for any region X on the screen being hit:

1. $P(X|\text{slit 1 open}) = p_1$ 2. $P(X|\text{slit 2 open}) = p_2$

3. $P(X|\text{both slits open}) = p_{12}$

Suppose we try the following model. There is a parameter Y which takes value 1 or 0 depending on whether the electron passed through slit 1 or 2. Then surely(?) we have

4. $p_1 = P(X|Y = 1)$ 5. $p_2 = P(X|Y = 0)$

6. $p_{12} = P(X|Y = 1 \text{ or } Y = 0)$

It follows at once that p_{12} is some sort of average (a convex combination) of p_1 and p_2.[2] Therefore the model requires that:

7. p_{12} lies in the interval with endpoints p_1, p_2.

In other words, p_{12} cannot be larger than both, or smaller than both. In the experiment, unfortunately, the interference pattern which appears when both slits are open shows many violations of this condition.

One reaction would be to stick with the model, and reject the probability theory and/or logic involved in the deduction of 7 from 4–6. That is the cry for logical revolution. But the other, and more orthodox, reaction is to reject the representation of 1–3 by 4–6. To put it in other words: we can reject the idea that the electron must have a definite position (in slit 1 or in slit 2) at the time of its passing the barrier.[3]

I have said enough for now, I think, to show that the easy and quick argument, i.e.

random variables on a classical probability-space satisfy Bell's Inequalities; since those are violated, we must reject classical probability theory (and/or logic),

is a *non sequitur*. It is wonderful enough that some phenomena may not admit any possible causal model—wonderful enough to establish the need for a theory diverging drastically from classical physics. The surface phenomena in the experiment are easily representable by a classical probability function. This does not *ipso facto* remove their mystery. The point is only, to put it bluntly, that classical logic and probability theory as such do not rule out miracles, nor telepathy, nor violation of Bell's inequalities.

2. GENERAL AND GEOMETRIC PROBABILITY MODELS

The new phenomena do not force violations of classical probability theory or logic. On the other hand, they do not fit the classical Common Cause models. What sorts of models do they fit? What sorts of models does quantum theory utilize; do they perhaps depart from classical logic *in fact*? Let us take a closer look first at what theoretical models must be like in general, and then inspect the curiously geometrized kind of probability models found in quantum theory. In this chapter we will have a preliminary look only, but one sufficient to make clear the basic ideas. All of this will be properly and precisely generalized in the next chapter.

A theory is, in essence, a set of models. These models are provided in the first instance to fit observed and observable phenomena. But of course, the description of these phenomena is in practice already by means of some models—a very modest sort, which we call 'data models' or 'surface models'. When the theory is devised, and the official theoretical models are constructed, certain parts thereof are offered as images of the empirical phenomena; I call these *empirical substructures*. They are meant to be isomorphic to the data models that encapsulate the phenomena to be 'saved' by the theory. This can be put in another way: the data or surface models must be isomorphically embeddable in theoretical models.

A great deal of thought has been given, since the advent of quantum theory, to the form that any possible surface model must take. Foundational work on the theory, especially in the 'quantum logic' tradition, sometimes looks as if one is engaged in deducing the whole of quantum theory from assumptions about what any observable phenomenon (any surface model) must be like, a priori. These deductions do go remarkably far, on the basis of assumptions that seem surprisingly weak and plausible.[4] I shall not discuss this further here. For the purpose of our present chapter, it suffices to stay with a quite naive characterization of the surface models.

The preceding chapter focused on an experimental situation of a quite simple structure. We were given several alternative measuring arrangements, a classification of possible outcomes, and some probabilities extrapolated from (imagined) observed frequencies. As the structure we are given the following:

A *surface model* consists of:

   (*a*) two sets of observable conditions: *PRC* is a set of realizable measurement choices, and *PRS* a set of possible outcomes;
   (*b*) the *surface state P*, a non-negative real-valued function *P* with domain part of *PRC* × *PRS* and range [0, 1], which assigns probabilities of outcomes conditional on measurement.

This structure is subject to certain minimal conditions which (as discussed) guarantee that *P* is mathematically extendible to a classical probability function. The numbers assigned by the surface state I call *surface probabilities*. Surface models are also referred to as experimental or data models.

What about theoretical models? The 'classical' case studied in the preceding chapter we can sum up as follows. There is a partition *Q* (of 'hidden states') whose members determine, with probability 1, the exact outcome of each measurement performed. The more general case must therefore be the one in which even the finest partition—of possible ways the situation can be—fails to lead us to such perfect predictions of the outcomes of measurements, if performed. States there may be, but this much information they cannot give. Thus we arrive at the general idea of a theoretical model for an experimental situation:

> The model provides a family M of *observables* (physical magnitudes) each with a range of possible *values*; a set S of *states*; and a *stochastic response function* P_s^m for each m in M and s in S, which is a probability measure on the range of m.

The number $P_s^m(E)$ is to be interpreted as the probability that a measurement of m will yield a value in E, if performed when the state is s. This explains to some extent what it shall mean for such a theoretical model to 'fit' an experimental model. But it does not quite—it only tells us the probabilities of surface phenomena, on the supposition of a measurement *and* of a state. The latter is again something theoretical, behind the phenomena. I propose to use the most stringent notion of fit that we can have here:

> A theoretical model *MT fits* an experimental model *ME* just in case *MT* has some state s such that the function P_s^m contains the surface state of *ME*, relative to some identification of the measurement setups as measurement of the physical magnitudes m.

An objection will at once occur: what if the theoretical model is really right, but the experiment is being done many times with the system now in state s, then in state s', etc.? My answer is that, if this is possible, then the theoretical model had better also contain a state s^* which gives the correct results for that case. Clearly, s^* will have to be or act like a *mixture* or *average* of the states s, s', ... But we should not pretend that we already know a priori how such 'mixed states' are related to other states or other measurements—that will all be part of the theoretical model construction.

With this very general notion of theoretical model, *any* surface state can in principle be fitted. At the risk of being boring, I repeat that the surface state is just part of a probability function, in the sense of classical probability theory, which is defined for all surface phenomena. To that extent, anyway, every theoretical model is 'classical'. But these models need not be classical in the sense of being deterministic or even causal models. We should now look for some construction that does furnish us with surface states, as simply as is possible, but without entailing even partial determinism.

For ease of discussion, let us continue to consider only quantities that have 0,1 as possible values. Then a state s must specify two probabilities for a given quantity m: $p_1 = P^m_s(1)$ and $p_2 = P^m_s(0)$. Obviously, p_1, p_2 are non-negative and sum to 1. Now there is an easy geometric representation of this, with a vector of length 1 (see Fig. 5.1). By Pythagoras' theorem $x_1^2 + x_2^2 = 1$, and obviously these squares are non-negative. So why not represent the state by the vector (x_1, x_2) such that $p_1 = x_1^2$ and $p_2 = x_2^2$? We can immediately introduce more quantities, such as m', by rotating the axes of the coordinate system. The vector which is (x_1, x_2) as described in the m-frame is (y_1, y_2) as described in the m'-frame, but of course $y_1^2 + y_2^2 = 1$ again because length is invariant. So these numbers y_1^2, y_2^2 can be the probabilities $P^{m'}_s(1)$, $P^{m'}_s(0)$. The geometric frame of reference depicts, therefore, the experimental arrangement which measures the quantity in question.

If the vector (x_1, x_2) lies along the m_1-axis, then $x_2 = 0$, so $x_1 = 1$. In this case the outcome of an m-measurement is determined with probability 1 to be the first ($m = 1$) value; similarly for the m_0-axis and x_2. Thus we also say that the pair of vectors $(x_1, 0)$ and $(0, x_2)$ represent *eigenstates* of m: the states in which the outcome of an m-measurement is determined with value 1. (The possible values of m—in this case, 0 and 1—are its *eigenvalues*.) Call these vectors $|1\rangle$ and $|0\rangle$ or, if we need to be more explicit, $|m_1\rangle$, $|m_0\rangle$. Then the probabilities x_1^2 and x_2^2 represented by vector $v = (x_1, x_2)$ can be expressed also geometrically as follows:

$$x_1^2 = \cos^2 (\text{angle between } v \text{ and } |m_1\rangle)$$

$$x_2^2 = \cos^2 (\text{angle between } v \text{ and } |m_0\rangle)$$

And we can also express this algebraically in a convenient calculation form, because:

If (x_1, x_2) and (z_1, z_2) are two unit vectors (as described in the same coordinate system) then $x_1 z_1 + x_2 z_2 = \cos (\text{the angle between them})$.

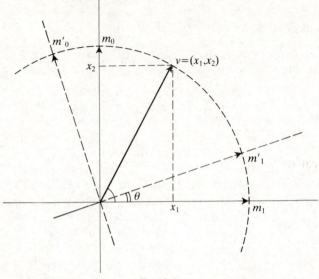

FIG. 5.1

Do we now know how to represent any surface state? Well, the quantities m and m' depicted above cannot be jointly measured (are 'incompatible'), as far as our representation is concerned; for we cannot select two frames of reference at once.

But it is clear that the choice of axes to represent quantities m and m' entails a certain relationship between the pairs of functions P_s^m and $P_s^{m'}$. We must proceed very delicately when we think about this, to avoid falling prey to false analogies with the earlier, causal models. Suppose the angle between the two sets of axes (in the plane) equals θ. If $s = |m_1\rangle$, that means that

$$P_s^{m'}(1) = \cos^2 \text{ (angle between } |m_1\rangle \text{ and } |m_1'\rangle)$$
$$= \cos^2 \theta = T(m' = 1|s)$$
$$P_s^{m'}(0) = \cos^2 \text{ (angle between } |m_1\rangle \text{ and } |m_0'\rangle)$$
$$= \cos^2 (\theta + \pi/2) = \sin^2 \theta$$

$$= 1 - \cos^2 \theta = 1 - T(m' = 1 | s)$$
$$= T(m' = 0 | s)$$

These two numbers are the *transition probabilities*. We can also write them so as to indicate the conceptual connection with conditional probability. So we can write $T(m' = 1 | s)$ as $T(m' = 1 | m = 1)$ since $s = | m_1 \rangle$, and therefore also as $T(m = 1 | m' = 1)$ since the quantity remains the same if we interchange m and m'. Such alternative ways of writing may be convenient, but must be used with great care to avoid philosophical confusion. The number so designated is the probability that an m'-measurement yields value 1, when performed on a system in state s, *on the supposition that* $s = | m_1 \rangle$, i.e. given that an m-measurement would have been *certain* to yield value 1. *It must not be confused with* 'on the supposition that an m-measurement (has or) would have yielded 1'; however we construe that, we court logical disaster if we drop the word 'certain'. (Given the controversies in our subject, it is impossible to emphasize this point too much.)

So the transition probability *is* a conditional probability, but with a special class of antecedents. Investigating this a little further, we can see what role it can play and also the disanalogy with a more unfettered classical use. Let us take another state s represented by a vector v in the same plane, and let the angles between v and $| m_1 \rangle$, $| m_1' \rangle$ respectively be ϕ and ψ, with $\phi + \theta = \psi$:

(1) $v = (x_1, x_2) = (\cos \phi, \sin \phi)$ in the m-frame

(2) $v = (y_1, y_2) = (\cos \psi, \sin \psi)$ in the m'-frame

So the probabilities are related as follows for the state s represented by vector v:

(3) $P_s^m(1) = \cos^2 \phi$ $P_s^m(0) = 1 - \cos^2 \phi = \sin^2 \phi$

(4) $P_s^{m'}(1) = \cos^2 \psi$ $P_s^{m'}(0) = 1 - \cos^2 \psi = \sin^2 \psi$

Remembering that transition probabilities are symmetric here, we can deduce:[5]

(5) $P_s^m(1) = P_s^{m'}(1)T(m = 1|m' = 1)$

$+ P_s^{m'}(0) [1 - T(m' = 1|m = 1)]$

$\pm 2\sqrt{[P_s^{m'}(1)P_s^{m'}(0)T(m = 1|m' = 1)}$

$\times (1 - T(m' = 1|m = 1))]$

of which the third term is sometimes called a 'probability interference' term. The equation may call to mind a classical theorem. Recall that $1 - T(m' = 1|m = 1) = T(m' = 0|m = 1)$, and so by the symmetry of T equals also $T(m = 1|m' = 0)$. Therefore (5) is the same as

(5*) $P_s^m(1) = P_s^{m'}(1)T(m - 1|m' = 1)$

$+ P_s^{m'}(0)T(m = 1|m' = 0)$

$\pm 2\sqrt{[P_s^{m'}(1)P_s^{m'}(0)T(m = 1|m' = 1)}$

$T(m' = 1|m = 0)]$

If you now thought of the state attribution $[m = 1]$ as an ordinary statement, which had $[m = 0]$ as its negation, and so forth, you would write (5*) in the form

(5!!) $P(A) = P(B)P(A|B) + P(\text{not-}B)P(A|\text{not-}B)$

$\pm 2\sqrt{[P(B)P(\text{not-}B)P(A|B)P(A|\text{not-}B)]}$

and so be under the impression that you had found a violation of equality

(6) $P(A) = P(B)P(A|B) + P(\text{not-}B)P(A|\text{not-}B)$

which is the classical 'theorem of total probability'. But of course this shows instead that in this context, although m' can only have value 0 or 1, the expression '$[m' = 0]$' does not stand for the negation of '$[m' = 1]$'. It stands instead for the state in which a measurement of m' is *certain* to have outcome 0.

Still, (5) and (5*) show that the probabilities for measurements of m are always constrained by the probabilities for measurements of m', together with the factor $T(m = 1|m' = 1)$ that relates the two observables. I say 'constrained', not 'determined', because square roots can be positive or negative, so we had to write $\pm$ in the equation. The connection with conditional probability will be pursued further in Chapter 6.

Proofs and illustrations

So far I have focused on the geometric representation of probabilities in the real plane, i.e. two-dimensional space with real-number coordinates. The possibilities of representation are enormously enriched by allowing more dimensions, and, even more liberally, complex-number coordinates. Here is a question to illustrate a use of three dimensions: can we have two observables A and B which are incompatible, and have only the two values ± 1, yet a state in which each possible value is equally likely to be found for either? Not if A and B are represented by axes in the same plane. So let us use three-dimensional space. Where X is either A or B, let X have only one eigenvector $|X_1\rangle$ for value $+1$ but two for value -1. Now can we choose a vector which makes a 45° angle (whose $\cos^2$ equals $\frac{1}{2}$) with both $|A_1\rangle$ and $|B_1\rangle$? Certainly, if we go outside the plane on which these two lie. These are all unit vectors, so their points lie on a sphere. Think of it as the Earth; let $|A_1\rangle$ be at the North Pole and $|B_1\rangle$ at Greenwich (latitude, let us say, 50°N). We look at the 45°N latitude and find a point Y on it. Draw a great circle through Y and Greenwich, and measure the separation. If it is less or more than $\frac{1}{8}$ of the circle (45°), move point Y until you get it right: then Y marks the state vector required. If it represents state s, then $P_s^A(1) = P_s^B(1) = 0.5$ and of course $P_s^A(0) = P_s^B(0) = 1 - 0.5 = 0.5$ too. This sort of pictorial argument will remain useful below.

In Chapter 6 I shall relate a fundamental theorem, due to Gleason, which determines exactly what combinations of probabilities can be represented in Hilbert space geometry.

3. ACCARDI'S INEQUALITIES

Not every experimental situation can fit the simple, causal models of the preceding chapter because Bell's Inequalities might be violated. But neither is it the case that every experimental situation can fit the simple geometric models of the preceding section. (This point is crucial to guarantee empirical significance!) We can conceive of violations of their basic features. Two of these features are common to all quantum-mechanical models. One is the rather obvious symmetry in transition probabilities, which we cannot expect very well a priori. The second is that for each m and b there must be a state in which an m-measurement is certain to have outcome b. One is tempted to say here: well, let the theorist conjecture all such states. But that is not generally possible; this feature limits what the theory can countenance as a genuine measurement of quantity.

Just as Bell's Inequalities are statistical invariants of causal modelling, so must there also be statistical invariants of geometrical probability modelling that describe *its* empirical limits. This point was made and developed by Luigi Accardi; I shall here report the initial results (Accardi and Fedullo 1982), which were stated in terms of transition probabilities.

Going back to the deduction of (5) in the preceding section, suppose that s is the eigenstate $|m_1''\rangle$ of a third observable, which also has just the two eigenvalues 0 and 1. For brevity write

$$T(m = 1|m' = 1) = p$$

$$T(m' = 1|m'' = 1) = q$$

$$T(m'' = 1|m = 1) = r$$

In a geometric probability model these are symmetric; i.e., $T(X = 1|Y = 1) = T(Y = 1|X = 1)$. This fact, together with the limitation to just two values, means that all transition probabilities are now determined. For example, for m and m' we have:

	$m' = +1$	$m' = -1$
$m = +1$	p	$1 - p$
$m = -1$	$1 - p$	p

There is also no difficulty at all in finding an angle $\theta_{mm'}$ such that $p = \cos^2 \theta_{mm'}$; in that case $1 - p = 1 - \cos^2 \theta_{mm'} = \sin^2 \theta_{mm'}$. Similarly for q and r. But now, all three angles $\theta_{mm'}$, $\theta_{mm''}$, and $\theta_{m'm''}$ must coexist between vectors in the same space! (Though not necessarily in the same plane.) Accardi deduces:

(7) If m, m', and m'' are represented by means of vectors based on the real number field, then

$$p + q + r - 1 = \pm 2\sqrt{(pqr)}.$$

This result is actually a corollary to a more general one, for, as is well known, quantum theory uses the geometry of vector spaces based on the field of complex numbers. That more general result is:

(8) (*Accardi–Fedullo*): If m, m', and m'' are represented by means of vectors based on the complex number field, then

$$-2\sqrt{(qpr)} \le p + q + r - 1 \le 2\sqrt{(pqr)}.$$

We see therefore that the admission of complex numbers allows for the whole range of possibilities, whose extreme limits appeared in result (7) about real number representation.

This can be illustrated by means of polarized light. Suppose filters F_A and F_B have orientations inclined to each other with angle θ_{AB}. We can then prepare a photon in eigenstate $|A_1\rangle$ by passing it through F_A. Any such photon has probability $\cos^2 \theta_{AB}$ of passing through filter F_B. The angles here are all co-planar, a rather simple situation. Yet if we thought that the filters did not change the photons sometimes, but only revealed preceding properties, then (as Accardi also points out) the numbers p, q, r would have to satisfy a typical Bell inequality—which is violated. This form of the Bell Inequalities (for the transition probabilities) is:

(9) (*Bell–Accardi*): $|p + q - 1| \le r \le 1 - |p - q|$

Of course, this is different from Wigner's formulation, which concerns probabilities of joint measurement outcomes.

We have now seen that there are strict theoretical limits to what (conceivable) empirical phenomena could be represented in the geometric probability models, just as there are for causal modelling.

Can there be phenomena that admit not even any complex geometric model? If such phenomena occurred, then quantum mechanics too would be in trouble. And indeed, it is not so difficult to envisage violations of condition (7), at least in the abstract. There are also beautifully picturable thought-experiments, devised by Mielnik (1968, 55)[6] and by Aerts (1986), in which such violations occur.

4. THE END OF COUNTERFACTUAL DEFINITENESS

Let us quickly take stock. In the preceding discussion of incompatible observables—parameters which cannot be jointly measured—we found no reason to call for a revolution in logic or probability theory. But this was in part because I did not ask for the imposition of such postulates as Value Definiteness and Veracity in Measurement. I do not think of this as a giving up, or as an admission or concession—I doubt that anyone ever cherished such principles. (You can't just introspect, catch a sense of comfort, and say oh, that is a principle which governed classical thought! Our breast harbours no such oracle of intellectual history.) The basic idea of measurement or observation has not had this connotation of passively received revelation since what Kant called his own Copernican revolution. A measurement situation puts nature to the question—in the Inquisitors' infamous sense—and so yields one of the possible values. Measurement reveals; it reveals something; but not always, in a straightforward way, a value already possessed independently.[7]

We turn now to some other arguments directed against logic and supposedly classical intuitions. Here we are much more closely concerned with Bell's Inequalities. John Bell himself, in his (1981) article on Bertlmann's socks, confronts us with more or less the following puzzle.[8]

Suppose we think of the two-generals situation as involving two particles travelling to the two receivers. There they will meet barriers, which depend on the setting. Armand chose setting 2 on the right, and his particle passed. Now, given that the right particle passed the barrier of setting 2, our perfect correlation entails that the left particle was not able to pass the setting 2 barrier there. However, Alfredo chose setting 1, and the left particle did pass. So we conclude:

1. The left particle was able to pass barrier 1 and was not able to pass barrier 2,

and write this in our experimental report.

Let us abbreviate 'the left particle was able to pass barrier i' as $A(i)$. That abbreviates 1 to [$A(1)$ & not-$A(2)$]. The following is a theorem of classical probability theory:

2. The probability of $A(1)$ and not-$A(2)$ *plus* the probability of $A(2)$ and not $A(3)$ *is greater than or equal to* the probability of $A(1)$ and not-$A(3)$.

But we continue the experiment, writing down many reports like statement 1. Then we extrapolate the probabilities from our data, and lo! theorem 2 is violated.

Is 2 really a theorem? The reader can check it with a little calculation, without trouble. So the experiment has violated our classical blend of logic and probability. Yet it did this by its usual violation of Bell's Inequalities.

There is no sleight of hand here, with omitted suppositions of measurement, as in the preceding section. Each experimental report, however, involved an inference from a measurement outcome to the existence of a property that the particle already had when it came to the barrier. This was the *modal* property of *being able to pass*—or else not being able to pass. Now this inference had its own ideas, so to say, about what measurement reveals. And it also relied on the conclusion that, if the right particle passed the barrier, this showed that the left particle could not have passed it *if it had been put to the same test*. So we have an inference to a 'counterfactual conditional': a conditional statement about what would have happened if something else had been done.

This subject of counterfactual conditionals has been much studied in contemporary logic, and we are in a good position to evaluate such arguments. The initial literature concerning this twist to Bell's results, however, proceeded without reference to recent logic. Mainly following Henry Stapp, some writers have given short and elegant derivations of Bell's Inequalities by relying on intuitions concerning counterfactuals (see Stapp 1971; Eberhard 1977; Herbert and Karush 1978). The main one is:

3. *Counterfactual Definiteness*: If a measurement can only have outcomes 0 or 1, then one of the following two statements is true:
   (*a*) if the measurement is (be, were) done, then the outcome will be (would be) 0;
   (*b*) if the measurement is (be, were) done, then the outcome will be (would be) 1.

This is an instance of a principle of the first successful logic of counterfactual conditionals, which was due to Robert Stalnaker (1968):[9]

4. *Conditional Excluded Middle*:

$$(A \rightarrow B) \text{ or } (A \rightarrow not\ B)$$

How does this lead to Bell's Inequalities?

Well, as Common Cause factor, we choose the conjunction of all such conditionals which are true as the particles leave the source. By 3, this conjunction implies logically what the outcome will be for each setting choice. That puts us immediately at the end-stage of the deduction of Bell's Inequalities, just at the point where one introduces Wigner's eightfold classification.

If Stalnaker's (1968) original logic of conditionals were an inalienable part of classical logic, we would now have grounds for a logical revolution.[10] But in the history of this chapter of logic, Conditional Excluded Middle had been immediately attacked by David Lewis, and Stalnaker had weakened it to a status in which it no longer licenses Counterfactual Definiteness (see van Fraassen 1974*b*). Thus, those arguments in the physics literature rested on a 'plausible intuition', which had however already been rejected in formal logic. One good example to bring out the reasons is the venerable example, Which is true, if Verdi and Bizet had been compatriots, Verdi would have been French? or Bizet would have been Italian?

This defence has one shortcoming: it leaves open that there may be *some* sense of the conditional, for which 3 and 4 hold. There are happily two other defences. One is that, in the last stage of the Bell's Inequalities deduction, in the preceding chapter, we needed to appeal to Hidden Autonomy. That is the assumption that the hidden Common Cause factor is stochastically independent of the settings. In our present application to this hidden variable made up of counterfactuals, that appears to require something of form

5. $P(A \rightarrow B|A) = P(A \rightarrow B)$

But by some simple principles of probability, and a principle common to both Stalnaker and Lewis (that $A \& (A \rightarrow B)$ is equivalent to $A \& B$), we derive

6. $P(A \rightarrow B) = P(B|A)$

which is another hotly debated, and generally rejected, principle in this area (see Harper *et al.* 1981 and Hajek 1989 for details).

Finally, probabilities are not necessarily attachable to any and all propositions, no matter how tenuously linked to fact.[11] Perhaps they do not attach to counterfactual conditionals, for example.

So we have seen that the devastatingly demure argument with which we began was full of hidden complexities, none of them more than tenuously connected to classical probability and logic as such. Yet, again we have learned something. The violation of Bell's Inequalities demonstrates empirically that we should not look to measurement outcomes to give us direct information about state, propensity, capacity, ability, or counterfactual facts. From fact to modality, only the most meagre inferences are allowed.

5. MODELS OF MEASUREMENT: A TRILEMMA FOR INTERPRETATION

There is a distinct threat of circularity for such a theory as quantum mechanics, which mentions measurement in its basic principles and yet itself covers measurements as a subclass of interactions. In this section we shall have a preliminary exploration of the threat.

To begin, we need to specify what a theoretical model, as introduced in Section 2 above, looks like for a compound system. Suppose that, for individual systems X and Y and the compound system $X + Y$, we have

Sets of states S_X, S_Y, S_{XY}
Families of observables M_X, M_Y, M_{XY}
Stochastic response functions PX_s^m, PY_s^m, PXY_s^m where m and s range over the relevant observables and states in each case.

Obviously there must be certain relations among these which reflect the compound–component relation. Specifically, there must be a way of thinking, for example, of a state of X or Y as some aspect or part of a state of $X + Y$. This relationship can be captured by two functions:

$$h_X : S_{XY} \rightarrow S_X \qquad h_Y : S_{XY} \rightarrow S_Y$$
$$h_X^{-1}(S_X) = S_{XY} = h_Y^{-1}(S_Y)$$

Similarly each observable on $X + Y$ must somehow concern each of X and Y, however trivially, and we can let the same mappings extend to this relationship as well:

$$h_X : M_{XY} \rightarrow M_X \qquad h_Y : M_{XY} \rightarrow M_Y$$
$$h_X^{-1}(M_X) = M_{XY} \qquad h_Y^{-1}(M_Y) = M_{XY}$$

So far these mappings carry no information, but we note that M_X and M_Y include the trivial observables which always have value 1; call these i_X and i_Y. Now when $h_Y(m) = i_Y$, we can think of m as really just representing its image $h_X(m)$, an observable pertaining to X alone. In this way the families M_X and M_Y pertaining to X and Y alone have each a 'copy' inside M_{XY}.

To enter the last link, which will give all this a use, let us first simplify the notation. When s is a state of $X + Y$, we think of its images $h_X(s)$ and $h_Y(s)$ as 'reduced states' pertaining to its components. Let us designate these more briefly as #s and s# respectively. Secondly, when we look at an observable m pertaining to $X + Y$, let us

designate m as %n if $h_X(m) = n$ and $h_Y(m) = i_Y$;
designate m as n% if $h_X(m) = i_X$ and $h_Y(m) = n$

Then we need to have the following alignment of stochastic response functions:

$$PXY_s^{\%n} = PX_{\#s}^n \text{ for all } n \text{ in } M_X$$
$$PXY_s^{n\%} = PY_{s\#}^n \text{ for all } n \text{ in } M_Y$$

for all states in S_{XY}.

We will see this general scheme instantiated in a specific manner when we come to the quantum-mechanical treatment of compound systems in Chapter 7.

So far so good, but the theory also covers, as processes described, those measurement interactions for which it makes predictions. Suppose particularly that Y is the measuring apparatus for m, and the measuring process is the evolution of $X + Y$ from state s to state s'. Since Y is the apparatus, its final condition must register the measurement outcome. Therefore the probability measure $PX_{\#s}^{m}$ must at least indirectly give us probabilities for what state $s'\#$ of Y at the end must be.

A word of caution here. It will be natural and tempting to think that our present discussion applies directly to quantum theory—in the abstract, anyway—with those states being the usual quantum-mechanical states. That should *not* be assumed. What I have said so far (or am about to say) concerns at best the picture we shall have of what quantum theory says *after it is interpreted in some definite way*. These different interpretations will display some connection between the usual quantum-mechanical states and the states discussed here, but may or may not identify them.

To continue then: for each Borel set E of possible values of m there must be a corresponding set of possible states $Y(E)$ of system Y, so that PX_{x}^{m} admits of the dual interpretation:[12]

(*1*) $PX_{x}^{m}(E)$ is the probability that an outcome in E will be found, given that m is measured on X, beginning with an initial state s of $(X + Y)$ such that $x = \#s$.

(*2*) $PX_{x}^{m}(E)$ is the probability that the final state s' of $(X + Y)$ will be such that $s'\#$ is in $Y(E)$, when this process began with a certain initial state s of $(X + Y)$ such that $x = \#s$.

But it is clear now that in this story the transition $s \rightarrow s'$ is indeterministic. Indeed, PX_{x}^{m} is now seen to be in effect a probability measure on the set of possible final states of $X + Y$ resulting from this process that starts with a given initial state $s(X, Y)$.

So—this story cannot fit a theory in whose models of measurement the total system develops deterministically. And now we have come to the paradigm puzzle about quantum mechanics: in its models of an isolated process, the quantum-mechanical state *does* develop deterministically, in accordance with Schroedinger's equation.

There are only three ways to reconcile quantum theory with the conclusion we drew.

1. If a measurement takes place in isolation, the process is not subject to Schroedinger's equation, but evolves indeterministically (there is an 'acausal' transition).
2. No measurement interaction is the temporal evolution of an isolated system; an interaction with an environment, *incompletely described*, is part of what a measurement is.
3. The quantum-mechanical state is only one aspect or part of the total state of the system at a given time; which observables have what values is a second aspect or part of what the system is like at that time.

All three of these solutions are found, in various versions and guises, in the literature on the interpretation of quantum mechanics. What we note here is that the disjunction of this trio is forced on us by the general discussion. For it established:

If every process in an isolated system is a deterministic state-transition, and the probability measure PX_x^m is not trivial (with values 0, 1 only), and if measurement requires the coincidence of characterizations *1* and *2* of that probability measure, then no process is the temporal evolution of an isolated system.

The possible reactions, to allow for measurement in quantum mechanics after all, are therefore just these three:

(*a*) Not every quantum-mechanical state transition in an isolated system is deterministic.
(*b*) A measurement is a process in a non-isolated system.
(*c*) The quantum-mechanical state of an isolated system does develop deterministically, but it is not the total state, and some aspect of that total state evolves indeterministically.

The modal interpretation that I shall advocate will take the third option.

6. INTRODUCTION TO QUANTUM LOGIC

There is no need for a logical revolution to appreciate the subject of quantum logic. That quantum-mechanical propositions have a logic of their own was suggested by von Neumann in his book *Mathematical Foundations of Quantum Mechanics* in 1932. Birkhoff and von Neumann (1936) gave the idea an especially elegant form. Herman Weyl's essay (1940) 'The Ghost of Modality', written in memory of Husserl, related it to modal logic—indeed, Weyl gave in rudimentary but prescient form the outline of the semantic analysis that would eventually unify modal, quantum, and intuitionistic logic.[13] We are now in a good position to understand this logic, for many of its features appear equally well in the general context of geometric probability models.

A theoretical statement about a system, whether real or imagined, typically says something about its state. As representative here, let us take

1. The system is in some state s such that $P_s^m(x) = 1$

for which we can use our previously introduced term *state-attribution* (since it fits the use in Part I of that term, in classical contexts). That form is of wider application than it may seem; for suppose observable m' is defined by saying that it has value 1 exactly when the value of m is in interval E and has value 0 if the value of m outside that interval. Then we can equate:

2(a). The system is in a state s such that $P_s^{m'}(1) = 1$

(b). The system is in a state s such that $P_s^m(E) = 1$

thus broadening state-attributions to sets of values.[14] Finally, we abbreviate 2(b) to

3. $[m, E]$.

Now we can go to the geometric probability model, to see how this proposition is represented there. Obviously, it is true whenever the system is in some state which is an eigenstate of m with respect to interval E (i.e. of m', with respect to value 1). In a three-dimensional example, with m having distinct eigenvalues 1, 2, 3 and $E = \{1, 2\}$, we see that statement 3 requires the vector which represents state s to lie in the plane containing vectors $|m = 1\rangle$ and $|m = 2\rangle$. If $E = \{1\}$, that vector must actually be $|m = 1\rangle$; that is, it is constrained to lie in a certain line through the origin. If E had N distinct eigenstates of m in it, the constraint would be to an N-dimensional subspace. So: the state-attributions are represented by subspaces.

The logic of state-attributing propositions can now be 'read off' the model. We say that state *s makes true* or *satisfies* $[m, E]$ if it is as described in $2(a)$–(b). The first logical notions are clear:

4. $[m, E]$ *entails* $[m', E']$ exactly if any state which satisfies the former also satisfies the latter; that is, exactly if $[m, E] \subseteq [m', E']$

5. $[m, E]$ is the *conjunction* or *meet* of $[m', E']$ and $[m'', E'']$ exactly if the former is satisfied whenever the latter two are satisfied; that is, exactly if $[m, E] = [m', E'] \cap [m'', E'']$

Happily, the intersection of two subspaces is always again a subspace. As a special case we note:

6. $[m, E \cap E'] = [m, E] \cap [m, E']$

which makes good intuitive sense. For the other familiar operations of logic, more care is needed.

What does 'or' mean? One classically correct answer is this:

7. The *disjunction* (P **or** Q) of two propositions P and Q is the logically strongest proposition which is entailed by P and also by Q

'Logically strongest' means that it itself entails as much as possible. Hence we gather from definition 7:

8. P entails (P **or** Q); Q entails (P **or** Q)

9. if P entails R and Q entails R then (P **or** Q) entails R

These are also in lattice theory the principles for the *join* operation. There is indeed such an operation on the subspaces (they form a lattice, in mathematical parlance):

10. If P and Q are two subspaces, there is a smallest subspace $P \oplus Q$ which contains both—the *orthogonal union* or *join* of P and Q.

11. $[m, E] \oplus [m, E'] = [m, E \cup E']$

Corollary 11 shows us that in the simplest case (each proposition is 'about' the same observable m) this construction agrees well with our intuitive grasp of 'or'. But even here, we note at once that a probability measure can give 1 to $E \cup E'$ without giving 1 to either part E or E'. Similarly in the geometric representation:

12. If vector x is in P and vector y in Q, then not only x and y but every superposition (i.e. linear combination) $ax + by$ also lies in $P \oplus Q$.

So the orthogonal union can be much larger than the ordinary union. $P \oplus Q$ may represent a true proposition, while neither P nor Q does so.

We can proceed similarly with 'not', though here it is easier because we automatically deal with one observable only. In analogy to definition 7, we write

13. The *negation* (not-P) of a proposition is the logically weakest proposition which is logically incompatible with P.

Here P and Q are logically incompatible exactly if no state can satisfy both. The strongest such proposition Q is of course always the self-contradiction. If $P = [m, E]$, then the weakest such proposition—which entails as little as possible—must be $[m, R - E]$. And indeed, we have a corresponding operation on the subspaces:

14. If P is a subspace, there is a largest subspace $P^\perp$—the *orthocomplement* of P—such that $P \cap P^\perp$ contains only the null-vector.

The following are consequences of this definition:

15. $P \oplus P^\perp = $ whole space

16. If $P \subseteq Q$ then $Q^\perp \subseteq P^\perp$

17. $(P \cap Q)^\perp = P^\perp \oplus Q^\perp;$ $(P \oplus Q)^\perp = P^\perp \cap Q^\perp$

18. $[m, E]^\perp = [m, R - E]$

Properties 17 are called DeMorgan's Laws; corollary 18 shows again that we have good intuitive contact with our pre-theoretic intuition.

But $P^\perp$ and P do not together exhaust the whole space, although $P \oplus P^\perp$ does. The vectors in $P^\perp$ are those which are orthogonal to the ones in P. Thus, if P is a line through the origin, and the whole space is three-dimensional, then $P^\perp$ is a plane through the origin. The 'Excluded Middle'—property 15—is therefore somewhat illusory, for it does not imply 'Bivalence'. A given state may satisfy neither P nor $P^\perp$. Of course, this derives from the fact that a probability function must assign 1 to the whole of its range, but may fail to assign 1 either to a given set E or to its complement $R - E$.

Birkhoff and von Neumann noted especially that the familiar *distributive laws* of classical logic are not obeyed here. Let P and Q be two orthogonal lines through the origin, containing vectors x and y respectively. Now let $a > 0$ and consider line S which contains superposition $ax + (1 - a)y$. Then we see that all three lines meet pairwise in the origin—the null vector:

(*a*) $(S \cap P) \oplus (S \cap Q)$ = the null space

(*b*) $S \cap (P \oplus Q) = S$

since $P \oplus Q$ is the plane that contains all the superpositions of P and Q, and therefore S. The distributive law, which fails here, holds in classical logic as

19. (*S* **and** *P*) **or** (*S* **and** *Q*) = [*S* **and** (*P* **or** *Q*)]

and this discrepancy is often pointed out as *the* failure of classical logic in quantum theory.

Let us follow this introduction with a more advanced presentation, to introduce the abstract concepts at work here.

A partially ordered set (*poset*) is a set K ordered by a relation $\leq$ which is:

reflexive, i.e. $x \leq x$ for all x in K;

transitive, i.e. if $x \leq y$ and $y \leq z$, then $x \leq z$, for all x, y, z in K;

antisymmetric, i.e. if $x \leq y$ and $y \leq x$ then $x = y$.

The notions of greatest lower bound (g.l.b.) and least upper bound (l.u.b.) in this partial ordering are generalizations of meet and join, to cover more than just finite combinations. Thus:

> the g.l.b. of a subset X of K in the ordering $\leq$ is the element VX such that $VX \leq y$ for all y in X, and if $z \leq y$ for all y in X also, then $z \leq VX$.

Dually for the l.u.b. of X. (Informally, the l.u.b. 'lies below' (or, in logic, entails) all and only those elements 'above' all members of X.) But there may not be any such elements in a given poset.

> *Definition*: The poset K, $\leq$ is a *lattice* if every finite subset of K has a g.l.b. and l.u.b. (with respect to $\leq$ in K), and is a *(sigma-)complete* lattice if this is so for every (countable) subset of K.

The lattice laws are simply the conditions required by this definition, as stated in the paragraph above.

There are many sorts of lattices. The g.l.b. or *join* of the whole of K, if it exists, is the unit element 1 and its l.u.b. or *meet* is the null or zero element 0. If there is an operator $^{\perp}$ on K which satisfies *mutatis mutandis* the description given in 14–18 above, the lattice is called *orthocomplemented*. If the distributive law 19 is satisfied (where **and** and **or** stand for meet and join), the lattice is of course called distributive. If it is both orthocomplemented and distributive, then orthocomplement is unique, and the lattice is called *Boolean*. It is then indeed a Boolean algebra, familiar to us from set theory and classical logic both.

In the present context, it is the lattice of subspaces that is of interest. I shall not make the notions of space and subspace precise until the next chapter, but shall here state some geometrical intuitions. The subspaces do indeed form an orthocomplemented complete lattice, partially ordered by set inclusion. It is also *atomistic*. (An atom is a non-zero element x such that $0 \neq y \leq x$ only if $y = x$. The property of atomism is that for each non-zero element y there is some atom x such that $x \leq y$, and indeed y is the l.u.b. of the set of such atoms.)

For a space with dimension greater than 1, this lattice is not distributive. It may be remarked that, if the dimension is finite, the lattice is still *modular*; i.e.,

$$(S \cap P) \oplus (S \cap Q) = S \cap (P \oplus Q) \qquad \text{if } Q \subseteq S$$

but this too fails for infinite dimensionality. What properties do hold with complete generality? Two subspaces are *compatible* if they are made up by orthogonal union of members of a set of mutually orthogonal subspaces. This can be stated in lattice-theoretic terms in various ways.[15] Jauch notes that the subspace lattice is always *weakly modular*; i.e.,

if $S \subseteq P$ then S and P are compatible.

Apart from the above, the only further property of the subspace lattice which is at all well known is the so-called Ortho-Arguesian Law, but it is already noxiously complicated.[16] The fact is that the set of laws which characterize exactly the family of lattices of subspaces is not axiomatizable. This was shown by Rob Goldblatt (1984). There are however representation theorems, due to Piron and others, which answer the question by less strict standards than that of the logicians' axiomatizability.[17] The only question is whether this fragment is especially important in some way. Von Neumann himself flirted later with needed extensions. (In terminology which will be explained later, the assertion that a system is in a particular mixed state, for example, is *not* a state-attribution in the above sense.) On some interpretations of quantum theory the entire focus is on this fragment; on others the fragment is too meagre to describe all significant features of a theoretical model. (I shall opt for the latter.) But the family of state-attributions, and its logical structure, does remain very important on all accounts, even if not paramount.

7. IS QUANTUM LOGIC IMPORTANT?

Now I have explained quantum logic, and the meaning of its operations; did you have to relinquish classical logic to understand me? Of course not.

You may well reply that I wrote all this in the metalanguage, in which language and semantic facts are described. So the logic of the language in which this chapter is written is classical. The language we spoke *about*—the little language of state-attributions—has a logic which is not classical. This is what we established.

Yet we established also that the language of state-attributions has an exact copy in our classical language. The proposition [*m*, *E*] we talked *about* has as copy in our language statement (2*b*) of the preceding section, which we can assert if we like. Thus, the little language of state-attributions, whose 'inner logic' is not classical, is a fragment of a larger language, whose logic is classical.

There is absolutely nothing revolutionary about a fragment whose inner logic is not that of the whole.

In the last section of the preceding chapter and the first of this one, I discussed an equivocation which could lead one to the conclusion that experimental results which violate Bell's Inequalities also violate classical probability theory. Because of the latter's close connection with logic, one might then also conclude that logic and/or the probability calculus must be revolutionized. It is not my idea to conjecture that anyone who advocates *that* conclusion was actually misled by an equivocation.[18] But I do want to insist that the very valuable concepts, techniques, and insights gained in the course of developing the quantum-logical approach to quantum mechanics can be appreciated well enough aside from any such non-standard interpretations. The interpretation I shall advocate, beginning with Chapter 9, will also focus on families of propositions whose internal logic is quantum logic. But this advocacy will still go well with a resolutely classical point of view with respect to logic and probability themselves.

PART III
Mathematical Foundations

The empirical content of the quantum theory can be summed up as: all empirical phenomena admit models of the sort that were sketchily introduced in the preceding chapter. Our next step is to describe more precisely and comprehensively the exact class of models that quantum mechanics provides. Happily, we have more than half a century of foundational research to draw on. Unhappily, it is not possible to present the results briefly while remaining rigorous and entirely neutral with respect to interpretation. These next two chapters will concentrate on what is specially required for our purpose: joint probabilities, symmetries, conservation laws, selection and superselection rules, composition of systems, interaction, and of course measurement.

I have tried to keep in mind two possible sorts of readers, with quite different needs. The first has little mathematical background; so I start off slowly, and often give proofs for special or simple cases, so that this reader can get to a working knowledge of foundations at an elementary level. The second is mathematically advanced; he or she can use these chapters as a handy reference, a key to the inevitable idiosyncrasies of terms and notation later on, and a guide to recent literature. Neither needs proofs for the main theorems, though there the one needs more faith than the other. For a less abstract introduction to foundations, oriented more towards the physics itself, I recommend Hughes (1989); for greater rigour and mathematically more advanced points of view, consult for example Redhead (1988) and Beltrametti and Cassinelli (1981a).

6

The Basic Theory of Quantum Mechanics

This chapter will treat of single systems, ignoring that they may consist of parts, and may interact with other systems.

1. PURE STATES AND OBSERVABLES

The custom in quantum theory to refer to physical quantities as *observables* reflects the initial conviction that each physical quantity must be associated with a realizable measurement (or experimental) arrangement. The stochastic response function is intuitively thought of as giving the probabilities of measurement outcomes. Later we shall look to see how measurement too can be modelled as a quantum-mechanical process, but for now we rely on that intuition. In the notation of the preceding chapter, we read $P_s^m(E)$ as the probability that a measurement of observable m yields a value in Borel set E. Before turning to the Hilbert space representation, I want to explain some concepts definable generally in terms of the stochastic response function, which will prepare us for questions to be raised later.

There are two concepts of *purity* for states. To begin, if s is a state, and m is an observable, let P_s^m be the function which assigns probabilities to outcomes of measurements of m, on the basis of state s. Then if s, t, and u are states such that

$$(1) \qquad P_s^m = bP_t^m + (1 - b)P_u^m$$

for a number $0 < b \leqslant 1$ and all observables m, we call s a *mixture* of states t and u. This convex sum of the stochastic response functions can also be countable, provided only that the 'weights' are non-negative and sum to 1. We call s *pure* if it is not a mixture of other states. When (1) holds we also call t and u *components* of s. These components provide an example of the following relation:

(2) t is *possible relative to* s iff P_t^m assigns probability 1 wherever P_s^m assigns the probability 1.

When equation (1) holds, we may say that t is 'classically' possible relative to s; we must leave open here whether we shall also find 'non-classical' cases. This relation yields a second concept of purity, but I will use a different word:

(3) s is *prime* iff for any other state t, t is possible relative to s only if s is also possible relative to t.

In other words, the set of certainties in such a state is maximal; it cannot consistently be increased.

In quantum-mechanical models, the same word 'pure' can be used for both concepts because the two defined extremes coincide.[1] It is also true in quantum mechanics that the complete class of states consists of the pure states and their mixtures (i.e. that decomposition of form (1) cannot go on for ever, but must eventually reach just pure states, at least if we allow countable sums).[2]

When we relate states and observables to each other via the stochastic response function, however, we cannot speak so categorically about what quantum mechanics assumes without straying into interpretation.

(4) States s and t are *empirically indistinguishable* exactly if $P_s^m = P_t^m$ for all observables m. Observables m and n are *empirically indistinguishable* exactly if $P_s^m = P_s^n$ for all states s.

Does empirical indistinguishability imply identity? Since any representation can without loss of adequacy be embedded in a 'larger' representation, the question makes no sense in general. After all, the set of states is a postulation, a theoretical construct used to model the phenomena. We can only ask: is there a 'smallest' model in which this is so?

But quantum mechanics is a theory, and a theory is identified by the set of models with which it provides us. Can we not just ask whether empirical indistinguishability implies identity in each of these models? This is the point where foundations and interpretation merge. The models provided in physics practice form an open-ended class, and are in certain respects only sketched. On the interpretation which I will advocate in the next Part, the question is answered *yes*. As far as the present chapter goes, that answer seems obviously compatible with the theory until we come to superselection rules; then it still is, though somewhat less obviously.

2. PURE STATES, OBSERVABLES, AND VECTORS

From now until we come to superselection rules, we will discuss only models in which all pure states are represented by vectors. In this section the observables are all represented by sets of vectors ('eigenvectors'). There is a simple rule, *Born's Rule*, for calculating probabilities in this representation. Though simple, the representation is neither unique nor elegant, it omits mixed states, and treats observables at best somewhat clumsily.

The models all use as pure state-space a separable Hilbert space, which is a vector space with inner product. I shall take some knowledge of vectors for granted, but will list and illustrate the basic definitions for Hilbert space in *Proofs and illustrations*. The inner product of vectors x and y I will denote $(x \cdot y)$; the norm or length of x, i.e. the square root of $(x \cdot x)$, as $|x|$. There will be three notational quirks. The first is that if b is a complex number, I write b^2 to denote its squared modulus, i.e. b^*b. This is to make my notation as far as possible neutral between real and complex numbers, a difference which does not usually intrude on general discussion. The second is that Σ sums over all indices when it appears without any. The third also concerns indices. These are usually written as subscripts, e.g. x_1, $x_2, \ldots, x_i, \ldots$ But if the term itself occurs as a subscript, or some other baroque form, it is convenient to write $x(1)$, $x(2), \ldots, x(i), \ldots$ instead. This is accurate, since x_i is just the value of some function—call it x—at the number i, after all.

Each vector represents a pure state; two vectors x and y represent the same pure state if they are parallel (i.e. if $x = ky$ for some number k). When $z = \Sigma a_i x_i$ we say that z represents a *superposition* of the states represented by the vectors x_i.

Suppose that observable m has a set S of possible values. We call S its *spectrum* and these values its *eigenvalues*.[3] To represent the observable, we choose a basis $\{x_i\}$ for the space and a mapping: $x_i \longrightarrow m_i$ of this basis on to S. The observable is represented by the basis so associated with it, and we call the members of that basis *eigenvectors* of the observable, sometimes writing $x_i = |m_i\rangle$ as convenient shorthand. (This is adapted from Dirac notation, which I shall not use in general.)

Now we are ready to state the rule for calculating probabilities of measurement outcomes in this representation. We begin with a simple case: observable m has exactly one unit eigenvector $|m_k\rangle$ corresponding to possible value m_k. Then:

> *Born's Rule* (simple case): If m is measured in the pure state represented by unit vector $y = \Sigma c_i |m_i\rangle$, the probability of outcome m_k equals the number $c_k^2 = (|m_k\rangle \cdot y)^2$.

This can be connected with the geometrical intuition as follows. Each vector y can be uniquely represented in terms of any basis $\{x_1, x_2, \ldots\}$ of unit vectors, using the inner product: $y = \Sigma(x_i \cdot y)x_i$. That is easy enough to see; for let $y = \Sigma c_j x_j$. Then

$$(x_i \cdot y) = \left(x_i \cdot \sum c_j x_j\right)$$

$$= \sum_j (x_i \cdot c_j x_j)$$

$$= \sum_j c_j(x_i \cdot x_j)$$

$$= c_i(x_i \cdot x_i)$$

$$= c_i$$

for the vectors x_i are orthogonal—hence $(x_i \cdot x_j) = 0$ if $i \neq j$—and of unit length. We also use the following terminology: if $y = \Sigma c_i x_i$, then $c_k x_k = (x_k \cdot y)x_k$ is the *projection* of vector y *along* basis vector x_k. The squared length of the projection along $|m_k\rangle$ is therefore exactly the probability mentioned in the Born Rule.

This rule referred to a unit vector, and applied only to the case in which there is just one basis vector for each eigenvalue. To cover all other cases is also straightforward. If the vector y you use to represent a pure state is not a unit vector, equate the probability to that calculated for the unit vector $y/|y|$. If the basis $\{|m_i\rangle\}$ that represents observable m contains duplications—e.g. $m_1 = m_2$—we say that m is not *maximal* or that its spectrum is *degenerate*. In that case we proceed a little more gingerly; here is the general statement.

Born's Rule: The probability of a measurement of observable m, as described above, on a system in state y, having value b as outcome, equals $\sum_i \{((|m_i\rangle \cdot y)^2 : m_i = b\}$.

We note in any case that we could have thought here of m as a function of a maximal observable, and reached the same conclusion. Suppose that the vectors $|m_i\rangle$ and $|m_j\rangle$ are sometimes distinct even when $m_i = m_j$. Take a maximal observable m' with the same eigenvectors, and define $f : f(m'_i) = m_i$. The probability of a measurement of m having outcome b equals the sum of the probabilities of a measurement of m' having one of the outcomes m'_i such that $f(m'_i) = b$. Hence we have reached a consistently generalized version of Born's Rule. Because of this, we can summarize this section in the following slightly more elegant form.

Summary of first form of representation

1. A maximal observable m with the discrete set of distinct possible values (*eigenvalues*) m_1, m_2, ... has an associated basis of unit vectors x_1, x_2, ... (called its *eigenvectors*). We write $x_i = |m_i\rangle$ to indicate the one-to-one correspondence.

2. A pure state is represented by a unit vector y, which can be expanded in terms of the orthonormal basis $\{x_i\}$ in the form $y = \Sigma c_i x_i$, where $c_i = (x_i \cdot y)$. The probability of a measurement of the maximal observable m yielding outcome value m_k equals c_k^2; that is to say, $(x_k \cdot y)^2$. Specifically, the probability of outcome eigenvalue m_k in eigenstate x_k equals 1. Any scalar multiple ky of y ($k \neq 0$) represents the same state as y, but the rule of calculation is given for the unit vectors.

3. Every observable m' with discrete spectrum can be equated with a function of some maximal observable m, in the sense that m' has eigenvalues $\{f(m_i)\}$ where m_1, m_2, ... are the eigenvalues of m; and m' has the same eigenvectors as m. The probability of a measurement of m' in state

y having outcome b equals the probability that a measurement of m in state y has one of the outcomes m_k such that $f(m_k) = b$.

Proofs and illustrations

Here I will give the precise definitions of the notions of vector space and Hilbert space, and then describe the simplest examples of such spaces.

To define a vector space, we begin with a number field. The two most common are the field of real numbers and that of complex numbers. Every complex number has the form $a + bi$, where a, b are real and i is the positive square root of -1. The *conjugate* of $(a + bi)$ is $(a + bi)^* = (a - bi)$, and I write b^2 as short for b^*b when b is complex, whenever that is not confusing.

A vector space S over number field F is a set with operations of scalar multiplication and addition. For each b in F and x, y in S, bx (the scalar multiplication of x by b) and $(x + y)$ are also in S. The operations are such that

(1) there is a unique zero vector $\varnothing = 0x$ for all x in S;
(2) $a(x + y) = ax + ay$; $(a + b)x = ax + bx$;
(3) $\varnothing + x = x$;
(4) $+$ is commutative and associative.

The simplest example is the space F^n over F whose members are the n-tuples $\langle b_1, \ldots, b_n \rangle$ of members of F. The operations are defined by $a\langle b_1, \ldots, b_n \rangle = \langle ab_1, \ldots, ab_n \rangle$ and $\langle b_1, \ldots, b_n \rangle + \langle c_1, \ldots, c_n \rangle = \langle b_1 + c_1, \ldots, b_n + c_n \rangle$. Of course, we are familiar with this from analytic geometry, where $\langle b_1 \ldots, b_n \rangle$ would be the set of coordinates of a point in an n-dimensional space.

Geometric spaces have an *inner product*. In the n-dimensional space F^n this can be defined by $\langle b_1, \ldots, b_n \rangle \cdot \langle c_1, \ldots, c_n \rangle = b_1^* c_1 + \ldots + b_n^* c_n$. The general properties of the inner product are these: for each x, y in S there is a number $(x \cdot y)$ in F such that

(5) $x \cdot y = (y \cdot x)^*$

(6) $x \cdot x \geq 0$ and $x \cdot x = 0$ if and only if $x = \varnothing$

(7) $$x \cdot (y + z) = (x \cdot y) + (x \cdot z)$$

(8) $$(x \cdot by) = b(x \cdot y)$$

Note accordingly that $(bx \cdot y) \neq b(x \cdot y)$ in general but $(bx \cdot y) = (y \cdot bx)^* = [b(y \cdot x)]^* = b^*(y \cdot x)^* = b^*(x \cdot y)$. The number $(x \cdot x)$ however is obviously real, and $(x \cdot y)^2 = (y \cdot x)^2$; when talking about the Born probabilities it is often convenient to rely on these facts, and I shall do so without comment.

In terms of the inner product, we define the *length* or *norm* $|x|$ of vector x:

$$|x| = \sqrt{(x \cdot x)}$$

Thus $|\varnothing| = 0$ and $|x| > 0$ if $x \neq \varnothing$. We call x a *unit vector* if $|x| = 1$. In the example of a two-dimensional space

$$|(b, c)| = \sqrt{(b^2 + c^2)}$$

which ties in neatly with the Pythagorean theorem and the geometric picture (Fig. 6.1). The inner product has also a more direct geometric significance in Euclidean space. First of all we call x and y *orthogonal*, writing $x \perp y$, exactly if $(x \cdot y) = 0$. To generalize this, suppose that the vector in the figure has unit

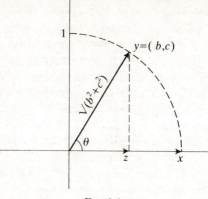

Fig. 6.1

length $\sqrt{(b^2 + c^2)} = 1$. Then $\cos \theta = b$ and $\cos (90° - \theta) = c$. Consider now the unit vector $\langle 1, 0 \rangle$:

$$\langle 1, 0 \rangle \cdot \langle b, c \rangle = b + 0 = b$$

But θ is the angle between $\langle 1, 0 \rangle$ and $\langle b, c \rangle$. In general, in a Euclidean space, if x and y are unit vectors, then $(x \cdot y)$ equals the cosine of the angle between them.

A *basis* of the space is a set B of mutually orthogonal vectors such that every vector is a linear combination of members of B. When the members of a basis are all unit vectors, it is called 'orthonormal', and I shall simply write 'basis' for 'orthonormal basis' except in contents where that might lead to confusion.

All bases for a space have the same cardinality, which is the dimension of the space. We call a space *separable* if it has bases which are finite or countably infinite. Henceforth we assume that all spaces considered are separable. A Hilbert space is a vector space with inner product, which is *complete* in the sense that, if a sequence of vectors converges, then there is a vector in the space which is its limit. To make this precise, call the sequence $x_2, x_2, \ldots$ of vectors a *Cauchy* sequence exactly if the sequence of norms $|(x_m - x_n)|$ converges to 0. The vector space is complete if for every such Cauchy sequence it has a limit which is a vector in the space.

In consequence, we can in Hilbert space make up infinite (countable) linear combinations of vectors. The infinite sum $\sum_{i=1}^{\infty} b_i x_i$ is the limit of the sequence of finite sums $\sum_{i=1}^{n} b_i x_i$ (and exists provided that sequence converges). If $x_1, \ldots, x_n, \ldots$ are vectors in a Hilbert space, then the smallest Hilbert space which contains all of them is called the *subspace* $[\{x_1, \ldots, x_n, \ldots\}]$ *spanned by* these vectors. A vector y is called a *superposition* of $x_1, \ldots, x_n, \ldots$ exactly if y belongs to that subspace. Note that the completeness property which defines Hilbert space guarantees that all the finite and infinite sums of vectors described above are superpositions of those vectors.

The n-dimensional spaces described above as examples are all Hilbert spaces. The simplest infinite-dimensional Hilbert space is F^{∞}, the set of countable sequences $\langle b_1, b_2, \ldots \rangle$ of numbers in the field F, such that Σb_m^2 is finite. (Recall that if b is complex then b^2 means b^*b.) The operations of scalar product, sums, and inner product are defined pointwise, just as above.

3. OBSERVABLES AND OPERATORS

The representation so far has been simple but clumsy. Still remaining with the pure states only, we turn now to the representation of observables by means of operators.

3.1. *Representation by Hermitean operators*

Let m be an observable with possible values a_1, a_2, ... (duplications such as $a_1 = a_2$ allowed in this list), and corresponding orthonormal basis $\{x_1, x_2, \ldots\}$. Define the following operator on all basis vectors:

$$(1) \qquad Mx_i = a_i x_i$$

and extend it to all the vectors in the space by linearity, i.e.

$$(2) \qquad M(by) = bMy \qquad M(\Sigma y_j) = \Sigma My_j$$

An operator M satisfying (2) is called *linear*. The one we are looking at has several other properties. First of all, it is *everywhere defined*: it operates on all vectors in the space. Secondly, its *eigenvalues* (numbers or such that $Mx = ax$ for some vector x) are all real. Thirdly, it may have the following two properties:

Bounded: there is a positive number b such that $|Mx| \leq b|x|$ for all x

Self-adjoint: $(x \cdot My) = (Mx \cdot y)$

M is *Hermitean*; this term, so common in the physics literature, though mostly avoided now by mathematicians, has not had an everywhere uniform usage. Generally it just means self-adjoint; sometimes it means bounded and self-adjoint. Unbounded self-adjoint operators cannot be everywhere defined. Like most other complexities too intimately connected with the infinite, I shall remark on them only in passing.

The Hermitean operator M represents the observable m very conveniently. For example, given a unit state vector $x = \Sigma c_j x_j$, the probability of value a_j being found is c_j^2, so the *expectation value* of the measurement outcome equals $\Sigma a_j c_j^2$. This number is again identifiable by an inner product construction:

$$(x \cdot Mx) = \left(\sum_j c_j x_j \cdot M \sum_k c_k x_k \right)$$

$$= \left(\sum_j c_j x_j \cdot \sum_k c_k a_k x_k \right)$$

$$= \sum_k c_k a_k \left[\sum_j c_j (x_k \cdot x_j) \right]^*$$

$$= \sum_k c_k{}^* c_k a_k$$

$$= \sum_k (x_k \cdot x)^2 a_k$$

which is just that number, the expectation value of the outcome of a measurement of this observable.

All of the above works equally well if some of the possible values a_j are the same. If for example $a_1 = a_2 = a$ but $a_1 \neq a_i$ for any i after 2, then the probability of a measurement finding value a, given state $x = \Sigma c_j x_j$, must equal $c_1^2 + c_2^2$. Let us define a new operator I_a^M as follows:

$$I_a^M x_i = x_i \qquad \text{if } a_i = a$$

$$I_a^M x_i = \varnothing \qquad \text{if } a_i \neq a$$

where $\varnothing$ is the null vector, and again extend the operator from the basis $\{x_1, x_2, \ldots\}$ to the whole space by linearity. Then if x is as above, we have

$$I_a^M x = I_a^M(\Sigma c_j x_j) = c_1 x_1 + c_2 x_2$$

The above calculation of $(x \cdot Mx)$, if now used for $(x \cdot I_a^M x)$, gives us the number $c_1^2 + c_2^2$, that is, that very same probability. So we have the results:

The expectation value of a measurement of m in unit state vector x equals $(x \cdot Mx)$, and the probability that this measurement will yield value b equals $(x \cdot I_b^M x)$.

There is another way to express this probability. Suppose $x = \Sigma c_i x_i$; then $I_a^M x = \Sigma\{c_k x_k : a_k = a\}$. Now the length of that vector is $\sqrt{(\Sigma\{c_k^2 : a_k = a\})}$. If we drop the square root sign, we have exactly the probability. So we can also say:

The probability that a measurement of M in state (unit

vector) x will yield value b equals $|I_b^M x|^2$.

This construction is easily generalized to sets of values:

$$I_E^M x_i = x_i \qquad \text{if } a_i \in E$$
$$I_E^M x_i = \varnothing \qquad \text{if } a_i \notin E$$

and then we deduce:

> The probability that a measurement of m in state (unit vector) x will give a value in E equals $(x \cdot I_E^M x)$, or equivalently $|I_E^M x|^2$.

The operators I_a^M, I_E^M are *projection operators* which are definable as: Hermitean operators I such that $II = I$. Their only eigenvalues are 1 and 0, because if $Ix = bx$ then $IIx = I(bx) = bbx$, and since therefore $b^2 x = bx$ it follows that b is 0 or 1. There is again a clear geometric intuition, for $|I_b^M x|$ is the length of the projection of x along the eigenvector $|b\rangle$ of M, if there is only one eigenvector corresponding to eigenvalue b.

Operators, like any other functions, can be combined by the same operations as their argument. Thus, for operators on any vector space, we can write

$$(A + B)x = Ax + Bx \qquad (\Sigma A_j)x = \Sigma A_j x \qquad (kA)x = kAx$$

These particular functions are linear, and so are, as it were, inherited from the structure of the vector space. (In Sections 3.2 and 3.3 I shall consider functions of observables more generally.) In the case we have been examining, therefore, we can write $M = \Sigma\{bI_b^M : b \text{ an eigenvalue of } M\}$. We call the equation of M with this convex combination of projection operators a *decomposition* of M.

Not all Hermitean operators on all Hilbert spaces have this simple construction as a sum of projection operators. The ones that do are exactly the ones we say have a *discrete pure point* spectrum, this spectrum being the discrete set a_1, a_2, a_3, ... There are observables, such as position, which are not to be represented in this way. I shall restrict the discussion to omit them (but see *Proofs and illustrations* for some further details about spectral decomposition of observables).[4]

Summary of second form of representation

1. Every observable m is represented by a Hermitean operator M. Vector x represents an eigenstate of m corresponding to eigenvalue a if and only if $Mx = ax$. The observable is non-degenerate (maximal) if all eigenvectors corresponding to any one eigenvalue are parallel (i.e., $Mx = ax$ and $My = ay$ implies $x = ky$ for some number k).

2. A Hermitean operator I which is its own square, i.e. is such that $II = I$, is a *projection operator*. Given observable M and Borel set E, we can define I_E^M to be the Hermitean operator I such that Ix is x if $Mx = ax$ for some value a in E and is the zero vector if $Mx = bx$ for some value b outside E. If the eigenvalues of M are a_1, $a_2, \ldots$ then $M = \Sigma a_i I_{a(i)}^M$.

3. The probability that a measurement of M in a pure state represented by unit vector y has an outcome in set E equals the probability that a measurement of I_E^M in that state has outcome 1. That probability is the number $(y \cdot I_E^M y)$, which is also equal to the number $|I_E^M y|^2$. In general, the expectation value of the outcome of a measurement of observable M in unit state vector y equals $(y \cdot My)$.

Proofs and illustrations. In quantum mechanics infinite-dimensional Hilbert spaces are certainly important, and unfortunately the limits of our discussion above are connected with dimensionality. Let us here note some of these. If the space is finite-dimensional and A is Hermitean (or unitary, if it is a space on the complex number field; see Section 6 below), then A can be described completely in terms of its eigenvalues and eigenvectors: there is an orthonormal basis of the space consisting of eigenvectors. This *may* be so if the space is not finite-dimensional (and my discussion is restricted to those operators for which it is so), but not necessarily.

In the infinite-dimensional case, a Hermitean or unitary operator may have no eigenvalues or eigenvectors at all—this is so for the traditional 'basic' observables of position and momentum. There is still always a spectral decomposition $A = \int r \, dE_r$ where the 'spectral family' $\{E_r\}$ is a set of projection operators too. The operator E_r is the projection I_E^A with E the set of all real numbers less than or equal to r. The equation means that

$$(x \cdot Ay) = \int r\, \mathrm{d}(x \cdot E_r y) \qquad \text{for all vectors } x \text{ and } y.$$

The set of numbers r on which E_r increases is the *spectrum* of A. This is then divided into the *point spectrum* (those points r at which E_r jumps) and the *continuous spectrum* (points r where E_r increases continuously). The *eigenvalues* of A are exactly the numbers in its point spectrum; the *possible values* are all numbers in its spectrum. Hence if the continuous spectrum is empty, then the possible values are just the eigenvalues, and we say that A has a *discrete* or *pure point* spectrum. In a more general way, the spectral decomposition establishes a one-to-one correspondence between the self-adjoint operators A (bounded and unbounded) and the *projection-valued measures* (sigma-homomorphisms of the real Borel sets to the projection operators) $E \longrightarrow I_E^A$.

What does this mean for interpretation? If r is a possible value, it can appear as a measurement outcome. However, only if r is an eigenvalue of A is there any state in which it is *certain* to be the outcome of an A-measurement. This introduces a complication for interpretation, and there is important recent work on this subject.[5]

3.2. *Projection operators and subspaces*

A projection operator clearly represents a very simple observable, one that has only 0 and 1 as possible values. The observables with pure point spectra are thus all definable in terms of these simple (*yes*/*no*) observables. That is why much writing on the foundations of quantum mechanics has focused on projection operators. John von Neumann referred to them as *propositions*: I_D^M represents the proposition that a measurement of m is certain to yield a value in D. Others have called them *questions*: to measure the observable represented by I_D^M is to ask a question which can only have the answers *yes* (1) and *no* (0). What I have called the *surface state* in preceding sections is clearly definable by their means:

$$P_x^m(E) = (x \cdot I_E^M x)$$

The reason such an operator I is called a projection operator is that it 'projects' each vector into a certain subspace. Define

(1) $$S(I) = \{x : Ix = x\}$$

Then $S(I)$ is a subspace. (That means: if $x_1, \ldots, x_n, \ldots$ are in $S(I)$, then so are their scalar multiples, and if the numbers $\{a_n^2\}$ sum to 1, then $\Sigma a_n x_n$ is also in $S(I)$.) If $S = S(I)$ we shall also designate I as I_S, and call I the *projection* on S.

The set $S^\perp$ of vectors y which are orthogonal to all members of S is called the *orthogonal complement* of S. If S is a subspace, then so is $S^\perp$. Also, if I is the projection on S, then

$$I^\perp \text{ defined by } I^\perp x = x - Ix$$

is the projection on $S^\perp$. Thus x is as it were split into two orthogonal parts, Ix and $I^\perp x$, which belong to the orthogonal subspaces S and $S^\perp$. These parts sum to x, so nothing is left out in this splitting. We call Ix the *projection* of x on subspace $S(I)$.

The smallest subspace spanned by a set of vectors $x_1, \ldots, x_n, \ldots$ is called $[x_1, \ldots, x_n, \ldots]$, the *span* of these vectors. Obviously a single vector x spans a very small subspace $[x]$, the one-dimensional 'ray' consisting of all scalar multiples of x. The projection on $[x]$ is accordingly called a one-dimensional projection operator, and I shall more briefly designate it as I_x. This sort of projection operator is explicitly definable:

(2) $$I_x y = [(x \cdot y)/(x \cdot x)]x$$

$$I_x y = (x \cdot y)x \text{ if } x \text{ is a unit vector.}$$

Thus the inner product $(x \cdot y)$, if real and positive, is the length of the projection of y along x when they are unit vectors.

Proofs and illustrations. This fits in very well with the geometric link between inner product and cosine in the figure above. There $z = (x \cdot y)x$ and has length $(x \cdot y)$ if x has length 1; but of course the length of z is also the cosine of the enclosed angle θ.

3.3. *Compatible observables and joint probabilities*

In Chapter 4 we noted Arthur Fine's (1982*b*, 1982*c*) proof which connected the Bell Inequalities with the existence of joint probability distributions. Both are characteristic of classical observables, but they reappear in quantum mechanics as an important special case. Intuitively, we think of observables as compatible if they are jointly measurable (with any desired precision). Here we shall give the quantum-mechanical characterization, but show how it is connected with classical features of the probabilities associated by Born's Rule.

Two projection operators I and J are *compatible* if they commute, that is if $IJ = JI$. Two observables A and B are *compatible* if I_E^A commutes with I_F^B for any two Borel sets E and F. When A and B are bounded this just amounts to $AB = BA$. This relationship is not transitive; it is possible that A is compatible with B and B with C although A is not compatible with C. This is true already for projections. In fact:

> *Theorem*: Projection operators I and J, which project on subspaces S and T, are compatible if and only if there are three mutually orthogonal subspaces S', S'', T' such that S is the orthogonal sum of S' and S'' while T is the orthogonal sum of S'' and T'.

Note that, when the condition is satisfied, then I is the sum of the projections on S and on S', while J is the sum of the projections on S' and T'; those three projections are also mutually *orthogonal*, in that applying one of them after another will turn a given vector into the null vector.[6]

More generally, if we have a family F of projections which all commute with each other, then there is also a family G of mutually orthogonal projections such that each member of F is a sum of members of G. The following theorem is due to von Neumann and Varadarajan:

> *Theorem*: A, B, ... are mutually compatible observables if and only if there is a single observable C such that A, B, ... are all Borel functions of C.

A Borel function f is an operation on real numbers such that, if E is a Borel set, so is $f^{-1}(E) = \{x : f(x) \text{ is in } E\}$. Given that f

is a Borel function, and $C = \Sigma c_i I^C_{c(i)}$ is an observable with pure point spectrum, $f(C) = \Sigma f(c_i) I^C_{c(i)}$ is again an observable with the spectrum $\{f(c_i)\}$. (For other observables a similar definition can be given via the spectral decomposition.) This has the following consequence for probabilities:

Theorem: For any state x, if $A = f(C)$ then $P^A_x(f^{-1}(E)) = P^C_x(E)$.

That is, the probability that a measurement of $A = f(C)$ will have as outcome a value r such that $f(r)$ is in E is exactly the probability that a measurement of C will have as outcome a value in E. The Born probabilities for measurement of $A = f(C)$ can therefore be calculated from those for C.

The converse of this statement holds also, rightly construed. To show this, we need to become more precise about the description of joint probability functions for several observables. Here I shall follow Fine (1982c), and Cassinelli and Lahti (1989b).

If p and q are probability measures defined on R, the function $pq(E \times F) = p(E)q(F)$, for Borel sets E and F, can be uniquely extended to a probability measure on R^2. The measure pq has p and q as first and second *marginals*; that is, $p(E) = pq(E \times R)$ and $q(F) = pq(R \times F)$. If q is concentrated on a countable set S, that means also that $p(E) = \Sigma\{pq(E \times \{k\}): k \in S\}$. The construction can again be extended similarly to any finite sequence of probability measures. Moreover, p and q are marginals of many probability measures r on R^2 besides the product measure pq; any such measure r is called a *joint distribution* of p and q. In the case of compatible observables A and B, we can define a joint distribution for their associated Born probability assignments: begin with

$$P^{A;B}_x(E \times F) = (x \cdot I^A_E I^B_F x)$$

and extend this to a measure on R^2. In that case P^A_x is obviously a marginal of the defined measure. Similarly for any finite sequence of compatible observables. In fact, something stronger can be said in this case, because we can also define observables which are functions of several commuting observables. Given two commuting operators $A = \Sigma a_i I^A_{a(i)}$ and $B = \Sigma b_i I^B_{b(i)}$ and a Borel function of R^2 into R, there exists an observable $C = \Sigma f(a_i, b_j) I^A_{a(i)} I^B_{b(j)}$. Clearly, A and B are also Borel functions of C: the probability that a measurement of C yields a value in E is the same as the probability that a measurement of A yields a number r such that, for some s, the couple $\langle r, s \rangle$ is in $f^{-1}(E)$. This construction is easily generalized to any finite number of mutually compatible observables, because the projections on their eigenspaces will all commute. If we now take specifically the Borel function f such that, for a given pair E and F,

$$f(r, s) = 0 \text{ if } r \text{ is not in } E \text{ and } s \text{ is not in } F$$
$$= 1 \text{ if } r \text{ is in } E \text{ and } s \text{ is not in } F$$
$$= 2 \text{ if } r \text{ is not in } E \text{ and } s \text{ is in } F$$
$$= 3 \text{ if } r \text{ is in } E \text{ and } s \text{ is in } F$$

then $E \times R = f^{-1}(\{1,3\})$ and $R \times F = f^{-1}(\{2,3\})$. So we argue that if $C = f(A, B)$ then

(a) the probability that a measurement of C has value 1 or 3 equals the probability that a measurement of A has a value in E;
(b) the observable I^A_E is the same as the observable $I^C_{\{1,3\}}$;
(c) P^C_x is a joint distribution for P^I_x and P^J_x where $I = I^A_E$ and $J = I^B_F$.

Could all this be done for incompatible observables as well? To answer this we must first define the requisite counterpart notion for arbitrary quantum-mechanical observables. Then there is a theorem which shows that the answer is *no*.

Definition: Observables $A_1, \ldots, A_n$ meet the *joint distribution condition* (briefly, (JD)) if and only if: for every state x there is a probability measure p_x on R^n such that, for any Borel function f of R^n into R, there exists some observable $M(f)$ such that

$$p_x(f^{-1}(E)) = P^{M(f)}_x(E)$$

$P^{A(i)}_x$ is the ith marginal of p_x

for all Borel sets E and any state x.

Joint Distribution Theorem (Fine): A finite family of observables meets (JD) if and only if they are all mutually compatible.

I will give a small but crucial part of the proof, to show how this works. Let A and B be two observables with pure point spectra $\{a_i\}$ and $\{b_j\}$, and suppose they meet (JD), and let x be a state. Let p_x represent their joint distribution on R^2, so that, for any Borel function f,

(1) for any Borel sets E and F, $p_x(f^{-1}(E)) = P_x^{M(f)}(E)$

It will be sufficient to show that, for any Borel sets E and F, the projection operators $I = I_E^A$ and $J = I_F^B$ commute. Let f be the Borel function that was defined above. Then we use (1) to conclude that there is some observable, call it C, which plays the role of $M(f)$ in (1) for this function. Since $E \times R = f^{-1}(\{1,3\})$ and $R \times F = f^{-1}(\{2,3\})$, we argue:

(2) $p_x(E \times R) = P_x^C(\{1,3\})$ and $p_x(R \times F) = P_x^C(\{2,3\})$

But these are also the marginals of p_x, hence equal the numbers $(x \cdot Ix)$ and $(x \cdot Jx)$. Since all this holds for any state x, we conclude that $I = I_{\{1,3\}}^C$ and $J = I_{\{2,3\}}^C$. Therefore I and J commute with each other. We conclude, since this was shown for arbitrary E and F, that A and B are compatible.

In the *Proofs and illustrations* for this subsection and the next I shall describe some further relevant results.

Proofs and illustrations. We can also define partial compatibility (partial commutativity), which can be utilized in discussion of 'crude' measurements. Let $\text{com}(A,B) = \{x \in H : I_E^A I_F^B x = I_F^B I_E^A x$ for all Borel sets E, $F\}$. This is a closed subspace; if it is H itself, then A and B are (totally) compatible, and if it is the null space they are totally incompatible. The theorem, due to von Neumann and Varadarajan, was generalized and connected with joint probability distributions by Gudder (1968b), Pullmanova (1980), and Ylinen (1985); see also Lahti and Ylinen (1987), Schroeck (1989), and Cassinelli and Lahti (1989b). Let I_{EF}^{AB} be the projection on the intersection of the subspaces on which I_E^A and I_F^B project; this acts like $I_E^A I_F^B$ exactly when those two commute.

Theorem: The function $E \times F \longrightarrow (x \cdot I_{EF}^{AB} x)$ extends to a (unique) probability measure on R^2 if and only if x is in com (A, B).

The more general form of this theorem for mixed states will be given in the *Proofs and illustrations* of the next subsection.

4. MIXED STATES AND OPERATORS[7]

We have so far only a representation for pure states. In the abstract notation we used to begin, if t and u are states, then there should, for example, be a state s such that

$$P_s^m = 0.25 P_t^m + 0.75 P_u^m$$

for all observables m. This state s is a *mixture* of t and u. But what represents it with respect to the Hilbert space? We should like ideally to have a uniform representation of both pure and mixed states, and this suggests that vectors have outlived some of their usefulness.

To arrive at such a uniform representation, we must do two things: explore projection operators somewhat more, and introduce a special function called the *trace*.

To apply the lessons of the preceding section, let us compare the pure state represented by unit vector x and the observable represented by projection I_x. The latter has only two possible values, 0 and 1. If measured in state x, we are certain to have outcome values 1, and if measured in a state y orthogonal to x, we are certain to find outcome 0. But we can go further: the probability of finding value b, if we measure arbitrary observable m in state x, could be expressed in terms of observable I_x. For obviously there is a one-to-one correspondence between the vector-rays and these one-dimensional projection operators. All we need to do is enter this correspondence into the rule for calculating probabilities.

There is a neat way to do exactly that, using the *trace* calculation; I shall describe it intuitively before giving the precise definition. The trace Tr is a *linear* map of operators into numbers, which is *order-independent* for products; i.e., $Tr(AB)$

$= Tr(BA)$. This mapping is such that, if I' is any projection operator, then

$$Tr(I_x I') = (x \cdot I'x)$$

But that means of course that the probability calculation we found at the end of the last section can be rewritten as

$$P_x^m(E) = (x \cdot I_E^M x) = Tr(I_x I_E^M)$$

Let us see now how we can extend this new representation of pure states, by means of operators, to one of mixed states as well.

Focus on the mixed state s described above, and suppose its component states t and u to be represented by orthogonal projections I_x and I_y respectively. Then for an observable represented by projection operator I', we must have:

$P_s^{I'}(1)$ = the probability of outcome 1 if I' is measured in state s

  $= 0.25$(probability of this outcome in state t)

   $+\ 0.75$(probability of this outcome in state u)

  $= 0.25\,Tr(I_x I') + 0.75\,Tr(I_y I')$

  $= Tr([0.25 I_x + 0.75 I_y] I')$

by the linearity of the trace function. So if we now define the new operator W to be $(0.25 I_x + 0.75 I_y)$, the above probability takes the form

$$P_s^{I'}(1) = Tr(WI')$$

which is the same general form as before. So operator W looks like a good candidate for the representation of that mixed state. It is also intuitively just right, because it is formed from the representations of the component pure states by the same linear combination as the probabilities themselves.

The linearity and order independence of the trace allows also a convenient calculation of expectation values. For suppose $M = \Sigma a_i I_i$. Then the expectation value of M in state x equals

$$(x \cdot Mx) = \left(x \cdot \sum a_i I_i x \right)$$

$$= \sum a_i (x \cdot I_i x) = \sum a_i Tr(I_x I_i)$$

$$= \sum a_i Tr(I_i I_x)$$

$$= Tr\left(\sum a_i I_i I_x\right) = Tr\left(I_x \sum a_i I_i\right)$$

$$= Tr(I_x M)$$

so we have exactly the same format for the calculation of expectation values as of probabilities. This is easily seen to work the same way if we turn to mixed state s and insert its representation W: the expectation value is then $Tr(WM)$.

It is clearly time now for official definitions. The above operator W is called a *statistical operator* or *density matrix*. We define:

(1) An operator A is of *trace-class* if the number $\Sigma(x_i \cdot Ax_i)$ exists and is the same for every basis $\{x_i\}$.
(2) If A is of trace-class, then $Tr(A) = \Sigma(x_i \cdot Ax_i)$, where $\{x_i\}$ is a basis.[8]
(3) A *statistical operator* (or *density matrix*) is a Hermitean operator W with discrete spectral decomposition

$W = \Sigma p_i I_{x(i)}$ where $\{x_i\}$ is a set of vectors and $\{p_i\}$ a set of positive real numbers that sum to 1.

Result: The statistical operators are exactly the positive Hermitean operators of trace-class, with trace 1.

A mixture can be made up out of any states you like.[9] Because of the linearity of the trace operation, weighted sums of statistical operators, *even if not mutually orthogonal*, yield corresponding mixtures of probability measures. For example, suppose x and y are any vectors whatsoever, not necessarily orthogonal, and $W = bI_x + (1 - b)I_y$ for $0 < b < 1$. Then if B is a base, and $I_E^M z = z'$:

$$Tr(WI_E^M) = Tr[(bI_x + (1 - b)I_y)I_E^M]$$

$$= \sum \{(z \cdot [bI_x + (1 - b)I_y]I_E^M z) : z \in B\}$$

$$= \sum \{(z \cdot [bI_x + (1 - b)I_y]z') : z \in B\}$$

$$= \sum \{b(z \cdot I_x z') + (1 - b)(z \cdot I_y z') : z \in B\}$$

$$= b \sum \{(z \cdot I_x z') : z \in B\}$$

$$+ (1 - b) \sum \{(z \cdot I_y z') : z \in B\}$$

and notice that $P_x^M(E) = Tr(I_x I_E^M) = \Sigma\{(z \cdot I_x I_E^M z) : z \in B\} = \Sigma\{(z \cdot I_x z') : z \in B\}$, and similarly for $P_y^M(E)$. Therefore the Born probability function for outcomes of measurements of M for state W is just the classical $b/(1 - b)$ mixture of the two corresponding probability measures for state x and state y.

We turn now to the relations in general between pure states and mixtures. Just as all probability used to be thought of as merely measuring ignorance ('it is one of these, but I don't know which'), so mixtures have been thought of as just representing ignorance of the system's real state which is pure. There is much to be said against this *ignorance interpretation of mixtures*, but I shall take it up in the next chapter. Undoubtedly it has on its side the fact that calculations with mixed states can typically be done in terms of the pure ones of which they are mixtures. In contexts of calculation, especially important are the orthogonal decompositions:

If W is a statistical operator, $\{y_i\}$ is a basis and $\{w_i\}$ is a sequence of non-negative real numbers such that $\Sigma w_i = 1$, then the equation

$$W = \sum w_i I_{y(i)}$$

is an *orthogonal decomposition* of W.

The numbers w_i, called *weights*, need not all be distinct; if they are not, this decomposition is not unique. These weights are obviously the eigenvalues of W. When some are equal, we call that also a case of *degeneracy*. When the states y_i are not all mutually orthogonal, the equation is a non-orthogonal decomposition.

From the point of view of interpretation, as opposed to ease of calculation, the orthogonal decompositions may or may not be especially important. Let us consider three other relations:

(4) x is a *component* of W exactly if x occurs non-trivially in a decomposition of W; i.e. if

$$W = bI_x + (1 - b)W'$$

for some $0 < b \leq 1$ and statistical operator W'.

(5) x is in the *support* of W exactly if x is in the subspace spanned by the eigenvectors of W with non-zero weights.

(6) x is *possible relative to W* exactly if for all observables M and Borel sets E, $P_W^M(E) = 1$ only if $P_x^M(E) = 1$.

The notion of relative possibility we saw already in more general form above. To begin the discussion of these three relationships, let us note that the last two coincide. For $I_{[S]}$ takes value 1 with probability 1 in state W which has support S, so to be possible relative to W a pure state must lie in $[S]$. But if x is in $[S]$ and T is any other subspace, then I_T takes value 1 with probability 1 in x if $[S] \subseteq T$, hence if T takes value 1 in W with probability 1. Since $I_E^M = I_T$ for some subspace T, the conclusion follows:

(8) Pure state x is possible relative to W if and only if x is in the support (image space) of W.

The relationship between a mixture and its components is rather more complicated, as will at once be clear from the following theorem:[10]

(7) *Theorem* (Hadjisavvas): x is a component of W if and only if x is in the range of $W^{1/2}$ (which is also the range of W in the finite-dimensional case).

To elucidate this, we shall first take a closer look at the support of W. Let $W = \Sigma\{p_x I_x : x \in S\}$ with all values p_x positive, $\Sigma p_x = 1$, and the set S orthonormal. The subspace $[S]$ is then the *support* of W. Let $S \cup T = \{y_i\}$ with S and T disjoint be an orthonormal basis, so we can always write $z = \Sigma c_i y_i$ for any vector z. If $y \in T$, then $Wy = \Sigma\{p_x I_x : x \in S\}$, which is the null vector because $x \perp y$ when x is in S and y in T. For the same reason $I_{[S]}y = \varnothing$. So the space is divided into the two subspaces $[S]$ and $[T] = [S]^\perp$. The operator W maps x either into $[S]$ or into $\{\varnothing\}$, so $[S]$ contains the range of W, all the 'images' of vectors under the mapping $x \longrightarrow Wx$. Accordingly we also call $[S]$ and $[T]$ the *image space* and *null space* of W respectively.

It is possible that $[S]$ is the whole space, so that $[T]$ is $\{\varnothing\}$. Is $[S]$ in general identical with the range of W? The answer is *yes* in the case of a finite-dimensional Hilbert space. For let $y = \Sigma\{c_x x : x \in S\}$ in $[S]$, and let $z = \Sigma\{d_x x : x \in S\}$. Then $Wz = \Sigma\{d_x p_x x : x \in S\}$, and this is y provided $c_x = d_x p_x$ for all x. Therefore we simply choose $d_x = c_x/p_x$ to show that this vector y in $[S]$ is itself in the range, i.e. is the image of z produced by applying W.

However, if the set S is not finite, then sum $\Sigma\{(c_x/p_x)x : x \in S\}$ may not exist, because the set of finite partial sums of which that is defined as the limit may not converge. It is easy to see how that could happen if the sequence $\{p_x\}$ itself converges to 0. Therefore it is not true in general that the range of W is the whole image space.

Now we can return again to the somewhat less tractable relation of being a *component*. In classical probability theory, any mixture (convex combination) of probability measures is again a probability measure. This is also true, *mutatis mutandis*, for mixed states in quantum mechanics. But consider now the 'converse' question: is there an observable Q such that

(9) If P, P' are probability measures, then $P = bP' + (1 - b)Q$, for $1 > b > 0$ and some probability measure Q, only if
   (a) $P(A) = 1$ implies $P'(A) = 1$;
   (b) $bP'(A) \le P(A)$ for all A in the domain?

Define $Q = (P - bP')/(1 - b)$. This is easily seen to be additive, and to assign 1,0 wherever P assigns 1,0. But it is not allowed to assign negative numbers, and it is guaranteed that it will not do so exactly if $P - bP'$ is nowhere negative, as (b) requires.

The similar conditions required for the component relation in quantum mechanics show at once that not every vector in the image space is in general a component. For in (4) we must have $W' = (W - bI_x)/(1 - b)$ giving non-negative probabilities by Born's Rule. With $Wy = p_y y$, this means that

$$Tr(W'I_y) = Tr(WI_y) - bTr(I_xI_y) = p_y - bTr(I_xI_y)$$

must be non-negative. Hence the 'transition probability' between x and y (squared cosine of the angle) must be no greater than p_y/b. Can we find such a number b for given x and W? We need $b \leq p_y/Tr(I_xI_y)$, but this series may be divergent if the decomposition is infinite. In that case there is no such number b. We conclude:

(10) In general, not all pure states possible relative to W (vectors in the image space of W) are components of W.

Summary of third form of representation

1. Each observable is represented by a Hermitean operator. Every state, pure or mixed, is represented by a Hermitean operator with trace 1; such an operator is called a *statistical operator* or *density matrix*. A statistical operator is a projection operator only if it projects on a one-dimensional subspace; if so, it represents a pure state.
2. If M represents an observable and W a state, then the expectation value of an outcome of an M-measurement in state W equals $Tr(WM)$. The probability that the outcome value is in the set D equals $Tr(WI_D^M)$.

Proofs and illustrations. The earlier calculation of probabilities and expectation values remains the most convenient for pure states. It is quite easy to see that they agree with the trace calculation. Given unit state vector $x = \Sigma c_i y_i$, the expectation of B in x is

$$(x \cdot Bx) = \sum c_i^* c_i b_i$$

if the vectors y_i are unit eigenvectors of B with $By_i = b_i y_i$. By the trace calculation, it looks like this:

$$Tr(I_xB) = \sum (y_i \cdot I_xBy_i)$$

$$= \sum (y_i \cdot I_x b_i y_i)$$

$$= \sum b_i(y_i \cdot I_x y_i)$$

$$= \sum b_i(y_i \cdot (x \cdot y_i)x)$$

$$= \sum b_i(x \cdot y_i)(y_i \cdot x)$$
$$= \sum b_i c_i^* c_i$$

which is of course the same. (It is important in both calculations that x is a unit vector. Otherwise $I_x y$ equals $[(x \cdot y)/(x \cdot x)]y$ instead.) If we know the decomposition of a given mixed state as $W = \Sigma p_i I_{x(i)}$, then we can use the old calculation instead; no need to use the idea of trace.

A second point that helps to move conceptually back and forth between the two sorts of state representations concerns what happens in projection. Suppose $z = I_S x$. Then how is I_z related to I_x? To put it another way, what operation on I_x corresponds intuitively to the projection of x on the subspace S?

Lemma: $I_S I_x I_S = I_z$

where $z = I_S x$

To prove this, consider any vector y and note that $I_S I_x I_S y = I_S I_x (I_S y) = I_S (x \cdot I_S y)x = (x \cdot I_S y)I_S x = (x \cdot I_S y)z = (I_S x \cdot y)z = (z \cdot y)z = I_z y$. Hence the two operators $I_S I_x I_S$ and I_z are the same everywhere.

There is also a useful way to describe the relation between a mixture and its components: by means of a matrix, i.e. a square array of numbers. Suppose first that $\{x_i\}$ is a base of unit vectors and $W = \Sigma p_i I_{x(i)}$. An operator is equally well described if we give its effect on every vector and for W that is clearly by the equation

$$Wz = \sum p_i I_{x(i)} z = \sum p_i(x_i \cdot z)x_i$$

Suppose now that $\{z_j\}$ is another basis of unit vectors, and for each i, $x_i = \sum_j a_{ij} z_j$. Then

$$Wz = \sum_i p_i \left(\sum_j a_{ij} z_j \cdot z \right) x_i$$

$$= \sum_i p_i \sum_j a_{ij}^*(z_j \cdot z) \sum_k a_{ik} z_k$$

$$= \sum_{jk} \left(\sum_i p_i a_{ij}^* a_{ik} \right)(z_j \cdot z)z_k$$

If we let $w_{jk} = \sum_i p_i a_{ij}^* a_{ik}$ then the square array $[w_{jk}]$ is called the matrix which represents W relative to basis $\{z_j\}$. If we look back now we see that in this terminology the matrix $[q_{jk} = p_j \delta_{jk}]$, where δ_{jk} equals 1 if $j = k$ and 0 otherwise, represents W relative to the original basis $\{x_i\}$. This matrix $[q_{jk}]$ is such that only its diagonal elements q_{jj} ($=$ the probabilities p_j) are non-zero, and is therefore called a *diagonal* matrix. We say then also that W is *diagonal in* basis $\{x_i\}$. Thus W has an orthogonal decomposition in a certain basis if and only if it is diagonal in that basis, which means that it is represented by a diagonal matrix relative to that basis.

Finally, we can state the complete theorem discussed in the previous subsection for compatible and partially compatible observables. (See Cassinelli and Lahti 1989*b*, Schroeck 1989, and the other references cited above.) The notation I_{EF}^{AB} denotes again the projection on the subspace which is the intersection of the subspaces on which I_E^A and I_F^B project. As stated before, $I_{EF}^{AB} = I_E^A I_F^B$ if *and only if* I_E^A and I_F^B commute. To illuminate the theorem due to Fine, described in the preceding subsection, we reflect as follows. Suppose now that G is a finite family of observables, and $G^* = \{I_E^A : A$ in G, E a Borel set$\}$. Now the members of G all commute with each other if and only if all the members of G^* commute with each other. This is the case, by the above theorem, if and only if all the functions $E \times F \longrightarrow Tr(WI_{EF}^{AB})$ can be extended to probability functions on R^2 for all components of all states W—hence for all pure states W. By Fine's theorem we see now that this is in turn equivalent to the existence of observables defined via Borel functions from the family G, which are statistically related to this family in the appropriate way.

5. GLEASON'S THEOREM AND ITS IMPLICATIONS

We now have a uniform representation of observables as Hermitean operators, and of states as Hermitean operators with trace 1. There is an overlap in representations: one and the same operator can represent both a state and an observable. This is perhaps a little curious but need not worry us much; for clearly, the state represented by W is just a state s such that the following holds:

Any state t is possible relative to s if and only if the observable represented by W has expectation value 1 in t.

The relationship of *relative possibility* was introduced in Section 1: that t is possible relative to s means that P_t^m assigns probability 1 whenever P_s^m does, for all observables m.

It is possible to ask a certain kind of completeness question for this representation. A state determines the probabilities of measurement outcomes. These are conditional probabilities: probabilities that measurement of m yields a value in D, given that m be measured. Suppose we are given an arbitrary but consistent description of such probabilities, and call it a 'putative state'. Does our representation of states have room for every such putative state?

Gleason's famous theorem answers this question affirmatively—relative to one precise way of taking the question. For to make the question definite, we must know what counts as a consistent description of such conditional probabilities. The way this is specified—as assumption of the theorem—is that all observables can be defined in terms of those represented by Hermitean operators, and that every Hermitean operator represents an observable. (In a later section we shall come up against the limits of this assumption.) The way in which operators represent observables is also taken for granted to some extent. The theorem is then that the possible probability assignments (those which respect these stipulations about the representation of observables) are exactly those determined by the density matrices in the usual way.

There are also two important corollaries which I shall take up. The first shows the impossibility of recasting quantum theory in classical mould, in a certain way. The second shows how states can be 'conditionalized' in a way that parallels the usual conditionalization of single probability functions. Both of these corollaries play an important role with respect to interpretation of the theory.

5.1. *Exposition of Gleason's theorem*

Recall that what a projection operator projects on is a subspace. If $y = \Sigma c_i x_i$, we call y a *superposition* of the states $\{x_i\}$. The set of all these superpositions (both countably infinite and finite sums) is the *span* of this set $\{x_i\}$ and is denoted $[\{x_i\}]$. This span is a subspace, and if the vectors x_i are mutually orthogonal unit vectors, they form therefore a basis of this subspace. Conversely, any subspace has such a basis.[11]

If S is a subspace, the set of vectors orthogonal to S is called $S^\perp$, the *orthocomplement* of S. If S and T are subspaces, then $S \cap T$, their *meet*, is also a subspace. The least subspace containing both S and T is called $S \oplus T$, their *join* or *sum* or *orthogonal union*. This is not the set-theoretic union of S and T, but the set of all sums of vectors $x + y$ such that $x \in S$ and $y \in T$.

The calculus of subspaces with these operations, and with $\subseteq$ as relation, is a familiar mathematical structure: it is a *lattice*. That means that $\cap$ and $\oplus$, when considered each of itself, act just like *and* and *or* in logic, or $\wedge$ and $\vee$ in Boolean algebra or set theory. Their countably infinite analogues $\cap_i S_i$, and $\oplus_i S_i$, taken by themselves, also exist and behave as usual. But the operations do not interact with each other in a Boolean way (see e.g. Beltrametti and Cassinelli 1981a; Kalmbach 1984). For example, the familiar *distribution law*,

$$s \cap (t \cup u) = (s \cap t) \cup (s \cap u)$$

$$s \cup (t \cap u) = (s \cup t) \cap (s \cup u)$$

does not hold for $\cap$ and $\oplus$. The orthocomplement again has some familiar features:

$$S^{\perp\perp} = S$$

$$S \cap S^\perp = \varnothing \qquad S \oplus S^\perp = H$$

$$S \subseteq T \text{ entails } T^\perp \subseteq S^\perp$$

where $\varnothing$ is the null space containing only the zero vector and H is the entire Hilbert space. It also obeys the usual rules of interaction with $\cap$ and $\oplus$, called DeMorgan's Laws:

$$(S \cap T)^\perp = S^\perp \oplus T^\perp; \qquad (S \oplus T)^\perp = S^\perp \cap T^\perp$$

But, owing to the failure of distributivity, we do not have an analogue in $\cap$, $\oplus$ to such Boolean equations as $S = (S \cap T) \cup (S \cap \bar{T})$.

All the usual Boolean equations do hold when the subspaces

S, T, R are mutually *compatible*. Two subspaces S and T are called compatible if $S \oplus T$ has a basis B, and if certain parts B_1 and B_2 of B are bases of S and T respectively. Let us illustrate this with a three-dimensional case. Suppose $\{x_1, x_2, x_3\}$ is a basis for S and so is $\{x_1, y_2, y_3\}$, but y_2, y_3 are not orthogonal to x_2, x_3. (For example, y_2 and y_3 are in $[x_2, x_3]$ but are formed from x_2 and x_3 by rotating both of them through some angle.) Then

> $[x_1]$ is compatible with $[x_1, x_2]$ and also compatible with $[x_1, y_2]$, but $[x_1, x_2]$ and $[x_1, y_2]$ are not compatible with each other.

So compatibility, though obviously reflexive and symmetric, is not transitive. It is not an equivalence relation but an 'overlap' relation.

I have discussed this calculus of subspaces because we can think of the measurement outcome probabilities as numbers assigned to subspaces. Recall that the probability that a measurement of m has as outcome a value in set D, given state x, equals

$$P_x^m(D) = (x \cdot I_D^M x)$$

But I_D^M is just the projection operator which projects on the subspace spanned by those eigenvectors of M, whose corresponding eigenvalues lie in D. Thus in our previous example, if $D = \{a_1, a_2\}$, where m has possible values $\{a_i\}$ and associated basis $\{x_i\}$, we have the following: $I_D^M x_i = x_i$ exactly if $i = 1, 2$. Hence if $y \in [x_1, x_2]$, then $y = cx_1 + dx_2$ and $I_D^M(y) = cI_D^M x_1 + dI_D^M x_2 = cx_1 + dx_2 = y$. So let us introduce the function P_x as follows: P_x maps the subspaces S into $[0,1]$ by the rule

$$P_x(S) = (x \cdot I_S x)$$

where I_S is of course the projector on S. This function P_x carries the same information as the whole family of functions $\{P_x^m : m \text{ an observable}\}$ — and indeed, no extra information, if every Hermitean operator (hence every projector operator) represents some observable. But while each function P_x^m is a classical probability function, it is clear that P_x is not. Instead P_x is a pasting together of probability functions, a condensed summary of them, and so it is a different kind of mathematical object. For example, a probability function is defined on a field of sets, which is a Boolean algebra; but the domain of P_x is the non-Boolean lattice of subspaces. Yet it has some familiar properties, and we use these to define its kind.

Definition: A (*quantum*) *probability measure* on a sigma-complete orthocomplemented lattice is a map P of L into $[0,1]$, such that:

(a) $P(\varnothing) = 0$

(b) $P(H) = 1$

(c) P is countably additive: for every countable sequence $\{S_i\}$ of mutually orthogonal elements in L, the series $\sum_i^n P(S_i)$ converges and $P(\oplus_i S_i) = \sum_i P(S_i)$

Here $\varnothing$ and H are the zero element and maximal element of the lattice; that it is sigma-complete means that the countably infinite meets $\cap_i$ and joins $\oplus_i$ also exist. We call S and T orthogonal exactly if $S \subseteq T^\perp$, or (equivalently, given the properties of the orthocomplement) exactly if $T \subseteq S^\perp$. The notion of measure has here been generalized in reasonable fashion, since it is a non-negatively real-valued, normalized, and countably additive function.

It is an immediate corollary that the class of probability measures on L is closed under mixtures (countable convex combinations). That is,

If $0 < w_i \leq 1$ and $\Sigma w_i = 1$ and $\{P_i\}$ is a countable set of probability measures on L, then the mixture $\Sigma w_i P_i$ is also a probability measure on L.

In preceding sections we represented pure states by the vectors x in a Hilbert space, each with its associated probability measure P_x on the lattice of subspaces. We have also seen how to make up mixtures. For this we used density matrices W and we had the probability measure P_W^m on the range of observable m, defined by

$$P_W^m(a) = Tr(WI_a^M)$$

Here too we can use our method of pasting together all the probability functions $\{P_W^m : m \text{ is an observable}\}$ into the function

$$P_W(S) = Tr(WI_S)$$

defined on the lattice of subspaces. These measures will typically be mixtures (but pure if $W = I_x$ for some vector x).

However, it is as yet an open question at this point in the exposition whether the above construction of probability measures really exhausts all the ones that exist, mathematically speaking. This question is answered by:

> *Theorem* (Gleason): If the dimension of separable Hilbert space H is ≥ 3, then, for every probability measure P on the lattice of subspaces of H, there is a unique density matrix W defined on H such that P derives from W; that is, $P(S) = Tr(WI_S)$ for all subspaces S.

The proof, given by Gleason in 1957, is long and non-trivial; it was only recently simplified by Cooke *et al.* (1985).[12]

Gleason's remarkable theorem shows us that we have, in the quantum-mechanical representation of empirical 'surface' probabilities, one of those gratifying cases in which the intuitive constructions genuinely exhaust all the mathematical possibilities.

5.2. *First corollary: no hidden variables*

One corollary to Gleason's theorem is Kochen and Specker's famous result (1967) that quantum theory does not admit of a certain sort of hidden variable interpretation (see also Fine and Teller 1978). (It should be added, however, that Kochen and Specker's proof gives a good deal more information.) The assumptions which define this sort look very weak, and are roughly as follows: (*a*) each observable has a sharp value which belongs to the spectrum of its representing operator; (*b*) two observables which have the same expectation value in every quantum-mechanical state are identical; (*c*) compatible observables are functionally related in the same way as their representing operators. (For further discussion, see Chapter 10 below.) Note that the second assumption implies that distinct observables cannot be represented by the same operator.

Suppose now that function v assigns to each observable a sharp value in its spectrum. By that implication of (*b*), this indirectly assigns the same value to each Hermitean operator which represents an observable. Let us assume this to include at least all projection operators; each then receives value 1 or 0. By assumption (c), it follows that no two orthogonal projections will each receive value 1. But, of any maximal set of mutually orthogonal projections, at least one will receive value 1, since the identity must receive 1 by (a) and is the sum of that set. Clearly, assignment v has indirectly yielded a quantum probability measure on the lattice of subspaces (since these correspond one to one to the projections in the right way). If we now finally add the supposition that the Hilbert space of pure states has dimension at least 3, then Gleason's theorem tells us that this measure derives from a statistical operator. Which statistical operator could this be? Let B be a basis, and let us extend v still further and say that it (indirectly again) assigns to vector x the same value that it assigns to $[x]$. Now clearly, v assigns 1 to one and only one member x of B, so that statistical operator must be I_x. But in that case v cannot assign either 1 or 0 to the observable I_y for any vector y which is neither parallel nor orthogonal to x. This contradicts the given.

This impossibility result is of prime significance for the interpretation of the theory.

5.3. *Second corollary: conditional probabilities and Lueders's Rule*[13]

Gleason's theorem has a very important second corollary, but it is fraught with interpretative difficulties. As a preliminary, I shall remind us of the subject of conditional probability as it already appears in a classical context. Then I shall raise the question: what is the exact formal analogy to this for quantum-mechanical states? The question has appeared in the literature almost solely in the guise of heavily interpretation-laden discussion of measurement (see e.g. Herbut 1969, Bub 1977). But it can be separated out and answered by itself.

Conditional probability has often been presented in terms of learning, in the sense of coming to know. The familiar formula goes something like this: if someone assigns probabilities to events X by means of function P, then $P(X|A)$ is the probability he assigned (would assign) if he learns[14] (were to learn) solely that A is the case. There are quite a number of things wrong with this formula, if taken as definition.

In the first place, it could at best define personal or subjective probability. If there is also physical probability, we need an account that covers that as well. Secondly, the formula is not applicable if A could never be one's sole, total new information, so the 'definition' would at best be partial. Thirdly, there are cases in which it is definitely false. (The second and third point have often been discussed in the literature on personal or subjective probability.) Worst of all, the above formulation, in terms of learning, and the equation $P(X|A) = P(X\&A)/P(A)$ have no logical connection with each other at all.

The equation (which holds provided $P(A) \neq 0$) does follow at once from the following account: $P(-|A)$ is the probability function P' which assigns 1 to A and agrees with P on the odds between events that imply A. Odds are probability ratios, so that means that $P'(X\&A)/P'(Y\&A)$ always equals $P(X\&A)/P(Y\&A)$, and this, plus the condition $P'(A) = 1$, yields the above equation at once.

A quantum-mechanical state is, as it were, a bundle of probability functions. Can such a state—the whole bundle, so to say—also be conditionalized in analogous fashion? We can approach this question now in the terms set by Gleason. It becomes then:

(Q1) For a given state W and subspace S, is there a state W' which gives (by the Born Rule) probability 1 to the observable I_S and agrees with W on the odds between observables I_T, I_U for T, $U \subseteq S$?

Before turning to the answer, let us remark that we are dealing here with the explication in quantum mechanics of the conditional or *transition probability* $T(A = 1|B = 1)$ discussed in general terms in the preceding chapter. The connection is this: if the state of the system is W, the 1-eigenspace of B is S and the 1-eigenspace of A is T, then $T(A = 1|B = 1)$ is the number

$$P_{W'}^A(1) = Tr(W'I_T)$$

where W' is the (unique) state related to W as described in (Q1). In view of this, it is quite easy to see what the answer to (Q1) would have to be if we restricted the discussion to pure states (see *Proofs and illustrations*).

It is possible to answer question (Q1) directly (see Beltrametti and Cassinelli 1981*a*). But it is also instructive to note that Gleason's theorem has an immediate corollary to answer question (Q1). For W induces a probability measure on the lattice of subspaces of the whole space. If we restrict that function to the subspaces of the subspace S, we shall have a new probability measure, on that lattice, provided we amend it by renormalizing. The latter means that, for $T \subseteq S$, we have the new assignment of probability

$$Tr(WI_T)/Tr(WI_S)$$

to T. This makes sure that S receives 1, and obviously the probability ratios for T, $U \subseteq S$ are preserved. Now Gleason's theorem steps in and says: there is a density matrix W' corresponding to this new probability measure, and it is *unique* if the dimension of S is ≥ 3.

The above discussion already tells us what W' is, restricted to subspaces of S. Only a little work is needed to see what it is on the whole Hilbert space in question. For clearly, we must have $Tr(W'I_V) = 0$ for all $V \subseteq S^\perp$, and with this additional information the following result can be obtained (see *Proofs and illustrations*):

Theorem: If P is a probability measure on the lattice of subspaces of H, there exists a probability measure P' on the same lattice such that $P'(T) = P(T)/P(S)$ for all subspaces $T \subseteq S$ of H. Moreover, if P derives from the density matrix W, then P' derives from density matrix $W' = I_S WI_S/Tr$ $(I_S WI_S)$, and from no other density matrices if S has dimension greater than 2.

This means that we can write unambiguously $P_W(T|S) = P'W(T) = Tr(I_S WI_S I_T)/Tr(WI_S)$ because $Tr(I_S WI_S) = Tr(WI_S I_S) = Tr(WI_S)$.

In case W is pure, there is a unit vector x such that $W = I_x$. In that case we have, by the lemma in Section 4 (*Proofs and illustrations*), that $I_S I_x I_S = I_z$ where $z = I_S x$. The conditionalization of this pure state on S is therefore represented by the vector $I_S x$, the projection of x on S. We note in passing that the divisor $Tr(I_S I_x I_S) = Tr(I_z) = |z|^2$ needs to be inserted because $z = I_S x$ is in general not a unit vector.

It may be objected that, while the abstract first part of the theorem takes a familiar enough form, what follows the 'Moreover' shows that this is deceptive. Why can't we just define $P(T|S) = P(T \cap S)/P(S)$? The answer is that this is indeed correct and deducible from the theorem, *provided* T and S are compatible subspaces. (In that case I_T and I_S commute, and $I_{T \cap S} = I_T I_S = I_S I_T$.) But the non-distributivity of the lattice defeats this idea in the general case. For suppose that T and S are orthogonal, but are not compatible with U, so $(T \oplus S) \cap U \neq (T \cap U) \oplus (S \cap U)$. Now if we condition on U in the classical way, we can get the numbers

$$p = \frac{P((T \oplus S) \cap U)}{P(U)} \qquad q = \frac{P(T \cap U)}{P(U)} \qquad r = \frac{P(S \cap U)}{P(U)}$$

with $p \neq q + r$. This is a violation of the additivity condition. Indeed, the example can be so chosen that $q = r = 0$; for let us choose three co-planar vectors s, t, u with $s \perp t$. Then $([s] \oplus [t])$ is the plane $[s,t]$ in which all of them lie; thus $([s] \oplus [t]) \cap [u] = [u]$. But $[s] \cap [u] = \emptyset = [t] \cap [u]$. And of course $P(\emptyset) = 0$, so the classical equation would demand $P([s,t]) = 0 + 0 = 0$ for any probability function P at all!

To put it another way: what we demanded of the 'conditionalized' state W' made very clear how it should behave with respect to observables that commute with I_S. But of course, it will also behave in some way or other towards non-commuting observables. Gleason's theorem then tells us that we really have no choice in the matter. The new state has already been fixed uniquely, and all we can do is ask submissively what is implied for all the rest.

What is the significance of conditionalization? We shall see later that in discussions of measurement and the EPR paradox it has taken on significance. This is certainly in part because of the mathematical relationship brought to the fore in our discussion so far—but also partly a matter of interpretation. To prepare us more thoroughly for this, we must take the discussion one step further.

Suppose A is a certain observable with eigenvalues a_1, a_2, ... Let S_i be the ith eigenspace, that is, the subspace of vectors x such that $Ax = a_i x$. These subspaces are a partition in the sense that they are mutually orthogonal and their orthogonal union equals the whole space. Now consider a given state W, and let

p_i be the probability that a measurement of A in state W will yield outcome a_i so that $p_i = Tr(WI_{S(i)})$. The question that was raised in discussions of measurement was:

(Q2) What is the state $W^1 = \Sigma p_i W_i$ such that, for each index i, the state W_i gives probability 1 to outcome value 1 for the observable $I_{S(i)}$ and agrees with W on the odds between observables I_T, I_U for T, $U \subseteq S_i$?

The answer can be seen at once if A has a simple spectrum so that each S_i takes the form $[x_i]$. For then we see immediately that

$$W^1 = \sum p_i I_{x(i)}$$

This is the equation von Neumann gave in connection with his Projection Postulate for the nature of measurement. By the corollary to Gleason's theorem that we just discussed, we see at once that the general answer (covering also non-maximal observables) is:

$$W^1 = \sum p_i I_{S(i)} W I_{S(i)} / Tr(WI_{S(i)})$$

which is $\Sigma I_{S(i)} W I_{S(i)}$ because $p_i = Tr(WI_{S(i)})$.

This is the corresponding equation behind what is known as Lueders's Rule, which Lueders proposed in 1951 as the proper generalized form of von Neumann's Projection Postulate.[15] I will not discuss here what significance von Neumann and Lueders attached to this—later we shall look carefully at the interpretation of quantum mechanics which connects all this with measurement. We may note however that, for classical probability theory, such 'multiple' conditioning (on a partition) is known as *Jeffrey Conditionalization* (see van Fraassen 1989, ch. 13).

Proofs and illustrations[16]

Here I shall first discuss the answer to (Q2) if it is restricted to pure states, an instructive special case; then I shall discuss a different approach to the question in its general form. As pointed out, we can understand (Q2) as being about transition probabilities. So let us ask: given pure state x and subspace S, can we find a pure state x' such that $I_S x' = x'$ and

$(x' \cdot y) = (x \cdot y)$ for all y in S? Since x can be written uniquely as $x = x_1 + x_2$ with x_2 in S and x_1 in $S^\perp$, it follows that $(x \cdot y) = (x_2 \cdot y)$ if y is in S. Therefore the answer is $x' = x_2$, i.e. the projection of x on S.

Question (Q2) was answered by appeal to the answer found to (Q1), and that in turn was a corollary to Gleason's theorem. But (Q2) can also be approached in other ways, which throw additional light on this notion of one state as the conditionalization of another state on some subspace. I shall briefly outline the results obtained by Herbut (1969),[17] but without the gloss in terms of measurement in which they were originally couched. I shall first state the main theorem, and then explain it.

Theorem: Let H be a separable Hilbert space, x an element of H, and S a subspace of H. The following three conditions are equivalent:
1. y is the unique element of S closest to x;
2. y is the unique element of S whose inner product $(y \cdot z)$ equals $(x \cdot z)$, for all elements z of S;
3. y is the projection of x into S.

The word 'closest' in condition 1 refers to the metric defined in terms of the inner product, so 1 means

1'. y is the unique element of S such that $|x - y| \le |x - z|$ for all z in S.

The projection of a vector x into a subspace S is the unique element x' such that $x = x' + x''$, with x' in S and x'' in $S^\perp$.

Although clause 2 obviously connects with questions about probability, the relevance of this theorem to the general question (Q2) is not immediately obvious. But the statistical operators on a Hilbert space are also elements of a certain other Hilbert space, to which this theorem can be applied. Herbut calls this the *operator Hilbert space*; let us designate it H^o:

H^o: The elements of H^o are the operators A on H with finite absolute norm; i.e., using * for adjoint, $Tr(A^*A)$ is finite. The scalar product $(B \cdot C)$ in H^o equals $Tr(B^*C)$. The linear combination $aB + bC$ is defined as usual for operators.

Returning now to question (Q1), and noting the connection

between inner product and probability, we can see that one answer to it will be formulable in terms of clause 2 in the above theorem, applied however to statistical operators (now treated as vectors in the operator Hilbert space). The noted equivalence of 1 and 2 leads then very quickly to the description of the new statistical operator, which is the same as in the answer we gave as corollary to Gleason's theorem.

There is a good deal of other useful information in Herbut's proof, and particularly in the Appendix, which investigates the relations between statistical operators and observables in general. I note some of it here. Let observable $A = \Sigma a_n I_n$ where I_n is the projection on its nth eigenspace. Then the probability

$$(1) \qquad P_W^A(a_n) = Tr(WI_n) = Tr(WI_n^2) = Tr(I_n WI_n)$$

because $I_n^2 = I_n$ and $Tr(AB) = Tr(BA)$. Hence this probability equals 1 if $W = I_n WI_n$. But moreover, the converse follows as well by a lemma due to Lueders, so we have

$$(2) \qquad P_W^A(a_n) = 1 \text{ if and only if } W = I_n WI_n$$

Next it is shown that the following three are equivalent:

$$(3) \quad (a) \quad W = I_n WI_n \qquad (b) \quad W = I_n W \qquad (c) \quad W = WI_n$$

and these equivalences quickly lead from (2) to

$$(4) \qquad P_W^A(a_n) = 1 \text{ iff } AW = a_n W$$

So we see that the case of probability 1 can be expressed for all eigenstates (mixtures as well as pure states) in the same form.

6. SYMMETRIES AND MOTION: SCHROEDINGER'S EQUATION

> At the still point . . . there the dance is,
> But neither arrest nor movement. And do not call it fixity
> . . . Except for the point, the still point,
> There would be no dance, and there is only the dance.
>
> T. S. Eliot, *Four Quartets*

We turn now to the dynamics of quantum-mechanical systems, and focus on the case of an isolated, conservative system. Here the temporal evolution has nothing to do with the absolute value of time, but only with the duration of the process. Given that the world is indeterministic, the outcome of any measurement is to some extent a matter of pure chance. But this does not rule out that this chance, the probability for measurement outcomes, itself evolves deterministically. Of course, if the measurement is actually *made*, that is an interference, and the system is no longer isolated. So we just ask here: how do the probabilities of measurement outcomes evolve, for as long as we do not measure, but leave the system isolated?

The idea that this evolution could be deterministic and a function of elapsed time only means that it is governed by a dynamic group (see Chapter 2, Section 3). The results of Part I, which were very general, will be recognizable here in new guise. Specifically, the very nice 'law of motion' which we found for classical probabilities (Chapter 3, Section 6), evolving in this way ($p \longrightarrow e^{tk}p$), will find here an analogue in Schroedinger's equation.

A symmetry of Hilbert space is a transformation which leaves all its structure the same. If U is such a transformation, and x, y are vectors, we must therefore have:

$$x \neq y \qquad \text{entails } Ux \neq Uy$$

$$z = kx \qquad \text{entails } Uz = kUx$$

$$z = x + y \qquad \text{entails } Uz = Ux + Uy$$

$$k = (x \cdot y) \qquad \text{entails } k = (Ux \cdot Uy)$$

for any scalar k. (As corollary, of course, it follows that if $k = |x|$ then $k = |Ux|$; i.e., U does not change the length of a vector.) Obviously, given these properties, U is a linear operator, and we define a *unitary operator* to be any one-to-one linear operator that preserves inner product (or, equivalently as it turns out, that preserves length). This is in fact equivalent to:

A unitary operator is a bounded linear operator, whose effect on an orthonormal basis is to produce another orthonormal basis for the space.

The eigenvalues of such an operator U must all have absolute value 1. For if $Ux = ux$ then $(x \cdot x) = (Ux \cdot Ux) = (ux \cdot ux)$

$= u^*u(x \cdot x)$, so $u^*u = 1$. Now that means that u is a complex number of form $(a + bi)$ such that $a^2 + b^2 = 1$. This in turn allows us to find an angle $0 \leq \theta \leq 2\pi$ such that $a = \cos\theta$ and $b = (\sin\theta)$, so $u = \cos\theta + i\sin\theta$. And, surprisingly enough, that is equal to the exponential $e^{i\theta}$. This is why unitary operators can be described somewhat metaphorically as imaginary rotations. We note in passing that only one of these eigenvalues can be a real number, namely 1, which labels the fixed points of U.

Turning now to evolution over time, we must stick robustly to the probabilities of measurement outcomes which are empirically attestable. To measure the observable I_x is to perform a test which has outcome 1 with certainty if performed on a system in state x, and outcome 0 with certainty if the state is orthogonal to x (and finally, outcome 1 with probability $(x \cdot y)^2$ if performed on some state y). Recall that for any pure state y the probability $(y \cdot I_x y) = (x \cdot y)^2$ is often called the *transition probability* between states x and y. (See Chapter 5 for a warning about the confusing connotations.)

Now suppose that the state $x = x(0)$ evolves over time into states $x(t) = T_t x$ for a certain dynamic group $\{T_t : t \in R\}$. We can prove quite simply that all the evolution operators T_t must be unitary. For consider: the probability of finding value 1 if we perform measurement $I_{y(0)}$ on this system at time $t = 0$ should be just the same as that of getting value 1 if at time t we measure $I_{y(t)}$ on it. So we must have

$$(x(0) \cdot y(0))^2 = (x \cdot I_{y(0)} x) = (x(t) \cdot I_{y(t)} x(t)) = (x(t) \cdot y(t))^2.$$

Therefore if T is the operator T_t we have $(x \cdot y)^2 = (Tx \cdot Ty)^2$. This is true if T is unitary, but it is also true if T is *anti-unitary*, namely if $Tcx = c^*Tx$ for any state x and complex number c, but otherwise like a unitary operator—in which case $(Tx \cdot Ty) = (x \cdot y)^*$. For after all, $(x \cdot y)^2 = ((x \cdot y)^*)^2$, so the two cases cannot be distinguished.[18] The point is of course that distinct vectors can give us the same surface probabilities, because these probabilities are squares of their coefficients.

However, these evolution operators form a one-parameter group, so that $T_{t+t'} = T_t T_{t'}$. Thus we also have $T_{2t} = T_t T_t$. If T is unitary, so is T^2. But in addition, if T is anti-unitary, then we have $(T^2 x \cdot T^2 y) = (Tx \cdot Ty)^* = (x \cdot y)^{**} = (x \cdot y)$. So T_{2t} is unitary no matter what T_t is. This reasoning holds for all values of t; hence T_t is unitary for all t. Therefore the dynamic group is a group of unitary operators. Assuming continuity of evolution, such a group can be described by means of another famous theorem.[19]

> *Theorem* (M. K. Stone): If $\{U_t\}$ is a one-parameter group of unitary operators and $(x \cdot U_t y)$ is a continuous function of t for all vectors x, y, then there exists a Hermitean operator H such that
>
> $$U_t = e^{-itH}$$
>
> and any bounded operator commutes with all the U_t if and only if it commutes with H.

Mathematically, this operator H is the infinitesimal generator of the group. In quantum mechanics it is called the *Hamiltonian* of the system, and the equation $U_t = e^{-itH}$ can be written, with $x(t) = U_t x$, in either of the forms

(1) $$x(t) = e^{-itH} x(0)$$

(2) $$i(d/dt)x(t) = Hx(t)$$

which are the familiar, equivalent statements of *Schroedinger's equation*. The observable represented by H is called the *energy* of the system.

I have focused so far on pure states. If $W(0) = aI_{x(0)} + bI_{y(0)}$ is a mixed-state subject to the same process, then it develops into $W(t) = aI_{x(t)} + bI_{y(t)}$. If $x(0)$ and $y(0)$ are orthogonal unit vectors, then they are here eigenvectors of $W(0)$ with corresponding eigenvalues a and b. The evolution U_t turns them into eigenvectors of $W(t)$ corresponding to the same eigenvalues. Let us abbreviate as follows: $x = x(0)$, $y = y(0)$, $x' = x(t)$, $y' = y(t)$, and $T = U_t$. Then $W(0)z = a(x \cdot z)x + b(y \cdot z)y$ and

$$W(t)z = a(x' \cdot z)x' + b(y' \cdot z)y' = a(Tx \cdot z)Tx + b(Ty \cdot z)Ty$$

Now notice that

$$TW(0)T^{-1}z = T[a(x \cdot T^{-1}z)x + b(y \cdot T^{-1}z)y]$$
$$= a(x \cdot T^{-1}z)Tx + b(y \cdot T^{-1}z)Ty$$

which therefore equals $W(t)z$ provided only the inverse T^{-1} of a unitary operator T has the property that $(Tx \cdot z) = (x \cdot T^{-1}z)$. And this is indeed true. Therefore we have found, by illustration, the form this law of motion takes for mixed states:

(3) $$W(t) = U_t W(0) U_t^{-1}$$

I assumed here that the evolution preserves the convexity structure of the set of statistical operators:

If $W(0) = \sum p_i W_i(0)$ then $W(t) = \sum p_i W_i(t)$.

This means of course that the evolution of a mixed state is derivative from the evolution of pure states, in the way that the ignorance interpretation would have suggested. Equation (3) was derived with that assumption, plus the previously derived fact that U_t is unitary (which followed from the assumption that the transition probabilities are preserved). There is also a theorem, due to Mackey and Kadison, that the assumption that the convexity structure is preserved in the evolution of statistical operators already entails that the evolution operators are unitary (see Beltrametti and Cassinelli 1981*a*, sect. 23.2).

7. SYMMETRIES AND CONSERVATION LAWS

What exactly can a conservation law be, for a probabilistic theory? In classical science, a conservation law states that a certain quantity—mass, say, or energy—remains constant in time, in an isolated system. In quantum mechanics the situation is not quite so simple at first sight. But in a certain respect it is also simpler, because the intimate relation between conservation and symmetry becomes crystal-clear here.

Recall from the preceding section that there exists a Hermitean operator H, the Hamiltonian, such that

(*a*) $U_t x = e^{-itH} x$;

(*b*) any observable commutes with all the operators U_t if and only if it commutes with H.

This H also represents an observable, the *energy* of the system.[20] But what is the significance of that relationship, commutation with the dynamic group, described in (*b*) above?

Our interest in a group of transformations such as $\{U_t\}$ generally concerns exactly what is invariant under them, what is conserved. And that is just what (b) tell us:

> *Theorem*: A quantity is conserved (remains invariant over time) exactly if its representing operator commutes with all the evolution operators.

The main part of proving this is to say what it means. Conservation means here that the probabilities for outcomes of measurement remain the same in time. Suppose M does commute with all the operators U_t, and let us begin with eigenstate x such that $Mx = mx$. Then

$$U_t x = (1/m)U_t mx = (1/m)U_t Mx = (1/m)MU_t x$$

hence $M(U_t x) = mU_t x$. In other words, the evolution of a vector in an eigenspace of M stays inside that same eigenspace. Hence if measurement of M is certain to have an outcome in set D at one time, this remains certain at any later time.

Consider next the superposition $y = \Sigma c_i x_i$ where $Mx_i = m_i x_i$. In that case $U_t y = \Sigma c_i U_t x_i$. As we have just seen, the vectors $U_t x_i$ belong to the same eigenspaces as the x_i, and occur with the same coefficients c_i. Hence all the measurement outcome probabilities for M are the same in the evolved state as in the initial state.

The theorem says 'if and only if'; so suppose M is not conserved. That means that for some y the probability of some measurement outcome, say m_k, is different in y and in $U_t y$. Looking at the above, we see that this must be due to c_k having changed to some other number d, so some of the x_i did not stay in their own eigenspace when acted upon by U_t. But if $Mx = mx$ and $MU_t x \neq mU_t x$, then M and U_t do not commute, as we can see from the second-last paragraph. Therefore the operator M does commute with all evolution operators if and only if all probabilities of measurement outcomes for M remain the same throughout time.

We have already implicitly introduced the nice geometric way to describe the same situation. The observable M corresponds to a partition of the Hilbert space into subspaces H_1, H_2, ..., the *eigenspaces* of M: $Mx = m_k x$ if and only if x is in subspace H_k. Now the statement that M commutes with the dynamic group means exactly that all these eigenspaces are invariant under the action of this group. Evolution of a state cannot lead out of, or into, such a subspace. The system's state-trajectory in time, if it is ever in such a subspace, remains in it, and never was out of it. In a terminology which we shall use again, these subspaces *reduce* each evolution operator.

Obviously H commutes with itself, and so as corollary to (*b*), and to this little theorem, we have the result that energy (by definition, the observable represented by H) is conserved over time.[21] It is one of those constant 'essential' characteristics left invariant by the evolution of state, of a system of the sort studied here. And that means: a system whose evolution in time is characterized by a dynamic group. And that means in turn: *one whose evolution in time has the time-translation symmetry*. So here we have, just falling out as a corollary, the deep connection between this temporal symmetry and conservation of energy.

An operator which commutes with the dynamic group, and thus partitions the space into time-invariant subspaces, is sometimes called a *selection operator*. The statement that this operator commutes with the dynamic group (equivalently, with H) is then called a *selection rule*. In the next section we shall take up a more drastic sort of selection—'superselection'.

As an illustration which will also help below, we shall look in some detail at another conservation law. This is the law of conservation of angular momentum, which follows from a spatial rather than a temporal symmetry. The formal structure is quite similar to what we have seen already. There is a group $\{V_\theta\}$ of unitary operators which represent physical rotation through angle θ, around a specific axis—say the z-axis of our spatial frame of reference. Then by Stone's theorem there is a special operator J, such that $V_\theta = e^{-i\theta J}$. This operator J, or more perspicuously J_z, represents angular momentum around the z-axis. Invariance of the state under rotation is thus logically linked with the conservation of angular momentum. In general we must say: *every physical symmetry leads thus to a conservation law*.

To make this illustration concrete, we need to look at the unitary representation of the spatial rotation group, and the associated observable of angular momentum. Suppose that the state is specified as a function not of time, but of spatial coordinates. Let r, s, ... stand for positions in Euclidean space, and let our states be given as functions $x(r)$, $x(s)$, ..., $y(r)$, $y(s)$, ... This is actually how Schroedinger originally constructed a specific example of Hilbert space, for his wave mechanics.[22] Now the symmetries of Euclidean space are represented among the symmetries of the Hilbert space built upon it. Consider for example the group of rotations R_θ around the z-axis in Euclidean space, and the functions V_θ defined by

$$V_\theta x(r) = x(R_\theta r)$$

These functions V_θ are a group of unitary operators which represent the group $\{R_\theta\}$. Obviously we can reason about it much as we reasoned about the group $\{U_t\}$, which in that same sense represents the group of translations of time:

$$U_b x(t) = x(t + b)$$

Again therefore Stone's theorem applies, and we can express V_θ in the form

$$V_\theta = e^{-i\theta J}$$

The operator $J = J_z$ is again Hermitean; it is called the *angular momentum* (around the z-axis), and Stone's theorem tells us that an observable commutes with J_z if and only if it commutes with all the V_θ.

A physical system may be invariant under rotation, in the sense that the evolution of a rotated state is exactly the same as the rotation of an evolved state.

Theorem: If a physical system is invariant under rotation (around an axis) then its angular momentum (around this axis) is conserved.

To see how this follows, note that we are here thinking of the state as a function of both time and space coordinates $x = x(r, t)$, so that $V_\theta x = x(R_\theta r, t)$ and $U_b x = x(r, t + b)$. The invariance explained above means that $U_b V_\theta = V_\theta U_b$ for any b and θ. So we see at once that U_b commutes with all the V_θ, and hence with J_z. But then J_z commutes with H and is conserved, by our previous result.

8. THE RADICAL EFFECT OF SUPERSELECTION RULES

The topic we now broach was a relatively recent introduction, but is very important for philosophical discussions. It will make us look back on the exposition so far, and see that some of it involved real oversimplifications. The selection operators partition the Hilbert space into subspaces which cannot be entered or left—but the superselection operators partition it into subspaces between which there can (also) be no genuine superpositions. This is a very drastic effect. As we shall see, it affects even such very basic ideas as that pure states are uniquely representable by single vectors.

A selection operator was an operator which commutes with the operator H or, equivalently, with all evolution operators. Its partition of eigenspaces 'reduces' the evolution operators in this sense: each such eigenspace is invariant under temporal evolution. Now, a *superselection* operator is defined as one which commutes (not just with H, but) with *all* observables.[23]

The first question is: does this definition pick out anything except trivialities? And the answer is: that depends on what observables there are. If all Hermitean operators on the Hilbert space represent observables, then only the identity operator, and its constant multiples, are superselection operators. That is trivial. But the famous paper by Wick *et al.* (1952) argued cogently that there must be non-trivial superselection operators. In that case, then, not all Hermitean operators represent observables. This possibility had always been considered in the abstract (though von Neumann argued against it), and now it had to be taken seriously.

As a special case, think of how we can represent rotation through 360° (say, around the z-axis). As far as ordinary space is concerned, this leaves everything the same. Now recall how rotation in general was represented in the preceding section. Rotation through angle θ is represented by unitary operator V_θ. These operators form a one-parameter group, and by Stone's theorem, $V_\theta = \mathrm{e}^{-\mathrm{i}\theta J}$ for all angles θ. Now as a matter of fact, this operator J has only integers and half-integers as eigenvalues. For the special case of 360° or 2π radians, we have

$$V_{2\pi}x = x \text{ if } Jx = kx \text{ for any integer } k$$

$$= -x \text{ if } Jx = (k/2)x \text{ for any odd integer } k$$

So we see that $V_{2\pi}$, although it represents the identity operator among rotations, is not itself the identity.

Measurement predictions are not affected by this, in the following sense: probabilities of measurement outcomes are the same for states x and $-x$, and hence for x and $V_{2\pi}x$. But in superpositions the difference shows up: those probabilities are not the same for $y + x$ and $y - x$, and so may not be the same for $y + x$ and $y + V_{2\pi}x$. Various experiments illustrate this fact (see Klein and Opat 1975, 1976).

The complete physical invariance under 360° rotation can be imposed in the form of a *superselection rule*:

$V_{2\pi}$ commutes with all observables.

By definition, this statement says that $V_{2\pi}$ is a *superselection operator*. We shall not expect this relatively trivial example to be of earthshaking importance, but let us see what consequences this rule has.

We have just seen that $V_{2\pi}$ has the two eigenvalues ± 1, so it has two eigenspaces—let us call them S^+ and S^-—which partition the Hilbert space. This partition *reduces* all observables in the sense we encountered before.

For all vectors x and observables A, the vector Ax belongs to S^+ (respectively, S^-) if and only if x does.

In other words, each of these two eigenspaces is invariant under the action of any Hermitean operator which represents an observable. (See *Proofs and illustrations* for the proof.)

The existence of a superselection operator therefore gives us a partition—consisting of the eigenspaces of this operator—which reduces every observable. These subspaces are also called the *coherent* subspaces. What exactly does this do to the states?

There are three interesting consequences. The first is that the principle of superposition is curtailed: what looks mathematically like a superposition of pure states may actually represent a mixed state. The second is that the representation of states—by density matrices, as established by Gleason's theorem—is no longer unique. The third, still more curious, is that even the representation of pure states *may not* be unique.

For the first consequence, let us consider a superposition of two vectors x and y which belong to different coherent subspaces—say x is in S^+ and y in S^-, and $z = ax + by$. All the curious features of quantum mechanics are related to the fact that *in general* vector z (or density matrix I_z) represents a new pure state. Certainly I_z is not mathematically the same as

$$W = a^2 I_x + b^2 I_y$$

which represents a mixed state. But because x and y belong to distinct coherent subspaces, we shall find that I_z and W do, *in this case*, represent the same state; for all probabilities of measurement outcomes are the same in both.

The proof will be found in *Proofs and illustrations*. Its idea is simple enough. If partition $\{S^+, S^-\}$ reduces observable A, then A really falls apart into two operators, one on each subspace. We might call these A^+ and A^-. The expectation value of the outcome of a measurement of A will be the sum of the expectation values of A^+ and A^-. But only the part of the state that lies in S^+ is relevant to the former, and only its part lying in S^- is relevant to the latter.

This is the sense in which no genuine superposition of states represented by vectors lying in distinct coherent subspaces is possible.

The second consequence, that representation of state is no longer unique, follows at once as a corollary. After all, density matrices I_z and W are distinct, but (however we choose a and b such that $a^2 + b^2 = 1$), I_z and W represent the same state.

We can always find a 'canonical' representation which is unique. A quick look will show that W does and I_z does not commute with the projections I^+ and I^- on the coherent subspaces. Given a density matrix W, we can make up a new one:

$$W^s = \sum \{IWI : I \text{ is the projection on a coherent subspace}\}$$

which represents the same state as W and does commute with all the projections on coherent subspaces (Beltrametti and Cassinelli 1981a, sect. 5.4). Indeed, we can choose this as the unique 'canonical' representation of state. But there is one remaining worry: do two vectors that lie in the same coherent subspace sometimes represent the same state?

This brings us to the third consequence: they may or may not.

For we make room for the existence of non-trivial superselection rules by saying that not all Hermitean operators represent observables. At that point, we no longer know which operators *do* represent observables. This question has no uniform general answer. The easy way to close the discussion at this point would be to say: every vector in a coherent subspace represents a unique pure state, as does equivalently the projection along that vector (one-dimensional projection). This requires the existence, for every two vectors x and y lying in a coherent subspace, of an observable M with different expectation in x and y. If I_x is an observable, that will do very well. But if we do not know what observables there are, we have admitted (at least in principle) the possibility that x and y, although distinct unit vectors, have no separating observable. The empirically purest case could then at best be represented by the two-dimensional subspace $[x, y]$ or projection $I_{[x,y]}$, and not by a one-dimensional subspace or projection operator.

Messiah and Greenberg (1964) explicitly acknowledged this possibility and introduced the term 'generalized ray' for it. The possibility should be taken seriously, in part because of its connection with quantum-statistical mechanics, as we shall see in a later chapter. But it would mean that a certain amount of theory needs to be rewritten. The projection $I_{[x,y]}$ does not have trace 1 but trace 2. So in this case the Hermitean operators which represent states are not singled out by having trace 1. Instead, we would first have to say that a pure state is represented by I_s if and only if the expectation of all observables is the same for all vectors in S, but not always the same for a vector in S as for a vector outside S. We would have to say that a mixed state is represented by a convex combination of such projection operators. Finally, a variant of Gleason's theorem would have to be proved for this, and some standard terminology revised. Most of this looks fairly straightforward, except that it would be risky to speculate about the exact variant of Gleason's theorem (i.e. how much uniqueness remains).

What happens to temporal evolution? Schroedinger's equation depicts a process which turns vectors into vectors, and hence (we would have said before) pure states into pure states. But now some vectors (those outside coherent subspaces) represent mixed states. So could the evolution of an isolated system now turn a pure state into a mixture?

The answer is *no* if the Hamiltonian H is an observable. For if it is, then it commutes with the superselection operator, and hence the latter represents a conserved quantity. This was shown in the preceding section, and means that temporal evolution cannot lead out of a coherent subspace. (The superselection operator is then also a selection operator and the eigenspaces of the superselection operator are invariant under the action of the dynamic group.) Notice the supposition: if H were not an observable, it could be one of those Hermitean operators that do not commute with the superselection operator. So then a pure state *could* evolve into a mixture.[24] But the assumption that the Hamiltonian H represents an observable, namely the energy, is very strong and appears to be usually taken for granted.

What we have now discussed for the superselection operator $V = V_{2\pi}$, which has only the two eigenvalues $+1$ and -1, can be generalized. First of all, we can imagine a superselection rule that effectively partitions the space into any number of subspaces. (We still refer to members of this partition as *coherent* subspaces.) Secondly, we can imagine that we have more than one superselection rule. In that case the situation is not very different if the superselection operators all commute with each other. For then they can all be defined as functions of a single superselection operator. But if they do not all commute, the situation is a little more complex.

In fact the Permutation Group, which we shall see at work in our discussion of the statistics of identical particles, is non-commutative. So this more complicated case is important; we shall postpone its complexities until the necessity arises.

Proofs and illustrations

Let us first prove that, if S^+ and S^- are the two eigenspaces of unitary superselection operators V (e.g. $V_{2\pi}$), then each is invariant under the action of a Hermitean operator M which represents an observable. Since V is unitary, it has an inverse V^{-1}; because V commutes with M, so does its inverse. Now let x be in S^+ and y in S^-. Then we can give a slightly roundabout proof that My is also in S^-, by showing that it is orthogonal to x in S^+.

$$(x \cdot My) = (x \cdot V^{-1}VM_y)$$
$$= (x \cdot V^{-1}MVy)$$
$$= (Vx \cdot VV^{-1}MVy) \quad \text{because } V \text{ is unitary}$$
$$= (Vx \cdot MVy)$$
$$= (+ x \cdot M(-y)) \quad \text{because } x \in S^+, y \in S^-$$
$$= (x \cdot -My)$$
$$= - (x \cdot My)$$

but since only 0 equals minus itself, this means that $(x \cdot My) = 0$, i.e. $x \perp My$. This reasoning holds for all vectors x in S^+, so My is orthogonal to all of S^+ and therefore belongs to S^-. Thus, if y is in S^-, so is My for any observable M — and *mutatis mutandis* for S^+.

Next we prove that, if x and y lie in S^+ and S^- respectively, the two density matrices

$$I_z, \text{ where } z = ax + by$$
$$W = a^2 I_x + b^2 I_y$$

represent the same state. (There is no *genuine* superposition of states belonging to distinct coherent subspaces.) It will suffice to consider an observable M and compare the expectation values $(z \cdot Mz)$, $(x \cdot Mx)$, $(y \cdot My)$. (For simplicity of calculation, first set $a = b = 1$ here.)

(1) $\quad (z \cdot Mz) = (z \cdot (I^+ + I^-)Mz)$
$$= (z \cdot (I^+Mz + I^-Mz))$$
(2) $\quad = (z \cdot I^+Mz) + (z \cdot I^-Mz)$
(3) $\quad = (z \cdot I^+I^+Mz) + (z \cdot I^-I^-Mz)$
(4) $\quad = (z \cdot I^+MI^+z) + (z \cdot I^-MI^-z)$
(5) $\quad = (z \cdot I^+Mx) + (z \cdot I^-My)$
(6) $\quad = (I^+z \cdot Mx) + (I^-z \cdot My)$
(7) $\quad = (x \cdot Mx) + (y \cdot My)$

Here step (1) follows because $I^+ + I^-$ is just the identity operator I_H; step (3) because projection operators are idempotent; step (4) because they commute with M (because V does); step (6) because projection operators are Hermitean. The rest follow because $I^+ z = x$ and $I^- z = y$. (If more generally we replace x and y by ax and by, the coefficients a^2 and b^2 appear.)

This is what is meant when one says that a superselection operator splits the Hilbert space into subspaces, between which no superposition is possible. The point is *not* that a certain algebraic operation $x + y$ is not meaningful. It is rather that a state so described is empirically indistinguishable from a mixture.

Kay-Kong Wan (1980) gave the following elegant and concise representation of quantum theory with superselection rules. It is however not the most general case, because it allows only for commuting superselection operators.

Postulate: Every quantum system with its associated Hilbert space H possesses the following two properties:

(*a*) *On states*: Let S denote the set of all pure states. Then $S \subseteq H$ and H may be decomposed into a direct sum of mutually orthogonal subspaces H_n, i.e.

$$H = \bigoplus_n H_n$$

such that

$$S = \bigcup_n H_n$$

There is no further such decomposition of H_n.

(*b*) *On observables*: Let Q denote the set of all observables. Then a bounded Hermitean operator A on H belongs to Q if and only if A leaves H_n invariant; i.e.,

$$\text{if } \chi_n \in H_n \text{ then } A\chi_n \in H_n$$

We make the following observations. Part (*b*) is in no way entailed by the theory of superselection; it is a postulate opting for the largest feasible choice of represented observables. (Unbounded observables can be defined.) Secondly, for any observable A we have

$$A = \sum_n P_n A P_n$$

where P_n is the projection on H_n. Thus these orthogonal projections are superselection operators. They are essentially the only ones: all are sums of these projectors. Thus the set of superselection operators is *Abelian*; i.e., any two of them commute with each other.

In a later chapter we will look at a family of non-commuting superselectors, the permutations. There we will see what may be denied in Wan's Postulate. The orthogonal coherent subspaces do not sum to the whole Hilbert space any more, and the postulate of Dichotomy (restricting elementary particles to fermions and bosons) eliminates many pure states and observables admitted by Wan.

7

Composite Systems, Interaction, and Measurement

THE three main issues in the philosophical foundations of quantum mechanics are measurement, the 'paradoxes' (Schroedinger's Cat, EPR, etc.), and the problem of identical particles. Each of these concerns the composition of several systems—sometimes interacting and sometimes not—which is a subtle matter in quantum mechanics. Dirk Aerts very aptly sums up these issues as the problem of the One and the Many, which has here taken on a new form of life. This chapter contains the technical background for those discussions, with only a little philosophical discussion to explain their putative significance.[1]

1. COMPOSITION

Some systems are composed of others, or may be so modelled—a solid or gas composed of molecules, molecules of atoms, atoms of elementary particles. Any physical theory must allow the representation of compound systems, and give some guidance as to how the state of the whole is related to the state of the parts. Whether there should be very strict constraints on this has long been a question in natural philosophy; a logical or philosophical *atomism* holds that the properties of the parts entirely determine those of the whole, while a *holism* allows for some independence of the whole. The question is sometimes logically finessed: could not the relations between the parts be themselves parts of the whole—or, being related in some fashion, be itself a property of the part? But since a physical theory begins with a specific form of representation for the states of systems in general, the question still remains whether the state of the whole, *as represented*, is determined by the states of the parts, *so represented*. We shall see here that the quantum-mechanical treatment is, in the above terminology, holistic rather than atomistic—one way in which today's *atom* is not yesterday's *atomos*.

Let us consider two systems X and Y, and the compound system $X + Y$. The pure states of X and Y are represented by the vectors of the Hilbert spaces H_1 and H_2, possibly the same.[2] Similarly, the pure states of $X + Y$ should be represented by the elements of a larger vector space H_{12}. This should have room at least for the case in which X and Y have nothing to do with each other physically, but are just joined 'in thought'. In that case, the state of $X + Y$ is just described by saying what states X and Y have respectively, say x and y. Let us denote this state of $X + Y$ as $x \otimes y$. Then the most conservative picture we can draw of H_{12} is this:

1. There is a map $\otimes$ of H_1, H_2 into H_{12} such that:
   (*a*) the set of vectors $\{x \otimes y : x \in H_1, \ y \in H_2\}$ spans the space H_{12}
   (*b*) $\otimes$ is bilinear: $x \otimes (y + z) = (x \otimes y) + (x \otimes z)$
   $$(x + y) \otimes z = (x \otimes z) + (y \otimes z)$$
   $$b(x \otimes y) = (bx \otimes y) = (x \otimes by)$$
   (*c*) $\otimes$ multiplies the scalar products:
   $$(x \otimes y) \cdot (x' \otimes y') = (x \cdot x')(y \cdot y')$$

The conservativism lies especially in (*a*), because it is the nearest we can come to demanding an atomistic (i.e. non-holistic) policy while admitting that H_{12} must also be a Hilbert space. Clauses (*b*) and (*c*) say in effect that the operations definitive of a vector space with inner product are carried over 'pointwise' from the old context to the new. Again this is conservative: it forbids emergence of new structure, underivative from the old.

It is a theorem that for each H_1 and H_2 there is essentially only one such H_{12} (up to isometry); and this is called their *tensor product*. We also call $x \otimes y$ the tensor product of x and y; in many texts the symbol $\otimes$ is simply omitted. I will provide the details at the end of this section; the properties displayed in 1 above suffice for intuitive discussion.

Despite our conservativism, it is clear that we now have states for $X + Y$ which are *not* of the form $x \otimes y$, namely such superpositions as $(x \otimes y) + (x' \otimes y')$. Perhaps the bilinearity condition 1(b) looks at first sight as if it reduces this to $(x + x') \otimes (y + y')$, but that is not the case. Indeed, by 1(b), the latter expands to

$$(x + x') \otimes (y + y') = [(x + x') \otimes y] + [(x + x') \otimes y']$$
$$= (x \otimes y) + (x' \otimes y)$$
$$+ (x \otimes y') + (x' \otimes y')$$

which has two more summands than $(x \otimes y) + (x' \otimes y')$. So we now have 'mingled' states which will give us 'tangled' statistics for the compound system. The fact is that our insistence on conservativism binds us, with mathematical necessity, to the emergence of new states for the wholes which are not simple patchwork combinations of states of the parts.

Something similar happens with the observables. We could ask about $X + Y$ the question 'What are the respective momenta of X and Y?' but also the quite different question: 'What is the total momentum of $X + Y$?' or 'What is the sum of the momenta of X and of Y?'. Thus we must have room for observables which properly pertain to the whole system, and the crucial issue will be how they are related to the observables pertaining to the subsystems. This issue has two parts: the first concerns the Hermitean operators and the second, the probabilities of measurement outcomes.

If A and B are Hermitean operators on H_1 and H_2 respectively, we can define their tensor product $A \otimes B$ by

$$A \otimes B(x \otimes y) = Ax \otimes By$$

and extend it to all vectors in H_{12} by linearity. The result is a Hermitean operator on H_{12}. In addition, if A_1, A_2 are Hermitean, so is $aA_1 + bA_2$ for real numbers a, b. Therefore we have operators that work on the state of the whole in an intuitive way. The operator M defined on H_1 can be identified with $M \otimes I_2$ (where I_2 is the identity on H_2) and similarly for M' defined on H_2 and $I_1 \otimes M'$. These represent the observables that really pertain to only one of the subsystems. Probabilities of measurement outcomes are now formally calculated as before, but we shall be looking quite a lot at how these probabilities can involve correlations between the parts of the compound.

To complete the discussion, I will now state the official, abstract definition of the tensor product and its consequences, although it will play a very small role in what follows. (See further Jauch 1968, sect. 11–7, or van Fraassen 1972*a*, app.) I shall use the Greek letters for vectors in tensor product spaces.

If H_1 and H_2 are two Hilbert spaces, we define $H_1 \otimes H_2$ to be the set of *conjugate linear mappings* from H_2 into H_1; that means the set of maps φ such that:

(i) $\varphi(x + y) = \varphi x + \varphi y$

(ii) $\varphi(ax) = a^* \varphi x$

One such map is $(x \otimes y)$ defined for x in H_1 and y in H_2 by

$$(x \otimes y)(z) = (z \cdot y)x \text{ for any } z \text{ in } H_2$$

The maps $(x \otimes y)$, now called vectors in $H_1 \otimes H_2$, span that set considered as a vector space. The scalar product in this 'tensor product space' is defined by

$$(\psi \cdot \varphi) = \sum_r (\psi(y_r) \cdot \varphi(y_r))$$

where $\{y_r\}$ is any base of H_2. The properties listed in the text as 1(*a*)–(*c*) follow quickly from this definition. From them in turn we can deduce:

2. *Tensor Product Theorem*

(*a*) If $\{x_i\}$, $\{y_j\}$ are bases of H_1, H_2, respectively, then $\{x_i \otimes y_j\}$ is a basis of H_{12}

(*b*) $|x \otimes y| = |x| \, |y|$

(*c*) $ax \otimes by = ab(x \otimes y)$

(*d*) $(x + y) \otimes z = (x \otimes z) + (y \otimes z)$

$\quad\quad x \otimes (y + z) = (x \otimes y) + (x \otimes z)$

(*e*) $\displaystyle\sum_{i,j} [a_i b_j (x_i \otimes y_j)] = \sum_i \left[a_i \left(x_i \otimes \sum_j b_j y_j \right) \right]$

$$= \sum_j \left[\left(\sum_i a_i x_i \right) \otimes b_j y_j \right]$$

$$= \sum_{i,j} (a_i x_i \otimes b_j y_j)$$

Here (a) is deducible from $1(a)$ and (b); (b), (c), (d) are readily deducible from $1(b)$ and (c); and (e) is a generalization of $1(b)$, to countable sums.

We can also use the official definition of tensor product space to prove another basic property which I did not list above:

3. *Uniqueness Property*: If the vectors x_i are mutually orthogonal, and $\Sigma x_i \otimes y_i = \Sigma x_i \otimes z_i$, then $y_i = z_i$ for each index i.

Of course the same is true if $\Sigma y_i \otimes x_i = \Sigma z_i \otimes x_i$, but I shall just prove the one case. By definition we have

$$\left[\sum (x_i \otimes y_i)\right](w) = \sum (w \cdot y_i)x_i$$

$$\left[\sum (x_i \otimes z_i)\right](w) = \sum (w \cdot z_i)x_i$$

which are two descriptions of the same vector in H_1, if w is in H_2. But in H_1 each orthogonal decomposition of a vector is unique, hence we conclude

$$(w \cdot y_i) = (w \cdot z_i) \qquad \text{for any vector } w \text{ in } H_2$$

But from that we infer that $y_i = z_i$.

Proofs and illustrations

There is an important consistency point about measurement probabilities and composition which was raised by David Finkelstein and discussed by Jeffrey Bub.[3] The law of large numbers of classical probability theory says that, if we make independent trials of the same type, of which the possible outcomes a, b, c, ... have probabilities p_a, p_b, p_c, ..., then the probability equals 1 that the limit of the relative frequency of outcome a equals the number p_a. The second occurrence of 'probability' refers here to the same function p, extended by product construction from simple trials to sequences thereof. If the Born Rule is a tenable assignment of probabilities for quantum mechanics, then the tensor product construction must connect it to a similar result.

This is indeed the case. Let A be an observable, and for a given state x in its domain consider the N-fold state

$$x^N = x \otimes \ldots \otimes x$$

the product taken N times. The corresponding observable on this compound state is

$$A^N = A \otimes \ldots \otimes A$$

also taken N times, but we are more interested in the mean value:

$$A(i) = I \otimes \ldots \otimes A \otimes \ldots \otimes I$$

$$\langle A \rangle = (1/N) \sum A(i)$$

where in the first equation of course A occurs in the ith place, and $i = 1, \ldots, N$. Now we assert that there is a unique number, namely

$$r = (x \cdot Ax)$$

the expectation value of A in state x, such that

$$\lim_{N \to \infty} |\langle A \rangle x^N - rx^N| = 0$$

The proof was formulated elegantly by Bub's student David Devine as follows:[4]

$$(x^N \cdot (\langle A \rangle - r)^2 x^N)$$

$$= (x^N \cdot [(1/N^2) \sum A(i)^2 + (1/N^2) \sum_{i \neq j} A(i)A(j)$$

$$- 2(r/N) \sum A(i) + r^2]x^N)$$

Noting that $A(i)x^N$ is just $x \otimes \ldots \otimes Ax \otimes \ldots \otimes x$, with Ax in the ith place, this expression simplifies to

$$= (1/N)(x \cdot A^2 x) + [(N1)/N](x \cdot Ax)^2 - 2r(x \cdot Ax) + r^2$$

$$= [(x \cdot Ax) - r]^2 + (1/N)[(x \cdot A^2 x) - (x \cdot Ax)^2]$$

which does indeed tend to $[(x \cdot Ax) - r]^2 = 0$ as N goes to infinity.

2. REDUCTION

If compound system $X + Y$ is in state W on $H_1 \otimes H_2$, what about its components X and Y? A state is supposed to give us probabilities of measurement outcomes, so we must require:

> the expectation of A in the state of X equals the expectation of $A \otimes I_2$ in the state of $X + Y$

where I_2 is the identity on H_2. Similarly for component Y and $I_1 \otimes B$, if B pertains to Y. When A is a projection operator, that expectation value is just a probability.

We deduce at once, from Gleason's theorem, that we can indeed assign such a state to X. For the expectation values of $I_S \otimes I_2$ in state W are perfectly good probabilities to be associated with the subspaces S of H_1. Therefore the theorem tells us that there exists a statistical operator, which we may call $\#W$ or $\#W[X]$ or $\#W[1]$, that represents such a state. I will call this state a *reduction* of W, or the state of $X + Y$ *reduced* with respect to X, or with respect to the Hilbert space H_1. This terminology comes from a standard procedure used in this context, called 'reduction of the density matrix'. Details concerning this procedure will be given in *Proofs and illustrations*. But we can already say quite a bit about these reduced states, simply on the basis of how we just introduced them:

1. If W is a statistical operator on $H_1 \otimes H_2$, then $\#W[1]$ is the statistical operator on H_1 such that $Tr(A \, \#W[1]) = Tr((A \otimes I_2)W)$ for all Hermitean operators A on H_1.

This is easy to generalize to a general system $X_1 + \ldots X_N$ with a state W on the tensor product space $H_1 \otimes \ldots \otimes H_N$ and its reduction $\#W[K, \ldots, M]$ with respect to the subsystem $X_K + \ldots + X_M$. Notice that 1 (which can serve as definition) makes best sense in a context where all Hermitean operators represent observables. If it is not so, the statistical operator representing the reduced state can still be so defined, but obviously its job of representation could be done equally well by some others—uniqueness is then lost, as usual.

2. A reduction of a mixture is a corresponding mixture of reductions; i.e.,

$$\text{if } W = \sum p_i\, I_{\varphi(i)} \text{ then } \#W = \sum p_i\#I_{\varphi(i)}$$

It will be convenient to abbreviate $\#I_\varphi$ to $\#\varphi$. Result 2 is an immediate corollary to definition 1, because trace is linear.

3. Reduction is transitive; that is,

$$\#(\#W[J, \ldots, M])\, [K, \ldots, M] = \#W[J, \ldots M]$$

$$\text{if } J \le K \le M$$

This is quite easy to see for a small compound. Suppose $X + Y + Z$ is in pure state φ in $H_1 \otimes H_2 \otimes H_3$. Then A has in $\#\varphi[Z]$ the same expectation as $I_1 \otimes I_2 \otimes A$ has in φ, and also as $I_2 \otimes A$ has in $\#\varphi[Y + Z]$. Of course the two numbers must agree, because, by the definition, $I_2 \otimes A$ must have the same expectation in $\#\varphi[Y + Z]$ as $I_1 \otimes (I_2 \otimes A)$ has in φ. The abstract point is that the tensor product construction, like the composition of physical systems, is (essentially) associative.

4. If $X + Y$ is in the 'perfectly correlated' state, $\varphi = \Sigma c_i x_i \otimes y_i$ with $\{x_i\}$ and $\{y_i\}$ bases of unit vectors for H_1 and H_2, then

$$\#\varphi[X] = \sum c_i^* c_i I_{x(i)}$$

$$\#\varphi[Y] = \sum c_i^* c_i I_{y(i)}$$

I will refer to this corollary as the *Special Reduction* result. It is a very important case because discussions of measurement, and also of many paradoxes in quantum mechanics, centre on such states of perfect correlation. It is exactly in this case that we can see at once what the reduced states are. This corollary can be proved as follows.

The expectation value of B on Y equals

$$(\varphi \cdot (I_1 \otimes B)\varphi) = \left(\varphi \cdot (I_1 \otimes B)\left(\sum c_i x_i \otimes y_i \right) \right)$$

$$= \left(\varphi \cdot \sum c_i x_i \otimes B y_i \right)$$

$$= \sum c_i (\varphi \cdot x_i \otimes By_i)$$

$$= \sum_i c_i \sum_k c_k^* (x_k \otimes y_k \cdot x_i \otimes By_i)$$

$$= \sum c_i c_i^* (y_i \cdot By_i)$$

which is just the average of the expectation values of B in the pure states y_i, weighted with the factors $c_i^* c_i$. But that is the expectation in the mixture $\sum c_i^* c_i I_{y(i)}$.

This Special Reduction result has two obvious but important sub-corollaries:

5. If $\varphi = x \otimes y$, then $\#\varphi[X] = I_x$, and $\#\varphi[Y] = I_y$.
6. The reduced states are in general not pure but mixed, even when the total system is in a pure state.

This last observation brings us back to the same holism we saw in our discussion of composition, now seen 'from below', so to say. The 'tangled statistics' in the total system make it impossible in general to assign a pure state to its parts. Schroedinger called this *the* peculiarity of quantum mechanics. It is also quite easy to see that in consequence, the (mixed) states of the parts do not determine uniquely the state of the whole.

7. Pure state $\Sigma c_i x_i \otimes y_i$ has the same reductions as mixed state $\Sigma c_i^2 I_{x(i) \otimes y(i)}$

This is easy to verify; use the Special Reduction result 4 for the one and use results 2 and 5 for the other. Von Neumann proved the following much more general proposition (1955, 426–9):

8. For any states W_1 on H_1 and W_2 on H_2, there is a state W on $H_1 \otimes H_2$ with reduced states W_1 on H_1 and W_2 on H_2 respectively; W is unique if and only if at least one of W_1 or W_2 is pure.

I will not prove this; but in *Proofs and illustrations* I will prove another result due to von Neumann (1955, 436–7):

9. Every vector in $H_1 \otimes H_2$ can be written in the perfectly correlated form described in statement 4.

As von Neumann remarks, this means that a pure state on

$H_1 \otimes H_2$ always effects a one-to-one correspondence between certain quantities represented on those spaces.[5]

Proofs and illustration

There is a very abstract approach to composition and reduction, which is beautiful; and also a workaday recipe for 'reduction of the density matrix' found in textbooks.[6] But I shall follow the intermediate route charted by Everett (1973), though without broaching his interpretation.

In what follows, bases will all consist of unit vectors, unless noted otherwise. If vector z is in H_1, and $\{x_i\}$ is a basis for H_1, then z has a description:

$$(1) \qquad z = \sum (x_i \cdot z) x_i$$

i.e., z is the sum of its projections along the basis vectors. Similarly, therefore, if in addition $\{y_j\}$ is a basis for H_2, and φ is in $H_1 \otimes H_2$, it can be described as

$$(2) \qquad \varphi = \sum (x_i \otimes y_j \cdot \varphi)(x_i \otimes y_j)$$

If we set $c_{ij} = (x_i \otimes y_j \cdot \varphi)$, then we can rewrite this equivalently by grouping the factors in two ways:

$$(3) \qquad \varphi = \sum_i x_i \otimes \sum_j c_{ij} y_j = \sum_i x_i \otimes v(i)$$

$$= \sum_j \left(\sum_i c_{ij} x_i \right) \otimes y_j = \sum_j u(j) \otimes y_j$$

Now these groupings tell us in effect what the reduced states are. Indeed, with φ and the bases as above, we have the result:

(4) *General Reduction*: With φ as above,

$$\#\varphi[1] = \sum_j |u(j)|^2 \, I_{u(j)}$$

$$\#\varphi[2] = \sum_i |v(i)|^2 \, I_{v(i)}$$

Note that $|u(j)|^2 = \Sigma_i c_{ij}^* c_{ij}$ and $|v(i)|^2 = \Sigma_j c_{ij}^* c_{ij}$; so all numbers and vectors involved can be read off from the description of φ

in terms of the given bases. Note also, however, that the vectors $u(j)$ and $v(i)$ are in general not unit vectors and do not form orthogonal bases. The reduced states are indeed explicitly described here but are not orthogonally decomposed.

Let us just calculate the expectation value of $I_1 \otimes B$ in state φ, and see how it relates to the expectations of B in the states $v(i)$.

$$(\varphi \cdot (I_1 \otimes B) \, \varphi) = \left(\varphi \cdot \sum x_i \otimes Bv(i) \right)$$

$$= \sum_i (x_i \otimes Bv(i) \cdot \sum_k x_k \otimes v(k))^*$$

$$= \sum_i \sum_k (x_i \otimes Bv(i) \cdot x_k \otimes v(k))^*$$

$$= \sum_i \sum_k [(x_i \cdot x_k) \, (Bv(i) \cdot v(k))]^*$$

$$= \sum (v(i) \cdot Bv(i))$$

Because $v(i)$ is not in general a unit vector, the expectation of B in $v(i)$ equals $(v(i) \cdot Bv(i))$ divided by the square of its length, i.e. $|v(i)|^2$. Thus we have deduced

The expectation value of $I_1 \otimes B$ in φ is the average of the expectations of B in $v(i)$, weighted by the factors $|v(i)|^2$

which is just what the General Reduction result says about the reduced state $\#\varphi[2]$. A similar deduction yields its assertion about the other reduced state.

Because it is sometimes useful to get to an informative corollary about orthogonal decompositions, let us take a look at matrix representation. Recall from the preceding chapter that Hermitean operator A is represented relative to basis $\{x_i\}$ by matrix $[a_{mn}]$ exactly if

(5) $\qquad Az = \sum a_{mn}(x_m \cdot z)x_n$ for all vectors z

The first thing we need to know is the matrix representation of the reduced states. For φ as above, let us calculate on H_2:

$$\#\varphi[2]z = \sum |v(i)|^2 \, I_{v(i)}z$$

$$= \sum |v(i)|^2 \, (1/|v(i)|^2) \, (v(i) \cdot z)v(i)$$

$$= \sum_i (v(i) \cdot z) \sum_k c_{ik}y_k$$

$$= \sum_{ik} c_{ik}\left(\sum_j c_{ij}y_j \cdot z\right) y_k$$

$$= \sum_{jk} \left(\sum_i c_{ij}^* c_{ik}\right) (y_j \cdot z)y_k$$

so $\#\varphi[2]$ is represented relative to basis $\{y_j\}$ by the matrix $[b_{jk} = \sum_i c_{ij}^* c_{ik}]$. In just the same way, we can deduce that $\#\varphi[1]$ is represented relative to basis $\{x_i\}$ by matrix $[a_{ij} = \sum_k c_{ik}^* c_{jk}]$.

Recall also from the preceding chapter that the operator has an orthogonal decomposition in that basis exactly if the representing matrix is diagonal. That means that only the diagonal elements a_{ii} are non-zero. Now we have the wherewithal to prove the following result, also due to Everett:

6. If φ is as above, and $\#\varphi[1]$ has an orthogonal decomposition in basis $\{x_i\}$, then the decomposition of $\#\varphi[2]$ into the states $v(i)$ is also orthogonal.

In that case we call the equation $\varphi = \Sigma|v(i)|x_i \otimes v(i)/|v(i)|$ a *canonical* description of φ. This is a perfectly correlated form. Since obviously $\#\varphi[1]$ must always have at least one orthogonal decomposition, φ can always be written in such a canonical form. This was von Neumann's result described at the end of our section.

To prove this we need only show that the vectors $v(i)$ are orthogonal to each other:

$$v(i) \cdot v(j) = \left(\sum_k c_{ik}y_k \cdot \sum_m c_{jm}y_m\right)$$

$$= \sum_k c_{ik}^* \sum_m c_{jm}(y_k \cdot y_m)$$

$$= \sum_k c_{ik}^* \, c_{jk}$$

but that is an element in the matrix which represents $\#\varphi[1]$ relative to the basis $\{x_i\}$. By supposition, that element is zero unless $i = j$. Hence the vectors $v(i)$ are mutually orthogonal.

We observe from this result that the decompositions $\#\varphi[S']$ and $\#\varphi[S - S']$—in which a compound system is split into two parts—must have image spaces of equal dimension. Otherwise no such canonical description could be true. At first this is surprising; if, for example, H_1, H_2, H_3 have the same dimension, then $H_2 \otimes H_3$ has a much larger dimension than H_1. So we could choose mixtures W_1 and W_2 whose respective sets of eigenvectors span H_1 and $H_2 \otimes H_3$. But the theorem then tells us that there is no pure state φ in $H_1 \otimes H_2 \otimes H_3$ such that $W_1 = \#\varphi[S']$ and $W_2 = \#\varphi[S - S']$. The large compound system S must itself then be in some mixed state.

To finish, let us connect the important relation of reduction to that other important relation among states: relative possibility. Recall from Sections 1 and 4 of the preceding chapter that state W' is possible relative to W exactly if the former assigns probability 1 to all that the latter gives probability 1. The representation of this relation is most easily summarized in two steps: the pure states possible relative to W are represented by the vectors in its image space (see Chapter 6, Section 4); and then W' is possible relative to W if it is a mixture of such pure states, i.e. exactly if the image space of W' is part of the image space of W.

(7) If φ is possible relative to W, then $\#\varphi$ is possible relative to $\#W$

This holds for corresponding reductions to any subsystem.

The proof has three steps. Let total system S have a state W on Hilbert space $H_a \otimes H_b \otimes H_c$ while its subsystem S' has state-space H_b. Now let P be any projection operator on that space H_b, and suppose that P has expectation 1 in $\#W[S']$. Then $I_a \otimes P \otimes I_c$ has that same expectation in W, by definition.

Now let W have orthogonal decomposition $W = \Sigma w_i I_{x(i)}$ and $\varphi = \Sigma c_i x_i$ where c_i is not 0 only if w_i is positive, so that φ is in the image space of W. Then $(I_a \otimes P \otimes I_b)x_i = x_i$ for each i, because $I_a \otimes P \otimes I_b$ has expectation 1 in each component of

W. But then $(I_a \otimes P \otimes I_c)\varphi = \varphi$. Therefore P has expectation 1 also in $\#\varphi[S']$.

Take now as special case the projection P on the image space of $\#W[S']$. Since it has expectation 1 in that state, it also has this expectation in $\#\varphi[S']$, by our argument. So if the latter is the mixture $\Sigma v_j I_{y(j)}$, it follows that $Py_j = y_j$ for each index j for which v_j is positive. So all those components of $\#\varphi[S']$ are in the image space of $\#W[S']$ on which P projects. Therefore also the image space of $\#\varphi[S']$, spanned by those components, is entirely part of the image space of $\#W[S']$. This is what was to be proved.

3. INTERACTION AND THE IGNORANCE INTERPRETATION OF MIXTURES[7]

It has sometimes been suggested that mixed states in quantum mechanics are like the probability distributions used in classical physics, which merely represent ignorance of the true state. That would mean that every system is always really in a pure state, though we may not know which. This *ignorance interpretation of mixtures* was explicitly advocated by Hans Reichenbach (1948) as the key to interpreting measurement and distant correlations. Hans Margenau and his students, on the other hand, advocated the contrary idea that pure states are only a special subclass, and that a system may well be in no pure state at all.[8]

As it turns out, this has much to do with interaction. But already before that, it is clear that this ignorance interpretation would have to answer several disturbing questions. Exactly what are the pure states that system X may be in if it is in mixed state $W = bI_x + (1 - b)I_y$? The first answer might be: x with probability b and y with probability $(1 - b)$. But if $b = \frac{1}{2}$, this orthogonal decomposition is not unique, and we have equally $W = \frac{1}{2}I_z + \frac{1}{2}I_w$ if $[z, w] = [x, y]$ and $z \perp w$. Now the system cannot have probability $\frac{1}{2}$ each for being in distinct states x, y, z, w! Here we face a choice: shall we say that all orthogonal decompositions are on a par, or that one is privileged?[9] If the latter, quantum mechanics is definitely incomplete, for it does not tell us which. If the former, why exactly this privilege for orthogonal decompositions? Recalling the preceding chapter, we could widen the set of possible pure states for X to be the set of all components (of possibly non-orthogonal decompositions) or even to the whole image space of W.

Suppose someone decided to advocate this last version: X is in mixed state W only if X is in a pure state x which is possible relative to W (is in the image space of W). We would then ask him or her how different mixtures with the same image space arise, and the answer would be: that is the part which reflects our ignorance of how to assign probabilities. A little more would be needed to give a full answer, surely! But there is a greater difficulty.

If X is part of $X + Y$ in pure state $\varphi = (1/\sqrt{2})(x \otimes x' + y \otimes y')$, so that $W = \#\varphi$, it must follow that X is in either x or x' and Y is in either y or y'. The ignorance interpretation so far has given us no reason to eliminate the combination (X in x and Y in y'), yet we predict with certainty that a measurement of $I_{x \otimes y'}$ will have value 0. Could the ignorance interpretation be amended, with a special addition for the case in which the mixed state of X is the reduction of a pure state of a larger system? Certainly, but now we wonder where the probability interference effects come from. If the real situation is either (X in x, Y in y) or (X in x', Y in y'), why are the correct measurement outcome predictions not the same as for $X + Y$ in the mixed state $W' = \frac{1}{2}(I_{x \otimes y} + I_{x' \otimes y'})$?

It is sometimes said that the ignorance interpretation *entails* the false consequence that if X is in state W then it is in state W'. The above rhetorical question is turned into a *reductio ad absurdum*. As defenders have pointed out, this is not logically valid. In general, if we are told that X and Y are in pure states x and y, we will represent $X + Y$ as having state $x \otimes y$. But if we must accept that knowledge of the states of the parts does not uniquely determine *the* state of the whole, this could be the place to do it. It does look like grasping at straws, though. If the motive for the ignorance interpretation was that it is simple and dissolves mysteries, this motivation has now been lost.

To gain a somewhat larger perspective on this subject, we should perhaps reflect on the concept of state. This has at least three sides to it: (i) a state is the basis for (statistical) prediction; (ii) a state is prepared by means of a physical preparation or filtering procedure; and (iii) a state evolves in time, under constraints strictly linked to the characteristics of the system and its environment. In this chapter we have found a fourth side: (iv) the states of the parts of a compound system are functionally determined by, but in general do not determine, the state of the whole. Only if we concentrate entirely on (i) will we be able to think of the ignorance interpretation as at all 'obvious'.

Point (iv) strongly qualifies (ii): a system could have been prepared in a certain state, by past interaction with another system, from which it is now wholly separated. Looked at in another way, the two systems that have interacted are still together one total system—their own states are only reductions of the total state. And the statistical correlations in that total state may have the parts thoroughly 'tangled' (as Arthur Fine very aptly calls it).

In fact, interaction of two systems can tangle, untangle, and tangle again. A simple and instructive example is given by Beltrametti and Cassinelli (1981*a*, sect. 7.5) of a dynamic evolution in which a subsystem evolves continuously back and forth between pure and mixed states.

4. THE QUANTUM-MECHANICAL THEORY OF MEASUREMENT

The empirical basis for quantum theory is provided by phenomena, and more specifically by reports of measurement outcomes. On the other hand, what goes on in such a measurement itself is also a physical interaction, and must therefore in principle admit its own quantum-theoretical model. Twentieth-century ears, accustomed to the paradoxes engendered by theories applicable to the very resources used to state them, immediately detect the threat of vicious circles, paradox, and inconsistency here.

4.1. *In search of a physical correlate*

Not every procedure that ends with writing down a number is a measurement. On the other hand, not everything traditionally expected of measurement may be essential to it. Perhaps the ideal of 'just looking' was always something only God could do: a way of interacting with the object which, while having no disturbing effect whatever, gathers information about what the object *was* like, useful for predicting what it would be like later on. Such an impossible ideal should not guide our thinking about real measurement.

Our initial target is the term 'measurement' in the Born interpretation rule: if a *measurement* of observable M is made on a system in state W, the probability of an outcome in E equals $Tr(WI_E^M)$. This is a conditional statement, and we need to explain under what conditions its antecedent is true, and also correlatively when its consequent is true. The criteria of adequacy for the explanation we give *are not* that we capture the full meaning of 'measurement', but only that we correctly single out a class of processes which (*a*) meet *minimal* criteria for designation of the term 'measurement', and (*b*) are such that the statement implied by Born's interpretation rule can *tenably* be maintained to be true.

We should look at this in two ways. On the one hand, there is the scientist, who only looks to see whether the theory is empirically adequate: do the phenomena fit the models? On the other hand, there is the believer, who describes even what this scientist does in quantum-mechanical terms. Pure cases of either are fictions, but everyone can partake of each to some extent.

For the first, the phenomena are described by statements which are typically measurement outcomes (observable M has value b), or more generally a frequency count of such simple outcomes. The quantum-mechanical states need to be such that, from these reported phenomena, it is possible to infer 'backwards' to such a state. For example, we see a pattern of spots on a screen, and in the model we make up those spots are the points of impact of electrons, which emerged from a certain source in some state or other. Empirical adequacy demands that there be at least one state W such that the pattern of spots fits a position probability-distribution linked, via Born's Rule, to state W. Putting together the beginning and end of this little discussion, we derive at least this: an *outcome* of a measurement is a value attribution, a proposition that some observable had a certain value. This observable pertains not to the object, but to the measurement apparatus; however, the outcomes must correspond in a certain statistical way to a possible initial state of the object.

Coming now to the imagined believer with the quantum-mechanical world-picture, we must be careful to ask of him only as much as makes sense from *his* point of view. I said above that what goes on in a measurement is also a physical interaction. This does not mean that we can give a description of a certain kind of physical interaction—in terms provided by physics—and then say: this is what the word 'measurement' *means*. I do not wish to enter here into a discussion on the philosophy of mind, or the so-called mind–body problem, but I do wish to discuss measurement without being naive about such issues. When I say that I measured the temperature of the tea with a mercury thermometer, I imply first of all that a certain physical interaction took place. But I also imply that I did something intentional and deliberate, which distinguished that episode from, say, measuring the temperature of the thermometer by means of a cup of tea. I designated a certain observable (temperature of the tea) as the observable being measured, and another one (height of the mercury) as the 'pointer-observable'. In doing so, I *presupposed* that the thermometer had certain physical characteristics which made it suitable to be chosen as an apparatus for such measurements. The first stage in our inquiry is therefore not at all to find out what 'measurement' means, but only this:

> Given systems X and Y, and observables A pertaining to X and B pertaining to Y, what must be minimally required of these in order to measure A on X by means of Y (chosen as apparatus) with B designated as pointer-observable? What must be minimally required of an interaction between X and Y as necessary condition for A being measured on X by means of Y with pointer-observable B?

These questions can be answered in purely physical terms. If we regard the answers as complete within the domain of physics—omitting such factors as that some person made certain choices or had certain intentions—then we can henceforth say, without fear of confusion: a measurement interaction *is* (as far as physics is concerned) that sort of physical interaction.

Of course, I myself have relied on an assumption here. I assume that the Born Rule, which assigns probabilities of measurement outcomes, should be read as pertaining simply to those physical interactions *qua* physical interactions. Whether or not some person actually had such intentions or made such choices, I take to be irrelevant to those probabilities. It is on the basis of this assumption that I can hope to use the term 'measurement interactions', in the context of this book, to refer solely to physical processes of the sort required as necessary conditions for measurements *by us*. For this assumption I offer neither apology nor warrant.

4.2. *Metacriteria for measurement*

The most general notion of measurement requires therefore only that we be able to infer from information of outcomes to information about the measured object system's initial state. This may be formally captured as follows. A measurement process of observable A on object system X is characterized by four factors: the Hilbert space of the measuring apparatus Y, the pointer-observable B, the groundstate W of the apparatus, and the evolution operator U. At this point we do not assume that U is unitary. Also, I shall not specify at this point exactly how apparatus and object system are to be coupled at the outset. The initial reduced state of the apparatus must be the groundstate W. Supposing that the initial state of the total system $(X + Y)$ is V such that $\#V = T$ and $V\# = W$, it is required that U changes that to a final state $U(V)$ whose reduction $W' = U(V)\#$ to the final state of the apparatus alone is such that

(M1) $P_T^A(E) = P_{W'}^B(E)$ for all Borel sets E.

But this is not enough to have a measurement. Imagine a process which changes $(X + Y)$ when it is initially in state $(x \otimes y)$ so that it ends in state $(y \otimes x)$. Then we can say that, for *every* observable A pertaining to X, and setting $B = A$, the condition (M1) is satisfied. Indeed, if this is acceptable, we could also set $Y = X$, and say that every observable pertaining to X is always being instantaneously measured by X itself. But surely, no single observable is truly measured here, let alone all the mutually incompatible observables pertaining to X! Should we conclude then that the apparatus must be a 'classical' system, or must be subject to superselection rules which separate the pointer-observable's eigenspaces? That has sometimes been suggested even on the basis of a conviction that a measurement apparatus must be macroscopic, and that all macroscopic observables are thus. Unless the problems force us to do so, we should not go so far. The reason is that we should not simply pre-empt the true believer who takes it that quantum mechanics, via Born's Rule, also gives probabilities for outcomes of structurally similar microscopic processes. A more modest requirement is:

> (M2) W' is a mixture of B-eigenstates; i.e., $W' = \Sigma p_i W_i'$ such that $Tr(W_i' I_{b(i)}^B) = 1$.

Does this remove the intuitive difficulties posed above?

There is still a difficulty arising from 'accidental' degeneracy. Consider the following argument that it is possible to measure incompatible observables jointly sometimes. Suppose A and B are observables on a Hilbert space, with disjoint associated orthonormal bases $\{|a(i)\rangle\}$ and $\{|b(i)\rangle\}$. But suppose that $a(1) = a(2) = a$, and $[|a(1)\rangle, |a(2)\rangle] = [|b(2)\rangle, |b(3)\rangle] = S$. For simplicity, also assume that the apparatus has the same Hilbert space, and that A and B themselves serve there as the two pointer-observables for A and B pertaining to the object system. Then if the latter's initial state is in S, we conclude by (M1) and (M2) that the apparatus final state is $W' = I_a^A$. But notice that $I_a^A = \frac{1}{2} I_{b(2)}^B + \frac{1}{2} I_{b(3)}^B$. So we also have

$$W' = \tfrac{1}{2} I_{b(2)}^B + \tfrac{1}{2} I_{b(3)}^B$$

Thus, with B itself designated as pointer-observable for the measurement of B on the object, we conclude that (M1) and (M2) are both satisfied for a measurement of B. Therefore two incompatible observables have been jointly measured. This shows that (M1) and (M2) are not jointly sufficient to characterize measurement quantum-mechanically. This is however a case of *accidental degeneracy*; it occurred because we chose that particular initial state. Part of the solution to this problem, at least, must lie in the fact that we should insist that the process must be such that (M1) and (M2) are satisfied for *any* initial state of the object. Not just any process which happens to meet conditions (M1) and (M2) on its initial and end states qualifies as a measurement:

The apparatus M must be a physical structure such that (M1) and (M2) must hold whenever this process is initiated, and regardless of the initial state of the object system.[10]

The general idea is obviously this: since we model the phenomena by choosing incompatible observables to represent quantities which cannot be jointly measured, the quantum-mechanical correlate of measurement must be a process in which the observables designated as measured are compatible. Our end in view requires us to characterize the correlate so as to satisfy this condition. On the other hand, there may well be many different kinds of measurement, and there may also be a variety of alternative accounts of measurement which form parts of different tenable interpretations of quantum mechanics. Hence I propose to proceed somewhat delicately, in two stages. First, I shall define a wide class of physical interactions—to be called *measurement setups*.[11]

To set the stage, consider two systems X and Y, with Hilbert spaces H_X and H_Y and with Y so constructed that, if Y in *groundstate W* is coupled with X in initial state T, the evolution of $(X + Y)$ through a certain time interval is governed by evolution operator U. The crucial factors here are W and U, whose specification must include the information that identifies H_Y and H_X; and only H_X is relevant to what X is like. Identify the *apparatus Y* with the couple $Y = \langle W, U \rangle$. If we now look upon the process governed by U as a measurement, we are selecting for attention a *measured* observable A defined on H_X, and a *pointer*-observable B_A^Y which pertains to Y. This correspondence is defined simply by our decision to look upon the situation in this way, so we can choose B_A^Y to have the same eigenvalues as A does, and also such that the correspondence A to B_A^Y is one-to-one. But that also means that we must make our notions independent of this correspondence, which I shall call *cmp*: $A, Y \longrightarrow B_A^Y$; and of course, we must recognize that the choice of correspondence presupposes that (M1) holds. ('cmp' is a mnemonic name for the function that chooses a corresponding pointer-observable for the measured observable in question.)

Definition: $S(Y, A) = \langle A, W, U \rangle$ is a *cmp-measurement setup* for observable A on system X, exactly if

$Y = \langle W, U \rangle$ is an apparatus with W as ground-state (the *associated* apparatus), and evolution operator U transforms the initial state $T \otimes W$ of $(S + M)$ in such a way that (M1) holds for the reduced final state W' of Y, for pointer-observable $B = B_A^Y$, for all possible initial states (statistical operators) T on H_X.

In any definition of a kind of measurement, we must now consider the possibility that such an apparatus Y will serve for non-trivial cases of joint measurement. Hence we impose the following metacriteria:

Metacriteria for ideal measurement: A class CS of cmp-measurement setups is a class of *ideal measurements* only if, for every associated apparatus Y:

(M2) the reduced endstate W' of Y is a mixture of joint eigenstates of all observables B_A^Y such that $S(Y, A)$ is in CS;

(M3) if $S(Y, A)$ is in CS then $B_A^Y x = \varnothing$ for all x which are orthogonal to every possible reduced end state of Y;

(M4) if $S(Y, A(i))$ is in CS for $i = 1, \ldots, n$ and f is an n-ary Borel function, then there exists an observable A such that $S(Y, A)$ is in CS and $f(B_{A(1)}^Y, \ldots, B_{A(n)}^Y) = B_A^Y$.

Despite the new phrasing, (M2) has not been strengthened, and the function of pointer-observables in (M4) is well defined because (M2) and (M3) entail that the pointer-observables are mutually compatible.

In a moment I shall discuss the motivation and scope of these requirements. But whenever we lay down exact identities, we depart from pure explication of practice; even (M1) can be faulted in that respect. While I shall come back to this point, I think we should add here that any defined class of measurements must include a class of ideal measurements as special cases. The classes of von Neumann measurements and von Neumann–Lueders measurements discussed below are among the most prominent in the literature; we shall see that they satisfy this metacriterion.

These metacriteria would be overly restrictive if they ruled out kinds of measurement generally recognized as such in the practice of quantum mechanics. But, in themselves, the criteria are not very restrictive. Suppose that for the class CS we choose the singleton $\{S(Y, A)\}$. Then (M2) says that the apparatus ends up in a mixture of eigenstates of B_A^Y, and our preceding discussion surely forces us to require at least that much. If $S(Y, A)$ does not satisfy (M3), that is because of the correspondence cmp. But suppose that, under that correspondence, $B_A^Y = B$. Now let us define B' to be the observable which is like B on the subspace S of H_Y spanned by the components of the possible reduced end states of Y, and maps the orthogonal subspace into $\varnothing$ (i.e. $B' = I_S B I_S$). Then the new correspondence cmp′ which sets $BY_A = B'$ will leave (M1) satisfied but will also satisfy (M3). Finally, since CS has only one member, (M4) is trivially satisfied.

So what is the importance of these metacriteria? An interpretation of quantum mechanics must follow upon a specification of what counts as measurement. In constructing an interpretation, it is advantageous not to insist too strongly on one's own ideas about this specification. It is better to make sure that the interpretation will still be tenable given any tenable specification of measurement. So if these metacriteria are to be accepted, then to have a good interpretation it will suffice to have one that works, provided the class of interactions proclaimed to be measurements satisfy those metacriteria.

Metacriterion (M4) has not yet been motivated. Intuitively, it seems unexceptionable. Imagine an apparatus with 15 dials, each recording a simultaneous-measurement outcome—for example, the object measured is a car, and the dials indicate mass, tyre pressure, colour, etc. A paper-and-pencil operation suffices to calculate the value of derivative characteristics: the average tyre pressure, the ratio of mass to average tyre pressure, and so on. The condition says then that there is in fact an observable on the object system, which is measured with the defined pointer-observable.[12]

Metatheorem on Joint Measurement: If CS is a class of ideal cmp-measurements, then the observables A such that $S(Y, A)$ is in CS are mutually compatible.

Proof. In view of (M2) and (M3), we see that all the

pointer-observables have all eigenstates in common, hence they are compatible. By Fine's joint distribution (JD) theorem (see Chapter 6, Section 3), they satisfy the JD condition. Since the distributions P_T^A are the same as the distributions $P_W^{B(Y,A)}$, (M4) entails that the family of measured observables A such that $S(Y, A)$ is in CS also satisfy the JD condition and are therefore, by the same theorem, also mutually compatible.

Proofs and illustrations. To define the physical correlate of measurement, we try to state what conditions a process must meet to be a candidate for something by which we perform a measurement, under the conventions that identify our choice of groundstate, evolution interval, and pointer-observable. The conditions I have laid down do *not* entail that the apparatus is macroscopic. Yet I have left open that we may add further conditions, and there certainly are important accounts of measurement in the literature (some of which I will discuss in later chapters) which impose and utilize additional requirements. As things stand here, the physical conditions can be met by microscopic processes, and that means that the Born Rule entails also predictions for humanly unobservable events. That is how I think it should be. Empiricism should not insist that the theory cannot predict anything about the unobservable; in my opinion (as I explained in Chapter I) empiricism should enter at a different level.

To deal with the problem of accidental degeneracy, I imposed conditions on the operator that governs the temporal evolution, and not on the character of initial and final state alone. Hence it is not ruled out that the end state may be the same in two measurements, on systems in the same initial state, of two incompatible observables. This is a little curious, though not troubling if the concept of measurement pertains to the type of process as a whole. There is however another approach to the problem which appears to rule this out as well. In addition, it appears to eliminate the need for a choice of pointer-observable by entailing a condition on that choice which selects it uniquely. This is the approach taken by Zurek (1981), though with a very different idea about what a general account of measurement must do.[13]

In a measurement, Zurek specifies three significant elements:

the object, the apparatus, and the environment. As in the von Neumann measurements which we shall discuss in the next subsection, the interaction that occurs must be unitary and must correlate the measured observable A pertaining to the object with the pointer-observable B pertaining to the apparatus, so that the final compound state will have the form $\phi = \Sigma c_i |a_i\rangle \otimes y_{a(i)}$. Normally, the further interaction of the object + apparatus system with the environment will immediately destroy any such correlation. To be a measurement, Zurek specifies, the result must be sufficiently stable to constitute a record, and therefore the interaction with the environment must leave the pointer-observable undisturbed.

This is a condition on the temporal evolution, hence on the Hamiltonian which governs the process, just as my extra condition was; but it extends to the process that includes apparatus–environment interaction. Let us as before call the apparatus and object Y and X respectively, and the environment E. The total system is $E + Y + X$, and the process is governed by a unitary operator U, determined by its Hamiltonian

$$H_{EYX} = H_E + H_Y + H_X + H_{EY} + H_{YX} + H_{EX}$$

To simplify, it has been assumed that all interactions are pairwise, and Zurek further simplifies by assuming that $H_{EX} = 0$ (the object is isolated from direct interaction with the environment) and that the interaction of apparatus and object effectively ceases when the correlated state is reached.

The crucial new condition, however, is that the pointer reading must not be disturbed by environmental action. This means that the pointer-observable B commutes with the interaction Hamiltonian H_{EY}. (See the previous discussion of conserved quantities, in the preceding chapter.) If this condition picks out B uniquely, then we are finished.

This uniqueness need only be established up to compatibility, since a measurement that registers, say, charge also registers, via a mere paper-and-pencil operation, the square of the charge, the charge $+10$, etc. Since compatibility is not transitive, it is so far not ruled out that H_{EY} also commutes with some observable B' incompatible with B. This is not true if H_{EY} is a maximal observable, but it is possible if H_{EY} is non-maximal, even for two maximal observables B and B' in the Hilbert space of the apparatus. If H_{EY} is known and maximal, however, the occurrence of an A-measurement can be identified through the occurrence of the appropriate final state of (apparatus + object) alone, since the pointer-observable is unique (see further Dieks 1988a, sect. 5). This should certainly be noted, but I shall not make Zurek's condition part of the definition of a measurement interaction.

4.3. *The class of von Neumann measurements*

So far our discussion has been very general. Most of the literature on measurement, beginning with von Neumann's formalization of the discussion, concentrates on relatively simple paradigms. Von Neumann himself described the class most usually (and most easily) addressed. Following our preceding discussion, let us leave tacit the function cmp which represents our choice of pointer-observables, and proceed somewhat less formally.

> A measurement setup $S(Y, A)$ is a *von Neumann measurement* if the groundstate of the apparatus Y is a pure state y_0 and the evolution operator is unitary and such that, if the initial state of the object system is a pure eigenstate of A, then its final state is the same eigenstate of A.

There are three special conditions here: pure groundstate, unitary evolution, and non-disturbance of eigenstates of the measured observable. It is quite easy to see that this class includes a class of ideal measurements as special cases:

> A von Neumann measurement is *ideal* if the pointer-observable B is non-degenerate with the same spectrum of values as A, and the initial state of $(X + Y)$ is $T \otimes T_{y(0)}$.

As usual, I restrict the discussion to observables with pure point spectrum. Then suppose that A has associated basis $\{|a_i\rangle : i = 1, 2, \ldots\}$ of eigenvectors (degeneracy still allowed at this point in our discussion) and B has corresponding basis $\{y_i = |b_i\rangle : i = 1, 2 \ldots\}$. If $a_i = a_j$, then not only $b_i = b_j$ but $|b_i\rangle = |b_j\rangle$, so the correspondence is then many-to-one. The requirements imply therefore:

(1) $$U(|a_i\rangle \otimes y_0) = |a_i\rangle \otimes y_i$$

(2) $$U\left(\sum c_i|a_i\rangle \otimes y_0\right) = \sum c_i|a_i\rangle \otimes y_i$$

(3) $$\left[\sum p_i I_{|a(i)\rangle}\right] \otimes I_{y(0)} \xrightarrow{U} \sum p_i(I_{|a(i)\rangle} \otimes I_{y(i)})$$

$$= \sum p_i I_{|a(i)\rangle \otimes y(i)}$$

(4) $$\left[\sum p_j(I_{\Sigma c(ij)|a(i)\rangle})\right] \otimes I_{y(0)} \xrightarrow{U} \sum p_j I_{\Sigma c(ij)|a(i)\rangle \otimes y(i)}$$

Here (2) follows from (1) at once by the linearity of U; and (3), (4) follow because of the result that, if both reduced states of a mixture are pure, then the compound state is uniquely determined as the tensor product.[14] Von Neumann pointed out that in such a case immediate performance of the same measurement again will yield the same result. Indeed, he described the object system as having been *projected* into an eigenstate of A, since

(2*)   With initial state $\varphi = \sum c_i|a_i\rangle \otimes y_0$, the final state is

$$\varphi' = \sum c_i|a_i\rangle \otimes y_i$$

and the reduced final object and apparatus states are

$$\#\varphi' = \sum c_i^2 I_{|a(i)\rangle}$$

$$\varphi'\# = \sum c_i^2 I_{y(i)}$$

If the observable $A \otimes I$ is measured on this system in that final state, the outcome will again show value a_i with probability c_i^2. But moreover, if we design a two-part apparatus, of which one part measures A and the other immediately afterward measures $A \otimes I$ from its groundstate z_0, we see the evolution:

$$t = 0: \quad \left[\sum c_i|a_i\rangle \otimes y_0\right] \otimes z$$

$$t = 1: \quad \left[\sum c_i|a_i\rangle \otimes y_i\right] \otimes z_0$$

$$t = 2: \quad \sum c_i[|a_i\rangle \otimes |y_i \cdot \otimes z_i]$$

in which perfect correlations are set up all round. We could also design the apparatus so that a third stage checks on whether the end results of the first and second stage agree. That requires that the third stage with groundstate v_0 is such that the evolution has the character

$$[x \otimes |y_i\rangle \otimes z_j] \otimes v_0 \longrightarrow [x \otimes |y_i\rangle \otimes |z_j\rangle] \otimes v(ij)$$

where $v(ij)$ is v_+ if $i = j$ and v_- if $i \neq j$.

Clearly, the outcome, if $t = 0, 1, 2$ were as above but with the third apparatus coupled on, is

$$t = 3: \quad \sum c_i[|a_i\rangle \otimes |y_i\rangle \otimes z_i \otimes v_+]$$

so the reduced final state of the third apparatus is then just the pure state v_+, indicating total agreement of the first two outcomes. This *repeatability* of the measurement is a feature which von Neumann regarded very highly. In the next chapter we shall see what role it played in his interpretation, but in Section 4.6 we shall also discuss it as a demonstrable feature of measurement classes.

Taking (2*) as the summary statement of what von Neumann measurements are like, we see also that it characterizes the ideal measurements of this sort only in context. If we forget the context, the problem of accidental degeneracy will at once return. In this case, where the measurement effects a correlation of the measurement and pointer-observable, the difficulty takes a striking form. This was brought out by the Einstein–Podolski–Rosen (EPR) paradox, in which a certain initial state $x \otimes y$ is turned by a certain process into a final state:

$$(5) \quad \varphi = \sum c_i(x_i \otimes y_i) = \sum d_i(x_i' \otimes y_i')$$

where $\{x_i\}$ and $\{x_i'\}$ are bases of *incompatible* (non-commuting) observables B and B' while $\{y_i\}$ and $\{y_i'\}$ are bases of two other incompatible observables A and A'.

The difficulty is this: if (2*) were the defining characteristic of a measurement process, and such a process had final state (6), then we would have to say:

A was measured (with B as pointer-observable);

A' was simultaneously measured (with B' as pointer observable).

But non-commuting Hermitean operators are meant to represent observables which cannot be subjected to simultaneous joint measurement.[15]

This difficulty disappears entirely when we insist that it is not the end state alone, but the character of the entire process, that is at issue. Imagine that there is a process governed by unitary operator U such that

(6) $\quad U(x_0 \otimes y_i) = x_i \otimes y_i$

(7) $\quad U(x_j \otimes y_0) = x_j \otimes y_j$

Then we use the fact that U is unitary, and hence preserves inner product, to derive

(8) $\quad (x_0 \otimes y_1 \cdot x_1 \otimes y_0) = (x_1 \otimes y_1 \cdot x_1 \otimes y_1) = 1$

(9) $\quad (x_0 \cdot x_1)(y_1 \cdot y_0) = 1$

and similarly $(x_0 \cdot x_2)(y_2 \cdot y_0) = 1$. But that is impossible, since the $x_j (j \neq 0)$ and also the $y_i (i \neq 0)$ are mutually orthogonal.

The preceding theorem on joint measurement applies to ideal von Neumann measurements as well. But this case is so simple that we can give a more direct proof, by considering a special sort of initial state:

Theorem on Joint Measurement: If two observables have a joint von Neumann measurement, then they are compatible.

To prove this, suppose that system Y is at one and the same time an A and an A'-measurement apparatus with the same groundstate y_0, so both measurements can be performed at once. Now consider a state $x = \Sigma c_i x_i$ with all the numbers c_i^2 distinct. Let $x = \Sigma d_i x_i'$ also, with no assumption being made about the coefficients d_i. The end state of the process is $U(x \otimes y_0) = \Sigma c_i(x_i \otimes y_i) = \Sigma d_i(x_i' \otimes y_i)$. Then by Special Reduction, system X is in final state $T' = \Sigma c_i^2 I_x = \Sigma d_i^2 I_x'$. Because the numbers c_i^2 are all distinct, T' does, in this case, have a unique orthogonal decomposition. Hence the two sets of operators I_x and I_x' are the same. Hence A and A' are both functions of a single observable, and therefore are compatible.

4.4. *Von Neumann–Lueders measurements*

Von Neumann himself had raised the question whether his account applies to non-maximal observables. We may equally ask whether the restriction to pure initial states, and the strong correlation between measured and pointer-observables in the final state, entail an important loss of generality. Lueders (1951) gave a more general account, which we already touched upon in Chapter 6, Section 5 in connection with the conditionalization of quantum states.

We keep the idealization that the apparatus has exactly one pure eigenstate of pointer-observable B for each eigenvalue of the measured observable A. We start again with the system X in state $x = \Sigma x_i^a$ with $Ax_i^a = ax_i^a$, where a ranges over the distinct eigenvalues of A and $\{x_i^a\}$ is an orthonormal basis for the space made up out of unit eigenvectors of A. The subspace spanned by $\{x_1^a, x_2^a, \ldots\}$ for a given a is the *a-eigenspace* of A, and obviously I_a^A is the projection on that subspace, hence $x = \Sigma(I_a^A x)$. Most generally, the initial object system state is a mixture T of such pure states. The general description of a measurement proposed is now:

> The evolution U of the measurement setup is unitary, the groundstate y_0 is pure, and the end state of the object system is the conditionalization of its initial state on the partition of eigenspaces of the measured observable.

There is nothing in the general notion of measurement which motivates us to focus on what happens to the object system during the process. It is rather a special characteristic of the von Neumann measurement that the initial object system is undisturbed when it is an eigenstate of the measured observable. This is also a characteristic of the Lueders measurement, because the conditionalization of a pure state on a subspace is just its projection on that subspace. The above description entails therefore:

> If the initial state of the object system is x_i^a (or, indeed, if it is any unit vector in one of the eigenspaces of A), then its final state is the same. If the initial total state is $x_i^a \otimes y_0$ and the pointer-observable is non-degenerate, then the final total state is $x_i^a \otimes y_a$.

Thus the von Neumann measurement is a special case. Moreover, if the initial total state is just $T \otimes I_{y(0)}$ with $T = \Sigma p_a I_{x(a)}$, then the convex structure is preserved and so the fact that the evolution is unitary tells us that the final state is $\Sigma p_a I_{x(a)} \otimes y(a)$.

The exact connection with conditionalization is as follows. Clearly, the vector $I_a^A x$ represents the conditionalization of the state which is represented by x, on the a-eigenspace of A. The squared length of that vector is the corresponding Born probability for the outcome a in a measurement. Recalling the formula for conditionalization and writing x_a for $I_a^A x$:

(10)   The conditionalization of the state represented by x, on the a-eigenspace of A, is represented by the statistical operator

$$W_a = I_{x(a)}/|x_a|^2 = I_{x(a)/|x(a)|}.$$

In the case in which the initial total state is $\Sigma x_i^a \otimes y_0$ and the pointer-observable is non-degenerate, we can find the final object system state by reduction:

(11)   Final state of $X + Y$ is

$$\sum I_a^A x \otimes y_a = \sum x_a \otimes y_a$$

$$= \sum |x_a|(x_a/|x_a| \otimes y_a)$$

(12)   Final state of X is $\sum |x_a|^2 I_{x(a)/|x(a)|}$.

But that is the conditionalization of the initial object state I_x on the partition consisting of the eigenspaces of A, as was announced.

Exactly the same happens with a mixed initial object state, namely, that the reduced final object state is the same conditionalization of the initial object state. The complete Lueders rule is:

(13)   If the object is initially in mixed state W when A is measured, the reduction of the final total state gives the reduced object state

$$W' = \sum_a Tr(WI_a^A)\dot{W}_a^A$$

where W_a^A is the conditionalization of W on the a-subspace of A.

The number $Tr(WI_a^A)$ is of course the Born probability for the a outcome. Recalling what W_a^A is, we see that the reduced final state is:

$$(14) \qquad W' = \sum_a Tr(WI_a^A)(I_a^A WI_a^A)/Tr(WI_a^A)$$

$$= \sum_a I_a^A WI_a^A$$

which is a simpler expression, and useful in what follows.

To follow this relatively intuitive discussion with precision, let us again put this in our canonical form:

A measurement setup $S(Y, A)$ is a *von Neumann–Lueders measurement* if the groundstate of the apparatus Y is a pure state y_0, and if the evolution operator is unitary, and such that the reduced final state of the object system X is the conditionalization of its initial reduced state T on the partition of eigenspaces of A. The measurement is *ideal* if the pointer-observable is non-degenerate and has the same spectrum of values as A, and the initial total state of $X + Y$ is $T \otimes I_{y(0)}$.

Much of our reasoning about von Neumann measurements carries over to this more general class. That jointly measurable observables are compatible in the ideal case is again made easy to show by the non-degeneracy of the pointer-observable.

4.5. *Unitary measurements*

The generalization by Lueders is still a special case, from the general point of view of the quantum-mechanical theory of measurement. But how special? In the next subsection, discussing repeatability, we will see that continuous quantities cannot be measured that way. Restricting ourselves here to discrete quantities, that is, to observables with pure point spectrum, we will see that the von Neumann–Lueders measurement actually all but exhausts the class of measurements by our criteria.

Intuitively, as we saw above, a measurement could be any process in which, at least in a statistical sense, the final characteristics of the 'apparatus' give a reliable indication of the initial state of the 'object'. We are idealizing to the extent of asking the final apparatus state to reproduce the relevant statistics exactly, and we have added the requirement that the apparatus end up in a mixture of eigenstates of the designated pointer-observable. There is a large and growing literature on such processes, with many important results to classify the conceivable varieties of measurement setups. I shall refer here mainly to the work of Beltrametti, Cassinelli, Lahti, and Ozawa, and shall follow the exposition given by Lahti (1988).

The most basic result is due to Ozawa (1984): a measurement setup exists for every physical quantity. This was already established in von Neumann's discussion for discrete observables; Ozawa extends it to observables with continuous or mixed spectra as well. In the remainder of this section I shall concentrate on discrete observables, but shall not assume that a measurement needs to take anything like the form described by von Neumann.

Suppose that the observable A to be measured has eigenvalues a with degeneracy $n(a)$, and orthonormal basis of eigenvectors $\{x_i^a\}$ as before. Let the pointer-observable B be nondegenerate on its space, with the same point spectrum, and corresponding orthonormal basis $\{y_a\}$. Let the apparatus groundstate be the pure state y_0. All vectors below will be unit vectors unless the context shows otherwise. So far, the assumptions involved are generally regarded as merely simplifying. But now assume:

1. The evolution of the total system is a mapping of total states into total states which preserves the convexity structure of the statistical operators, and the extreme points.

This means that, if the evolution changes total initial state V_i to V_i', then (a) if V_i is pure, so is V_i', and (b) it changes the mixture $\Sigma p_i V_i$ into mixture $\Sigma p_i V_i'$. This has the following consequence (Beltrametti *et al.* 1989):

2. The evolution is a continuous linear extension of a mapping B of form

3. $$B(x_i^a \otimes y_0) = z_{ai} \otimes y_a$$

    where $\{z_{ai} : i = 1, \ldots, n(a)\}$ is a set of unit vectors which are mutually orthogonal in the second index.

That is, z_{ai} is orthogonal to z_{aj} if $i \neq j$, but possibly is not orthogonal to z_{bi}.

It is obvious where the von Neumann–Lueders measurements are located in this general scheme: they are the ones in which 3 takes the special form

4. $$V(x_i^a \otimes y_0) = x_i^a \otimes y_a$$

Let us call this operator V a *Lueders map*. It is unitary; but actually, that is not very special here. From the same source we have the theorem:

5. The map $x_i^a \otimes y_0 = z_{ai} \otimes y_a$ extends to a unitary operator U if and only if $\{z_{ai}\}$ is a set of unit vectors which are mutually orthogonal in the second index. The operator U can be decomposed as $U = U'V$, where U' is unitary and V is a Lueders map.

The class of measurement setups we are presently considering therefore all have the appearance of von Neumann–Lueders measurements which have been allowed or made to evolve unitarily a bit beyond a 'proper' end stage. This appearance may be a fiction; but the end would have been the same if that is what had happened. The change to the state of the object system in this case is as follows:

6. If the initial state of the object system is $x = \Sigma c_{ai} x_i^a$, then its final state is $\Sigma p_a^2 I_{z(a)}$, where $p_a^2 = \Sigma |c_{ai}|^2$ and $z(a) = p_a^{-1} \Sigma \{c_{ai} x_i^a : i = 1, \ldots, n_a\}$.

But are these measurement setups all measurements in our sense, with only the object system state somewhat different from what Lueders considered right? No, for the reduced state belonging to the apparatus is now not generally diagonal in the basis of eigenstates of the pointer-observable, and so this class as a whole fails to meet our metacriterion.

4.6. *Repeatable measurements and correlations*

Von Neumann–Lueders measurements have two characteristics, not unrelated to each other, which played an important role in von Neumann's thinking. One was that, at the end, the measured and pointer-observables are strongly correlated with each other. Thus, in the simplest case the end state is:

$$(1) \qquad\qquad \sum c_i \, (|a_i\rangle \otimes |b_i\rangle)$$

As we saw above, this means that, if we construct an apparatus which measures $A \otimes B$ but only registers agreement $(+)$ and disagreement $(-)$, and apply it to the end state of the first measurement setup, we will get the $(+)$ outcome with certainty. Secondly, if we were to construct an apparatus which simply measures A again, or, if you like, $A \otimes I$ on the total system $X + Y$ starting in the end state of the first measurement, we will get the same outcome twice. This point should be made precise, and that can be done in several ways. The first is again to think of ourselves as constructing a further measurement apparatus which checks for agreement $(+)$ and disagreement $(-)$ between the first and second A-measurement outcomes. It will show $(+)$ with certainty. The second, and more probative, way takes a little explanation.

It is convenient here to focus on the notion of *instrument*, which was originally introduced in the operational quantum mechanics due in various forms to Davies, Ludwig, and Mielnik. In the measurement process the compound state evolves in a way largely constrained by conditions pertaining to the reduced initial and final states: W changes to W', let us say, and initial object state T changes to T'. To characterize the latter change in a very general way, we define a function called the *instrument* associated with this measurement process. I shall designate it J (which depends on A and U):

For any initial state T of the object and any Borel set E, the function gives us $J(T, E)$ which is the final state T' of the object, conditionalized on the proposition that the value of A is in E with certainty. (More precisely: conditionalized on the subspace on which I_E^A projects, the E eigenspace of A.)

Many commonly discussed properties of measurement interactions are really best discussed in terms of the associated instruments.

We must emphasize here the exact relation between T' and J. If we know what T is *and* we know the value of $J(T, E)$ for *all* Borel sets E, then we can deduce what T' is, and conversely. The instrument is at this point for us a merely formal notion, whose interest lies in the uses to come.

To study repeatability, we ask what happens in sequential measurements, in which the system in state T is subjected to an A-measurement, and at the end the same system (now in state T') is subjected to the same measurement again. The final state at the end of the two-stage process is characterized by the family of transformed conditionalized states:

(2)　　　　　$J(J(T, E), F) : E, F$ any Borel sets

due to the successive applications of the same instrument. We now say that the measurement is *repeatable* if the second stage made no difference to Born probabilities, i.e. if

$$(3) \qquad J(J(T, E), F) = J(T, E \cap F)$$

with as corollary, of course,

$$(4) \qquad J(J(T,E),E) = J(T,E)$$

for arbitrary T, E, F. We may equally say that the instrument is repeatable.

Obviously, not all measurement processes have this property. The class of measurements discussed by von Neumann were said by him to be repeatable in this sense. The *instrument J* associated with the von Neumann–Lueders measurement is easily characterized in view of our preceding discussion, as the function

$$(5) \qquad J(T, E) = \sum_{b \in E} I_b^A T I_b^A$$

for state T and Borel sets E; and this is repeatable. For let $J(T, E) = T'$; then

$$(6) \qquad J(J(T, E), F) = J(T', F)$$

$$= \sum \{I_d^A \, T' I_d^A : d \in F\}$$

$$= \sum \left\{ I_d^A \left(\sum_{b \in E} I_b^A T I_b^A \right) I_d^A : d \in F \right\}$$

$$= \sum \left\{ \sum_{b \in E} I_d^A I_b^A T I_b^A I_d^A : d \in F \right\}$$

Now $I_d^A I_b^A = I_b^A$ if $b = d$, and it is the null operator which turns every vector into the zero vector if $b \neq d$. So $I_b^A I_d^A = I_d^A I_b^A = \delta(d, b) I_b^A$ where $\delta(d, b) = 1$ if $d = b$ and $= 0$ if $d \neq b$, and we continue the calculation:

$$= \sum \left\{ \sum_{b \in E} (d, b) I_b^A T \delta(d, b) I_b^A : d \in F \right\}$$

$$= \sum \{I_b^A T I_b^A : b \in E \cap F\}$$

$$= J(T, E \cap F)$$

as was to be shown. Repeatability, emphasized by von Neumann, is therefore mathematically demonstrable for this class of measurement interactions. The theory of measurement processes described within quantum mechanics already entails that, if a von Neumann–Lueders measurement is twice repeated, the statistical predictions for outcomes are exactly as for a single application.

But notice that the von Neumann–Lueders measurement was defined for an observable with discrete spectrum. Ozawa (1984) also proved the following:

(7) An observable admits of a measurement (precisely: a measurement setup with a positive evolution operator) which is repeatable if and only if the observable is discrete.

Thus, continuous quantities like position and momentum (unlike their arbitrarily finely discretized versions) do not admit of repeatable measurement. (See especially the discussion in Lahti 1988.)

In the cited paper by Beltrametti *et al.* (1989), repeatability is studied in general, and in relation to strong correlations. The general notion of a correlation coefficient is taken over from classical probability theory in a fairly straightforward way (see Beltrametti and Cassinelli 1981*a*, sects. 3.1 and 7.3). The coefficient of correlation between two observables A and B in state W is proportional to the expectation value of $A \otimes B$ in W *minus* the product of the expectation values of A in $\#W$ and B in $W\#$. If it equals 1, the correlation is perfect, and that is what we find in the end state of a von Neumann–Lueders measurement. Being interested in such perfect correlation only here, for discrete observables A and B with corresponding eigenvalues, we can explain it quite simply:

(8) A and B are perfectly correlated in state V exactly if

$$Tr((I_a^A \otimes I_b^B)V) = 0 \text{ whenever } a \neq b.$$

Let us recall the outcome of a general measurement of a discrete observable A from the last section. Recall the notation used:

(9) The map $x_i^a \otimes y_0 = z_{ai} \otimes y_a$ extends to a unitary operator U if and only if $\{z_{ai}\}$ is a set of unit vectors which are mutually orthogonal in the second index.

(10) If the initial state of the object system is $x = \sum c_{ai} x_i^a$, then its final state is $\Sigma p_a^2 I_{z(a)}$, where $p_a^2 = \Sigma|c_{ai}|^2$ and $z(a) = p_a^{-1}\Sigma\{c_{ai}x_i^a : i = 1, \ldots, n_a\}$.

We can now ask whether there is a perfect correlation in the total final state between the pointer-observable eigenvalues on the one hand, and on the other, *either* the eigenvalues of the measured observables, *or else* the states $z(a)$ which we know to occur always in an orthogonal decomposition of the final object state. These questions are answered by some revealing theorems:

(11) Given that the operator U is as in (9), then the projection operators $I_{z(a)}$ and I_a^B are perfectly correlated in the total end state (for all eigenvalues a which did not receive an initial Born probability zero), if and only if $\{z_{ai}\}$ is an orthonormal set.

(12) Given that the operator U is as in (9), then the projection operators I_a^A and I_b^B are perfectly correlated in the total end state (for all eigenvalues a which did

not receive an initial Born probability zero), if and only if $\{z_{ai}\}$ is an orthonormal set, with z_{ai} in the a-eigen-space of A.

(13) In both cases, the reduced final state of the apparatus is a mixture of eigenstates of the pointer-observable.

Note that in case (12) there is no difference to speak of between that sort of measurement and a von Neumann–Lueders measurement. In addition:

(14) A unitary measurement is *of first kind*, i.e. the Born probabilities for A are the same in the initial as in the final object system state, if and only if the perfect correlation condition of (12) holds.

First kindness is of course a cruder version of repeatabilility, and is implied by repeatability.

As announced in the preceding section, we have therefore found that the class of von Neumann–Lueders measurements looks practically exhaustive as far as measurements of discrete observables go.

4.7. *Approximation to measurement*

There is a sense in which incompatible observables can be measured: they can be measured jointly but crudely. Such a crude measurement is however just an ordinary measurement of compatible observables.

If all eigenspaces of A are subspaces of eigenspaces of A', let us call A' a *coarsening* of A. What I just called a crude measurement is the measurement of a coarsening. It is easily seen that A' might well be a coarsening of two distinct and incompatible observables. And more generally, A' may belong to a family of mutually compatible coarsenings of incompatible observables. Finally, in this derivative sense any two incompatible observables are jointly crudely measurable because the identity I is a coarsening of every observable.

This subject has more interesting aspects, of course; the Heisenberg uncertainty relations pertain to how much coarsening is necessary for crude joint measurement. We should also insist that we are not just playing with terminology; that I_E^A

takes value 1 can also be expressed by saying that A takes value in E, or that A takes 'unsharp' value E. Undoubtedly, measurement of I_E^A gives us some information about the initial object system state and pertaining to A.

But are all approximate measurements in practice to be construed this way? If I take a ruler and measure the table edge in centimetres, recording results to one significant decimal, even requirement (M1) is not fulfilled. This example we can plausibly handle as above, I think: the results I write down in metres, recording one significant decimal, do satisfy (M1). We may even think of this as a von Neumann measurement. But correlations are treacherous. There is the table, there is the ruler, there is the person who records. We expect something like this to happen:

$$|1.1\rangle \otimes y_0 \otimes z_0 \longrightarrow |1.1\rangle \otimes y_{1.1} \otimes z_0$$
$$\longrightarrow |1.1\rangle \otimes y_{1.1} \otimes z_{1.1}$$

But what if this person has a hangover, and writes down '1.2'? This is an extreme case, but in practice there are such errors.[16] It is exactly the part of statistics developed in the eighteenth and nineteenth centuries for this purpose which will translate a distribution in a finite set of readings into a probability distribution in which we have greater confidence. This latter statistic is what we take to be the proper summary of outcomes—that is, the place where we should see whether (M1) is satisfied. But this does not remove the problem of principle.

With the possibility of such errors, we cannot say that the Born Rule predicts the measurement outcomes. For in the case in question, the rule would say: given that a measurement is performed, the outcome 1.1 occurs with certainty. Since therefore the hangover scenario is not a measurement, do we have *any* prediction about what will happen? *Not via the Born Rule!* But who would doubt that we do use the theory to make predictions for such imperfect attempts at measurement?

This is the problem of principle: (i) with a strict criterion for measurement, we end up with *no* predictions for the processes we usually refer to as measurement, but (ii) with a permissive criterion for measurement, we shall imply that incompatible observables can be jointly measured. That is quite a dilemma.

Here is another way to pose it. In theoretical discussions of measurement we usually assume that the object + apparatus is an isolated system or, in Zurek's version, that the environment leaves the apparatus undisturbed in relevant respects. In practice, this will usually not be so. But then, since the only probabilities for events that the theory itself gives us came by Born's Rule, it follows that it usually makes no predictions at all. This is an apparent consequence which we cannot accept.

It is becoming increasingly difficult here to keep interpretation at bay. Let me advance an answer. What we discuss theoretically are models. It may well be that nature contains phenomena which fit some of these models exactly. But the phenomena *we* refer to do not typically fit exactly the models *we* construct. In the possible worlds in which quantum theory is true—that is to say, in its theoretical models—Born's Rule gives the probabilities of pointer-observable values for processes which fall in specific classes of measurements, and never for joint measurements of incompatible observables. Interpretation of the theory must make sense of these possible worlds. Acceptance of the theory, however—and now I speak as an empiricist—involves the decision to let the theory function as expert predictor (probability assigner) for the phenomena as *we* classify them. We can let it function that way to our chosen degree of approximation, with our chosen confidence intervals, and so forth. So *in use* the Born Rule is extended, *modulo* such qualifications, even to instruments subject to earth tremors and observers with hangovers. This answer is not open, I think, to any who want to insist that we can let quantum-mechanical probabilities guide our own expectations *only* on the basis of the belief that a certain quantum-mechanical statement (summed up as: 'this process is a measurement') is true.[17]

5. PREPARATION OF STATE

Von Neumann's interpretation of measurement, which we shall study in the next chapter, entailed that measurements are also state preparations. The system X on which we measure observable A emerges in an eigenstate of A. This assertion is an interpretation; it is not implied by anything we have found out about measurement so far. Recall from Section 3 above that attributing a mixture of eigenstates of A to X is *not* the same as saying that X is really in one of those pure states though we don't know which! But however all this may be (and we shall return to it at length), we do face another foundational question here: Just what sorts of interactions could be preparations of state? Can there be state preparation at all? Is there any process that simply prepares an ensemble of systems in state $|a\rangle$?

This cannot just be a matter of interpretation. What we are asking for is the possibility of a state preparation machine, call it M, with groundstate u. One suggestion could be that, if we couple it to a system X in state x, we have the evolution

$$u \otimes x \longrightarrow u_1 \otimes x \text{ if } x \in S$$

$$u_0 \otimes \varnothing \text{ if } x \in S^\perp$$

Hence, if $y = z + z^\perp$ with $z \varepsilon S$ and $z^\perp$ in $S^\perp$, then $u \otimes y \longrightarrow u_1 \otimes z$. This process will steer system X either into a state in subspace S or else into the null vector, which represents no physical state at all (destruction).

An intuitive example would be a Stern–Gerlach apparatus with two exits (for spin-up, spin-down), and one of the exits blocked by an absorbing wall.

But if this process is really a quantum-mechanical process, then it must be governed by a unitary evolution operator. Suppose for simplicity that X just has a two-dimensional Hilbert space with $S = [x_+]$ and $S^\perp = [x_-]$. Then the process is governed by operator U:

$$U(u \otimes x_+) = u_1 \otimes x_+$$

$$U(u \otimes x_-) = u_2 \otimes \varnothing$$

To be unitary, we require of U that $(Uy \cdot Uz) = y \cdot z$, i.e. that it preserves inner products. We calculate first:

$$(u \otimes x_+) \cdot (u \otimes x_-) = (u \cdot u)(x_+ \cdot x_-) = 0$$

$$(u_1 \otimes x_+) \cdot (u_2 \otimes \varnothing) = (u_1 \cdot u_2)(x_+ \cdot \varnothing) = 0$$

which looks fine. But now let U act on the superpositions in the following inner product:

$$(u \otimes (ax_+ + bx_-)) \cdot (u \otimes (cx_+ + dx_-)) = (u \cdot u)(a^*c + b^*d)$$

Calculating the dot product of what U turns these two vectors into, we find:

$$[(u_1 \otimes ax_+) + (u_2 \otimes b\varnothing)] \cdot [(u_1 \otimes cx_+) + (u_2 \otimes d\varnothing)]$$

$$= [(u_1 \otimes ax_+) + (u_2 \otimes \varnothing)] \cdot (u_1 \otimes cx_+)$$

$$+ [(u_1 \otimes ax_+) + (u_2 \otimes d\varnothing)] \cdot (u_2 \otimes d\varnothing)$$

$$= (u_1 \cdot u_1)a^*c + (u_2 \cdot u_1)0 + (u_1 \cdot u_2)0 + (u_2 \cdot u_2)0$$

$$= (u_1 \cdot u_1)a^*c$$

Since b and d appear in the first equation, are unconstrained, and do not appear in the second, the two results will in general disagree. Therefore U is not unitary.

It follows now that elementary quantum theory by itself denies the possibility of a state preparation process in the sense suggested. We need, or so it seems, a little miracle: something like von Neumann's Projection Postulate, which 'really destroys' selected components in a superposition. To add such a principle is to deny universal validity to Schroedinger's equation.

There are however several options. We can introduce the little miracle, and leave it unexplained. We can be more sophisticated, and note that in quantum theory, with superselection rules, there is a destruction of those components in a superposition which lie outside the coherent subspaces. That gives us state preparation processes—but note well that it works only when there are relevant superselection rules. That is certainly not true throughout, and so it seems that at least many sorts of states might not allow this sort of preparation.

Thirdly, we can say that, *in the sense needed*, state preparation is possible even if Schroedinger's equation has unrestricted validity. Of course, that sense cannot be the 'strict' sense thought up above. Yet again, there are several alternatives. One is to add some constraint to those which define repeatable measurement, but of the same general sort, to define 'preparatory measurement' (see e.g. Lahti 1988). A second alternative is to exploit infinite dimensionality. I shall now show that, on the general empiricist view adopted here, to exploit infinity is both helpful and all right. For let us suppose we have the above machine M with its groundstate u, but let $\{u_i\}$ be a basis of its Hilbert space H_M. We now consider two cases, first the one in which the Hilbert spaces are infinite-dimensional, and then the finite-dimensional case. We shall observe first of all that the impossibility proof above for strict state preparation does not generalize to the infinite-dimensional case. Then we shall see how the result can be carried over, in a sense, to the case of finite dimension.

First, then, let object system X have Hilbert space H of same dimension as H_M, and assume this is infinite. H has basis $\{x_j\}$, and we are asked to steer X into subspace $[x_j : j \geq k] = S_{k+}$. Let operator U be such that

$$U(u \otimes x_j) = u_j \otimes x_{j+k}$$

This can be a unitary operator; we showed this for $k = 0$ when we first discussed measurement, and this case is not essentially different. (To help the imagination, reflect on the fact that V defined by $Vx_j = x_{j+1}$ is a unitary operator in the infinite-dimensional case, a point which often appears in textbook exercises.) If the dimensionality were finite, however, the operator U could not be unitary; but here we assume it is infinite. Now notice that

$$U\left(u \otimes \sum c_j x_j\right) = \sum c_j u_j \otimes x_{j+k}$$

and the reduced state for X has no terms x_j for $j < k$. Therefore this reduced state places X entirely in the subspace S_{k+}.

What if X has a finite-dimensional Hilbert space? The empiricist approach adopted in Chapters 4 and 5 to the subject as a whole gives no objective status to this supposition. The phenomena involving X may require a Hilbert space of dimension *no less than* a certain number for their representation. But whatever can be represented in a given Hilbert space, H can also be represented in a space H' of higher dimension, provided only H' has a subspace isomorphic to the smaller space. Since the empirical adequacy of the model requires only that all the relevant phenomena can be represented in it, there is no objective upper bound to the dimensionality of 'the right' Hilbert space.

To apply this, let us go back to the two-dimensional case in which $H = [x_+, x_-]$. We now choose infinite-dimensional Hilbert space H' with basis $\{x_j\}$, and we identify

$$x_- = \sum \{c_j x_j : i < k\}$$

$$x_+ = \sum \{c_j x_j : i \geq k\}$$

These are orthogonal as required, and if we now let U and u be as in the second-last preceding paragraph, we see that

$$U\left(u \otimes \sum c_j x_j\right) = \sum c_j u_j \otimes x_{j+k}$$

which is orthogonal to all states of form $u' \otimes x_-$. Thus, x_- is not even possible relative to the new state of X found by reduction at the end of this process. It has disappeared entirely.

Of course, if we conclude that a given system X can undergo such a process, we cannot also claim that all phenomena pertaining to X can be represented in a finite dimensional-Hilbert space.

PART IV
Questions of Interpretation

The preceding chapters admittedly included some explicit though modest attempts at interpretation. Throughout, I took a theory to consist essentially in the family of models it provides for the representation of phenomena. I argued that certain sorts of phenomena logically require models at odds with determinism or even causality, at least in the strong sense of Common Cause explanation. But I also argued that no departures are needed from classical logic or probability theory. None of this is entirely uncontroversial. But there are questions of interpretation that go well beyond all this.

The three focal questions I shall address are: What is really going on in measurement? Is the quantum-mechanical description of nature incomplete? What is the identity of identical particles? In my opinion there is likely to be more than one tenable answer. Some I shall discuss critically; perhaps none is as yet free of all intuitive difficulties. I shall argue that diversity of interpretation does not entail lack of understanding—perhaps the contrary. But I shall advocate a specific approach to these questions: that of the *modal* interpretation or, more specifically, of what I have elsewhere called the Copenhagen variant of the modal interpretation.

8

Critique of the Standard Interpretation

WHEN von Neumann codified the mathematical foundations of quantum mechanics in 1935, he also gave it an interpretation. Undoubtedly, he took that interpretation to be implicit in scientific practice. If there is such a thing as the mainstream understanding of the theory during the fifty years that followed this work, it is von Neumann's. As I shall try to show, it involved two principles, one tacit and one explicit. The first is that all quantum-mechanical description can be given in terms of state-attributions; the second his famous Projection Postulate, the 'acausal' state transition in measurement. Our first task will be to enquire how the two principles are related to each other, and whether they are forced on us by the theory.

1. WHAT IS AN INTERPRETATION?

The interpretation of quantum mechanics is a lively philosophical issue, and controversial. Stances on this issue included Einstein's realism, Bohr's and Heisenberg's versions of the Copenhagen interpretation, von Neumann's postulate of 'acausal' collapse of the wave function, and the 'ensemble' interpretation of states. These views did not constitute specific, rigorously developed interpretations, such as we now have (notably, those which emerged in the detailed foundational work of Mittelstaedt, and of Ludwig; the quantum-logical interpretation developed by Putnam, Bub, Demopoulis, Friedman, and Stairs; the 'operational' theory of theories due to Foulis and Randall; the 'perspectival' interpretation of Kochen; and the 'modal' interpretation which I shall elaborate below). To understand such answers, we need to understand the question. So we must first ask: what is an interpretation of a theory? And this question in turn must be preceded by: what is a theory?

In Chapter 1 we saw that there are a number of answers to this latter question. According to the semantic view, to present a theory is to present a family of models. This family may be described in many ways, by means of different statements in different languages, and no linguistic formulation has any privileged status. Specifically, no importance attaches as such to axiomatization, and a theory may not even be axiomatizable in any non-trivial sense.

There are two important relations that a theory may have to reality. The first is *truth*: this means that one of the models is an exact copy of reality: each part or element of the model represents something real, and those real things are related in just the way that the model represents. The second is *empirical adequacy*: at least one of the models is such that all actual observable phenomena are correctly represented in it. This means also of course that there are two distinct, important forms of (unqualified) assent to a theory. The first is belief that it is true, and the second, mere belief that it is empirically adequate.

Ideally, belief presupposes understanding. This is true even of the mere belief that a theory is true in certain respects only. *Hence we come to the question of interpretation:* Under what conditions is this theory true? What does it say the world is like? These two questions are the same. The reason they are often difficult to answer is, in my opinion, that scientific discussion is so thoroughly focused on the question of empirical adequacy alone. As a result, philosophers who read scientific discussions of theories, even at a foundational level, tend to be disappointed and to accuse scientists of positivism and worse. The question of interpretation—what would it be like for this theory to be true, and how could the world possibly be the way this theory says it is?—does indeed go beyond almost all discussions in science. But it is also broached by scientists, just because the theory about the phenomena can rarely be well understood except as part of what it says the world as a whole is like.

But what answer could we give to such a question? Suppose I attempt to answer the question: what is the world like according to quantum mechanics? The question asks for the *content* of the theory, but an answer will strive for a certain completeness, and so almost inevitably will extrapolate. So, almost inevitably, my answer will be an interpretation of that theory. You could then *classify* my answer in various ways, by asserting for example that it disagrees with Bohr, or Einstein, or von Neumann at certain points. You could also *dispute* it, by pointing to some aspect of quantum mechanics which my answer leaves obscure, or fails to take into account. Even worse, you could *accuse* it of, in effect, producing a different theory by showing that quantum mechanics, under my interpretation, makes predictions different from or in addition to those of the theory itself.

Obviously, attempts to interpret are very much like, if not the same thing as, attempts to introduce hidden variables—to construct a hidden variable (h.v.) theory that subsumes quantum mechanics. The difference is that h.v. theorists might welcome the result that their construction yields different predictions from standard quantum mechanics. In that case, we definitely cannot speak of an 'interpretation' of quantum mechanics, but have before us an alternative theory, which, if successful, (*a*) agrees with quantum mechanics within limits of experimental error on previously found results, and (*b*) gives true predictions at variance with those of quantum mechanics for new experiments. This is the only sort of h.v. theory that could excite scientists as such very much, because it concerns empirical success. An h.v. *interpretation*, on the other hand, would be one that yields exactly the same predictions as quantum mechanics itself. In that case, the virtue that could be claimed would be this: the introduced h.v. are, as Feyerabend put it, 'empirically superfluous', but they show how the world could be the way quantum mechanics describes.

To put it paradoxically, any adequate interpretation must be an h.v. interpretation with empirically superfluous hidden variables. I mean this as follows. Suppose we agree that there can, in logical principle, be more than one adequate interpretation of a theory. Then it follows at once that interpretations go beyond the theory; the theory + interpretation is logically stronger than the theory itself. (For how could there be differences between views, all of which accept the theory, unless they vary in what they add to it?) So an interpretation introduces factors not found in the theory originally—and what else does 'hidden variables' mean? The empirical superfluousness is required to ensure that no new or different predictions are forthcoming—else we have an alternative theory rather than an interpretation.

2. TWO FORMS OF INDETERMINISM

Born's interpretation of the quantum-mechanical state, which consists in his rule for calculation of probabilities, became so thoroughly accepted that it must be counted part of the theory itself. At the time, it was put forward in opposition to, for example, Schroedinger's attempt to think of the state as a wave in a real medium as opposed to a 'probability wave' (i.e. a composite of probability functions which has some of the mathematical characteristics of a wave). But later questions of interpretation are always about how one might go beyond Born. These questions arise, in fact, as soon as we try to understand what Born's Rule says:

Born: The probability that a measurement of yes–no observable P will yield value 1, if made in state W, equals $Tr(WP)$.

This statement gives us the probability of the occurrence of an *event*, conditional on the occurrence of a certain measurement process in a given *state*. The indeterminism of quantum mechanics consists in the first instance in the fact that this probability may be neither 0 nor 1, even if the attribution of state W reflects no 'mere' ignorance on our part.

But what is this event? During the measurement process we see an initial situation change into a final situation. The initial situation can be analysed into two parts: the system of interest is in state W, and the environment (presence of and interaction with the measuring apparatus in its groundstate) has a certain character—call it IN. In the final situation, this system's state will have evolved into, say, W', and the environment will have a new character—call it OUT—which includes e.g. that the pointer on the apparatus now sits at the number 1. So we have a transition:

$$\text{IN}, W \longrightarrow W', \text{OUT}$$

This transition could be indeterministic in two ways. Recall that we have a probability for OUT. Another possible transition—one that could have occurred but did not—may be described as

$$\text{IN}, \; W \longrightarrow W'', \; \text{OUT}'$$

and it too had a probability.

> *Indeterminism, Form 1* (*indeterministic output*): Given the initial situation (IN, W), the final state W' is completely determined (so $W'' = W'$), and the probability we are given is really the probability of character OUT, given that the system is then in state W'.

> *Indeterminism, Form 2* (*indeterministic state transition*): Given the final state, the outcome character is completely determined (so if OUT $\neq$ OUT$'$ then $W' \neq W''$), and the probability we are given is the probability of the transition from state W to W', when the initial situation has character IN.

There is a third possible form, with the final state and outcome character more independent of each other, but then several probabilities are involved. Born gave us only one probability, so it would seem that we must restrict ourselves to these two forms. Now, which of these correctly typifies quantum mechanics?

This is a question of interpretation, and a number of different answers have been given.[1] Some fall very squarely in form 1 and some in form 2. The latter includes von Neumann's famous acausal state transition (Projection Postulate). The term 'collapse of the wave packet' usually refers to this, though sometimes it is used loosely to refer to whatever transition occurred from the possible to the actual—as *any* indeterministic theory must say there is. 'No-collapse' interpretations, such as the modal interpretation to be proposed below, take form 1.

3. WHAT HAPPENS IN MEASUREMENT? VON NEUMANN'S ANSWER

The interpretation of quantum mechanics must begin with a discussion of measurement. The reason is twofold: like all empirical theories, it is held accountable with respect to the phenomena reported as measurement outcomes; but it also purports to be the most basic physical theory, and hence covers the processes designated as measurement interactions. So we face, among other issues, a serious consistency problem. The first and basic principle of interpretation, Born's Rule, is stated in terms of measurement—and measurement is one of the processes in the theory's domain of application. The threat of inconsistency, or of a vicious circle, is therefore very real.

Most of the quantum-theoretical theory of measurement belongs to the mathematical foundations. From the preceding chapter I shall summarize only briefly the characteristics of von Neumann measurements, on which von Neumann's interpretation focuses. Let us take a system X in pure state $W = I_x$, where x is a unit vector in the appropriate Hilbert space H_X. Next, for the environment we take a measurement apparatus Y, in pure groundstate I_y (y also a vector, in the state space for Y), and assume that $X + Y$ in the initial situation is in pure tensor product state $I_{x \otimes y}$. Now IN is specified by saying that $X + Y$ is an isolated system in pure state $x \otimes y$. An observable is measured: let us call it A, and assume it to have a basis of eigenvectors $|a_i\rangle$ in H_X. There is a 'pointer-reading' observable: let us call it B and assume it to have associated basis of eigenvectors $|b_i\rangle$ in H_Y. And the measuring process, which is the evolution of this isolated system during a certain interval, has character

(S1) $$U(|a_i\rangle \otimes y) = |a_i\rangle \otimes |b_i\rangle$$

and therefore, if $x = \sum c_i |a_i\rangle$, also

(S2) $$U(x \otimes y) = \sum c_i |a_i\rangle \otimes |b_i\rangle$$

Now if IN is the character described by saying that, initially, isolated compound system $X + Y$ is in state $\varphi = x \otimes y$, then OUT is similarly described by the information that, at the final time, $X + Y$ is in state $\varphi_t = \sum c_i |a_i > \otimes |b_i\rangle$. We recall that, by reduction of the density matrix, this implies that

(a)　X is in $W_X = \sum c_i^2 I_{|a(i)\rangle}$

(b)　Y is in $W_Y = \sum c_i^2 I_{|b(i)\rangle}$

but that (*a*) and (*b*) together do not give all the information present in the full assertion:

(*c*) $X + Y$ is in final state $\varphi_t = \Sigma c_i |a_i\rangle \otimes |b_i\rangle$.

We must also emphasize here, as we did in the preceding chapter, that this description of measurement is not complete until we add that the process is not a measurement of *A* if it merely 'accidentally' takes the form (S2). The real condition is that the Hamiltonian (more generally, the group of evolution operators) which governs this process is such as to guarantee this form for groundstate *y* with *any* initial state *x* of the object. Only by reading our description in this strong sense is the measured observable A uniquely identifiable.[2] So far, the purely quantum-theoretical description.

But how shall we connect this with Born's Rule? There will be a connection only if the actual outcome of the measurement is included in the description of what happens at the end. We should be able to say that there is (also?) some true assertion of form:

(*d*) Pointer-reading observable B has value b_k

so this must at least be consistent with (*a*)–(*c*). How shall we construe (*d*)?

Von Neumann saw this problem very clearly, and in answering it he could reasonably think that he merely formalized what the developers of the theory had been saying informally. Indeed, if we look for instance at the famous Einstein–Podolsky–Rosen (EPR) paper, we find the authors writing in just the way von Neumann could have taken as paradigmatic—and most of the replies to EPR went along with it. Von Neumann formulated (even if he did not originate) the following answer to the question of how we shall construe an observable having a value:

(*e*) *Von Neumann's interpretation rule*: An observable *B* pertaining to system *Y* has value *b* if and only if *Y* is in a corresponding eigenstate of *B*.

That means, *Y* is in a state *W* such that $Tr(WI_b^B) = 1$. So far, the only interpretation we placed on this was Born's: in such a state, the probability of a measurement of *B* having outcome *b* equals 1. Can von Neumann consistently add that B *has* value b under exactly these conditions?

One possible objection is that a state is *nothing more* than a cluster of probability functions, a sort of mathematical summary of probabilities of measurement outcomes. If we say that, we land in an infinite regress. How can we equate the meaning of 'B has value b' with 'the probability that a measurement of B would yield value b equals 1'? The assertion that the measurement *does* yield value b, which is itself equivalent to the statement that e.g. pointer-observable C has value c, comes then to mean that another measurement would have outcome c with probability 1, and so on *ad infinitum*. If actual $\neq$ possible, we cannot everywhere equate the meaning of statements about what *is* the case with other ones that are about *what would be if*.[3] But I have tried from the beginning (since I knew this point was coming) to speak of states in a way that allows (for the possibility) that a state merely determines, and is not identical with, the corresponding cluster of conditional probabilities. As analogue, imagine FBI or Immigration Department files: everyone is there represented by some vital statistic (birthplace and date, social security number, etc.), but of course the person is only identifiable through, and is not identical with, these vital statistics. We may conclude that von Neumann's interpretation rule, although it is not logically forced on us, is a consistent addition.

But we have a second problem. Suppose that, in the measurement process described by (a)–(c), we have a non-trivial superposition (not every coefficient c_k equals 0 or 1). Then if the apparatus Y is completely described at the end of measurement by the reduced state W_Y, we note that, since $Tr(W_Y I^B_{|b(i)\rangle})$ $\neq 1$ for any index i, von Neumann's interpretation rule asserts that B does *not* have any value! That contradicts (d), the very assertion we are trying to construe.

Therefore von Neumann adds that the assertion that Y is in state W_Y is at best incomplete. Indeed, he asserts that

(f) At the end of the measurement there is some index k such that X is in state $|a_k\rangle$ and Y in state $|b_k\rangle$

This is consistent with (a) and (b) on one interpretation of mixed states: the ignorance interpretation. And that interpretation appears to be correct in many circumstances—but is problematic exactly when the mixed states are arrived at by reduction from a compound system in a pure state. Thus the real question is: is (f) consistent with (c)—the attribution of φ_t to $X + Y$?

Von Neumann regarded the two as *not* consistent. Hence he added his famous Projection Postulate:[4]

- (g) At the end of measurement process described by (S2) there is a further transition, an *acausal transition of state* φ_t to some state $|a_k\rangle \otimes |b_k\rangle$.
- (h) The transition from φ_t to $|a_k\rangle \otimes |b_k\rangle$ has probability c_k^2, i.e. the Born probability of the A-measurement outcome a_k for initial state x, i.e. the squared length of the projection of φ_t on the corresponding eigenspace.

The acausal transition is also commonly called a *collapse of the wave packet*, a term which derived from Schroedinger's wave mechanics formalism. Von Neumann did not ask how this transition occurs, restricting himself to questions of consistency. He added two arguments to support his interpretation. One is to the effect that the Projection Postulate adds no new empirical predictions. The other is to the conclusion that the phenomena of immediately repeated measurement require the Projection Postulate.[5] There is a certain tension between these two: if the phenomena demanded this Postulate, then there would have to be an inadequacy with respect to empirical predictions without it! Hence it seems that, if either argument is successful, the other must be wrong.

Before looking at von Neumann's defence of his Postulate, let us note that he interprets quantum mechanics as exhibiting indeterminism in form 2 (indeterministic state transition). Principle (e), which I call *von Neumann's interpretation rule*, insists that what the state is totally determines the truth value of any attribution of values to observables. Accordingly, the indeterminism noticed in Born's Rule has to consist in an indeterministic state transition. Once (e) is given, one *must* postulate an acausal transition of state.[6]

In this section, I have outlined only one answer to the questions about measurement. It is clear that alternatives must begin at (e), where von Neumann's interpretation rule augments Born's. If we agree with von Neumann there, the next cross-roads come, as it were, *after* the Projection Postulate. That is, once we agree with (e), we can only try to show that the 'acausal transition'—so called because it falls outside the process (S2) described by Schroedinger's equation—arises in some intelligible fashion, and does not lead either to inconsistency or to new, false empirical predictions.

Hence I shall turn first to von Neumann's defence, then to other defences of the Projection Postulate, which appeal to the idea of special 'macro-observables' or the macro–micro level distinction. In the next chapter I shall propose an alternative to von Neumann's interpretation rule.

4. VON NEUMANN'S FIRST DEFENCE: CONSISTENCY OF MEASUREMENT[7]

Von Neumann had two defences for his interpretation of measurement. The problem with them is that one makes the Projection Postulate empirically vacuous—hence purely a matter of interpretation—but the other gives it empirical import. This ambiguity in von Neumann's own thinking (are we merely interpreting, or adding a postulate demanded by the phenomena?) has plagued much subsequent discussion of the 'collapse of wave packet'.

Imagine that the complex system $X + Y$ of the preceding section, which was isolated (say in a hermetically sealed box out in space), is now looked at by an observer Z. Clearly, this breaks the isolation—for instance, Z opens the box and looks to see what the pointer on apparatus Y indicates. This too is a measurement.

As von Neumann noted, we can retrospectively analyse the situation in two ways. We can regard $(X + Y)$ as the subject on which a measurement is performed by interaction with Z. The 'pointer' on Z will now indicate the value z_k correlated with value $(a_k; b_k)$ of observable $A \otimes B$ on $X + Y$ for some index k. Alternatively, we can regard $(Y + Z)$ as the apparatus, with X as the subject of measurement. If C is the pointer-reading observable on Z, then $B \otimes C$ is the pointer-reading observable on $(Y + Z)$. However—here is the essence of von Neumann's consistency argument—the probability is in each case c_k^2 for the kth outcome. However we divide up the world here, we make essentially the same predictions.

The argument was clarified by Groenewold (1952, 1962) to show that, despite the talk about observables, no subjective element need have been introduced. Groenewold asks us to consider a system X at t_a in state x, subjected to a measurement by apparatus $M_1 = Y$ in interval (t_a, t) as in our schematic discussion above, and then to another measurement of observable C by apparatus M_2 on X during interval (t, t_b). Then we calculate the probabilities of outcomes of the second measurement in two ways:

(i) assuming that at t *n*o acausal transition occurs, but the measurement by M_2 is made on X in state W_X;

(ii) assuming that at t, in accordance with the Projection Postulate, X transits to pure state $|a_k\rangle$ with probability c_k^2, and the measurement by M_2 is made on X in the resultant state.

Again, it is easy to see that the probabilities of outcomes for the second measurement are the same in both cases, because $W_X = \sum c_i^2 I_{|a(i)\rangle}$. We can go further and introduce a third way of describing the situation:

(iii) assuming that M_2 measures observable $C \otimes I$ on $X \otimes M_1$ in state $\varnothing_t$ at t.

And again, we must arrive at the same probabilities of second-measurement outcomes, just because W_X is by definition the state for which probabilities for any observable C are the same as the probabilities for $C \otimes I$ on state $\varnothing_t$.

Thus we have a complete consistency of predictions, for measurements of observables pertaining to the system X, regardless of where we draw the line demarcating observed system from observing environment.

But, as Margenau (1936, 1950, 1963) clearly pointed out, this consistency proof is also an empirical redundancy proof. Since predictions of measurement outcomes for, say, assumptions (i) and (ii) are the same, the two situations described cannot be empirically distinguished by measurements performed on system X alone. Margenau concluded, therefore, that rejection of the Projection Postulate does not rob quantum mechanics of any predictive power. This conclusion must be qualified, as we shall see below.[8]

If we grant the point, then the retort would have to be that what is robbed is not prediction but explanation or interpretation—what does it *mean* to say that the measurement had a certain outcome? And indeed, as I argued above, the Projection Postulate is practically forced on us by von Neumann's response to the request for interpretation. Thus it focuses scrutiny on von Neumann's interpretation rule. Margenau himself saw this very well, and stated explicitly that, once the Projection Postulate is rejected, we face anew the question: what *happens* in a measurement?

A final word about the question of whether the addition of the Projection Postulate makes new predictions, going beyond the Born Rule. As David Albert has emphatically pointed out, there certainly are observables for which the predictions are different, depending on whether or not the acausal transition has occurred.[9] If the transition is from φ_t to $|a_k\rangle \otimes |b_k\rangle$, then a measurement of the observable represented by the projection on subspace $[\varphi_t]$ is certain to yield value 1 before the transition, and not afterward. Even if we do not know exactly when the transition occurs, we can surely carry out many measurements, whose outcomes will support or disconfirm this prediction.

That is a telling point in this context. However, we shall see below in Section 8 that the best elaboration of the Projection Postulate story involves exactly the denial that all Hermitean operators represent observables. Albert's main point, that there must be testable predictions in this sort of interpretation, may stand nevertheless, since we shall also see in Section 8 that this denial too appears to be empirically testable.

5. VON NEUMANN'S SECOND DEFENCE: REPEATABLE MEASUREMENT

The second defence listed above, concerning immediately repeated measurements, may well have been more influential than the first. If the wave packet does not collapse—if the state-vector is not projected into an eigenstate of the relevant observable —why should the immediate repetition of the same measurement yield the same result? It does; so surely(?) the state of the object system must have been changed from one in which the outcome was uncertain to one in which it was certain? Yet this defence does much more badly than the first, when properly scrutinized.

There have been two episodes in the discussion of this defence. The first belongs roughly with the discussions in the preceding section. It consists of von Neumann's analysis of the Compton–Simon effect, and responses essentially given in Margenau's and Groenewold's papers discussed above. The second episode was the purely quantum-mechanical analysis of measurement processes and the conditions under which they have the characteristic of repeatability demanded by von Neumann. I will take up each in turn.

Von Neumann's second defence may be read either as resting on a challenge to explain, or as claiming a predictive success. On the second reading, it says: unless the Projection Postulate is added, or made part of the interpretation, the interpreted theory fails to predict this salient fact of repeatability about repeated measurements. On this reading, then, von Neumann's Project Postulate has empirical import over and above the theory as interpreted by his interpretation rule. Let us see what we can make of this.

The sort of illustration that immediately occurs to one (and is often enough given) actually does not fit. Suppose for example that a beam of unpolarized light arrives at a vertical polarization filter. The photons go through with probability $\frac{1}{2}$, so the beam intensity is reduced from its initial value r to $r/2$. If we have two such filters, close together, they still reduce the beam intensity only by a factor of $\frac{1}{2}$. Does that not show the correctness of above reasoning?

No, for no measurement has been made twice. What observable plays the role of pointer-reading observable here? The answer is *none*, because no detection of the photons was described. We could put photographic plates behind each filter, but that would destroy the experiment because the photons would be absorbed before they can reach the second filter; similarly if we use charged particles and a magnetic field. If we detect the presence of the particle in one beam, the detection device will disturb the very features of the particle that help to determine its reaction to the second field. This point will be only a minor quibble, however, if we can formulate a reasonable design for repeated measurements.

So von Neumann looked for a more sophisticated method in which the first measurement is guaranteed not to disturb what the second measurement measures, and *yet* the whole setup counts as a twice-repeated measurement of a single observable (von Neumann 1955, 212–14). Is this possible?

In an experiment by Compton and Simon, light is scattered by electrons and the scattered light and scattered electrons are intercepted and have their energy and momentum measured. It was concluded from this experiment that the mechanical laws of collision hold. But von Neumann reformulates the conclusion as follows. If we assume that the laws of collision are valid, the position and central line of the collision may be calculated from the measurement of the path of *either* the light quantum *or* the electron after the collision. It is an empirical fact that the two calculations always agree. But the two measurements do not occur simultaneously; the measurement apparatus may be arranged so that either process may be observed first. So we have two measurements, M_1 and M_2, the second after the first; beforehand, their outcome is only statistically determined, but after M_1, the outcome of M_2 may be inferred. From this, plus the fact that M_1 and M_2 are *in effect* (i.e. via calculation) measurements of the same observable (say, a coordinate of the place of collision or of the direction of the central line), von Neumann infers that, if an observable is measured twice in succession, the second measurement 'is constrained to give a result which agrees with that of the first'. And since the outcome of the second measurement can be predicted with certainty, von Neumann infers that, after the first measurement, the measured system must be in an eigenstate of that observable.[10]

But what we have here is two measurements on distinct objects—an electron and a photon—which have interacted and whose states have become entangled. That is, we have a compound system $X + Y$ which has state $\varphi_t = \Sigma c_i |a_i\rangle \otimes |b_i\rangle$. In that case it is certainly true that, if we measure $A \otimes B$, we must get a pair of values $(a_i; b_j)$ such that $i = j$. What we calculate from them is not much to the point. It is only in the sense of a paper-and-pencil operation afterward that one observable is measured twice on the same (complex) system. This does not support the idea that the (component) system, subject directly to measurement, transits into some eigenstate.

Indeed, Groenewold's discussion already shows us exactly what we should say about immediately repeated measurements. In the discussion of his version of the consistency argument, let $M_1 = M_2$, so the same observable is measured twice over. We are not allowed to destroy the isolation, or the immediate succession, by interfering with any kind of look at the midpoint, time t. In the absence of such a look at the midpoint, however, we can check only the outcome of the second measurement. But, as he showed, our prediction of *that* outcome is the same, whether or not we suppose that an acausal transition occurred at time t. The second argument therefore has a valid case which establishes no more than the first—consistency—and derives its intuitive appeal from a mistaken picture that allows 'free looks' in the middle of an isolated process, without disturbance.

To the second episode in this discussion was the recent formal development of the quantum theory of measurement. It is now possible to demonstrate that repeated measurement does not provide a telling criterion for interpretation. Instead, it turns out that the issues concerning repeatability are already settled on the level of quantum theory itself, before we enter upon interpretation. In the measurement process we see both a change of state in the total system (object + apparatus) and, derivatively, changes of state—assigned by reduction—in each component. We have here two sorts of mappings—evolution and reduction—and the natural question to ask is (as mathematicians put it): can the diagram be completed? (This refers to Fig. 8.1.) To complete the diagram, an arrow also marked 'evolution' should be drawn along the bottom, to go from initial to final object state. That arrow purports to represent a well-defined function—and indeed, that function exists. The function J so described, which still depends on the measured observable A and the evolution operator U of the total system, because that derivative evolution does so, is the *instrument* of this measurement.

Imagine a setup in which the same measurement is executed twice in immediate succession. We must insist that the total process be isolated—no interference from outside, not even a

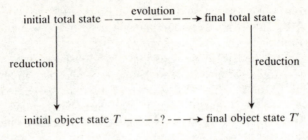

Fig. 8.1

quick look or photograph is allowed—and we must assume that the measurement apparatus resets itself (to its groundstate) without any effect on the object. The holism of quantum mechanics now makes it impossible even to think that the order in this two-stage procedure is immaterial—at least, not in general. But if the total two-stage derivative evolution

$$J(J(T, E), F)$$

is in fact no different, regardless of the order, and gives indeed the same predictions as the one-stage procedure would to someone interested in whether the outcome value is in both E and F, that is if

$$J(J(T, E), F) = J(T, E \cap F)$$

then we call the measurement (and the instrument) *repeatable*. What this implies is that, if we were to measure further whether there was value agreement between the first final state and the second final state with respect to the pointer-observable, then we would be sure to get a *yes* answer.

Not all measurement processes need have this property. The class of measurements discussed by von Neumann were said by him to be repeatable in this sense, and he explained their repeatability by means of the Projection Postulate. For that postulate said that the system really emerged in an eigenstate of the measured observable, and therefore the second stage had no effect at all. So the burning question for us is: is that explanation really needed? In the preceding chapter we already found that the answer is *no*. The instrument associated with the above-described von Neumann–Lueders measurement is repeatable.

We have now come to the conclusion that neither the Projection Postulate nor any other principle of interpretation is needed to explain repeatability. The theory of measurement processes described within quantum mechanics already entails that, if a von Neumann–Lueders measurement is twice repeated, the statistical predictions for outcomes are exactly as for a single application.

Proofs and illustrations

Margenau was of course quite right to point out that a particle subjected to measurement is usually absorbed at the end. This makes repetition of a measurement on the same particle rather difficult. But R. H. Dicke (1989) has devised a method of sequential measurement in which the information about what happens in the intermediate stages is encoded, preserved, and collected only at the end of the whole sequence. This is a process which, while remaining isolated, consists of a series of measurement interactions with the results encoded, stored, and eventually inspected together—though of course in the only sense in which that is allowed by quantum theory, and which does not necessarily fit classical requirements.

The technique is illustrated with the design of an experiment in which a photon is subjected to a sequence of measurement of circular polarization. The energy of the photon itself serves as 'apparatus register' on which the measurement results are encoded, and which can be 'read' at the end. I shall here just sketch the general design. The observable S to be measured has eigenvectors $|k\rangle$, $k = \pm 1$. The apparatus has state function $y(q)$ with groundstate y_0. The unitary operator U which governs the evolution

$$U\left(\sum c_k |k\rangle \otimes y_0\right) = \sum c_k |k\rangle \otimes y_k$$

is due to the interaction Hamiltonian $H_{int} = -CSq/\delta t$, which is the dominant term in the total Hamiltonian $H_{app} + H_{obj} + H_{int}$. Thus,

$$U = \exp[-i\mathrm{H}_{\mathrm{int}}(\delta t/\hbar] = \exp[iCSq/\hbar]$$

so that, in the above equation, $y_k = y_0 \exp[iCkq/\hbar]$. Now the variable $p = (\hbar/i)(\delta/\delta q)$ conjugate to q can serve as 'pointer' or 'register'. Its eigenvalues are $p_k = Ck$.

In the sequential measurement, the constant C is replaced by $C2^n$ for the nth measurement, and H_{app} is a function of p. The former change allows the sequence of results $S_k, S_j, \ldots$ to be encoded as a sum in the 'register', and the latter makes p a constant of the motion in intervals between the measurements, so that the record is stable.

The complete evolution of the compound state $\varphi = \Sigma c_k|k\rangle \otimes y_0$ is into the final state

$$\varphi' = U_n T \ldots U_3 T U_2 T U_1 \varphi$$

where $U_n = \exp[i2^n CSq/\hbar]$ and the Hamiltonian time displacement $T = \exp[-i(H_{obj} + H_{app})(\Delta t/\hbar)]$ operates during the intervals (assumed equal here) between the measurement stages. There are 2^n possible histories superposed in the total process, correlated with the terminal values

$$p_{kji} \ldots = C(2S_k + 4S_j + 8S_i + \ldots)$$

of the variable p. In other words, the final state φ' has the form of a superposition of tensor product states for the compound (object + apparatus), where on the apparatus side we see the indicator states

$$y_{kji} \ldots (q) = y_0 \exp[ip_{kji} \ldots q/\hbar]$$

with $p_{kji} \ldots$ as above.

In the experiment designed as illustration, a photon passes through a series of photon spin–energy correlators. These correlators each consist of two half-wave plates, one rotating and one fixed. Such a plate reverses the sign of circular polarization. If the disc is rotating with angular velocity $-\omega$, the energy $\pm 2\hbar\omega$ is transferred from disc to photon, with the sign determined by the spin state. If the photon was in a spin eigenstate, that is not affected (as is proper to a measurement) because the two plates together reverse the sign twice. But the photon energy is shifted by $\pm 2\hbar\omega$ depending on the spin state.

6. R. I. G. HUGHES'S ARGUMENT FROM CONDITIONAL
PROBABILITY

R. I. G. Hughes has argued in his recent book that the Projection Postulate can be understood as a conditionalization of the quantum-mechanical total (object system + apparatus) state, and cannot be removed from the theory (see Hughes 1989, sect. 8.8, 9.3, 9.4, 10.3). The argument occurs in the development of another interpretation, so may be taken out of context here. I want to show, however, that the argument cannot play the role of a defence of von Neumann's interpretation rule.

We have already seen how von Neumann–Lueders measurements are related to conditionalization. Such a measurement of observable A transforms the reduced initial object system state W into its conditionalization on the partition of eigenspaces of A; that is,

$$W \longrightarrow W' = \sum p_a W_a$$

where W_a is the conditionalization of W on the a-eigenspace of A, and $p_a = Tr(WI_a^A)$. If W is pure, then W_a is just the projection on that subspace. It is tempting to read the numbers p_a as measures of ignorance; really, at the end of the measurement the object system has transited into an eigenstate of A, though we don't know which, but it is W_a with probability p_a. And this temptation can now be reinforced in two steps. The first is to say that the apparatus is in some pointer-observable eigenstate—for example the one corresponding to value a. The second step is to say that, if we were given the information which value that was—for example a_0—then we should update our assignment of the state to the total system by conditionalizing on the subspace $H_x \otimes H_{a(0)}$ where $H_{a(0)}$ is the a_0-eigenspace of the pointer-observable. Then, because of the perfect correlation between apparatus Y and object system X, the reduction of *that* conditionalized total state to H_x would be $W_{a(0)}$.

This is seductive, but there are two assumptions at work, closely connected. The first is that, just because the pointer-observable has value a_0, it follows that the state of the apparatus is an $a(0)$-eigenstate of that observable. That, of course, is von

Neumann's interpretation rule ((e) in Section 3 above) at work. The second is that, if we find that the measurement outcome is a_0, we should think that the state has changed by conditionalization on $H_x \otimes H_{a(0)}$. That second assumption entails the first, so they are not independent. But the second is perhaps derived from a deeper assumption, namely that conditionalizing the quantum-mechanical state one attributes to a system is just updating one's opinion on the new information available. Even that deeper assumption, if we let it stand without controversy, needs the first to get off the ground. For what is the new information? Only if it is the information that the compound system is now in $H_x \otimes H_{a(0)}$ can we conditionalize on that. But the new information is only that the pointer-observable has value a_0—we need von Neumann's interpretation rule to translate that into: the new apparatus state is an a_0-eigenstate, or, equivalently, the new total state is in $H_x \otimes H_{a(0)}$. Hence these reflections support very clearly my suggestion that the crucial step beyond Born, taken by von Neumann, is his interpretation rule—after which the Projection Postulate is forced upon us. Hughes's argument illuminates this by connecting it with conditional probability, but does not add to it.

This will be clearer if we connect the argument further with the notion of instrument in the preceding section. The initial and final states of the system were called T and T' there, and the *instrument* of the process was the two-place function J which takes the pair $\langle T, E \rangle$ into the conditionalization of T' on the E-eigenspace of A.

But note that $J(T, E)$ is not T' for any E except ones such that the E-eigenspace of A contains the image space of T'. That this instrument is repeatable implies only that, if we design a process in which the A-measurement is twice repeated by the same apparatus on the same system (all in isolation), and we add a further apparatus which checks whether the same value was obtained twice, so to say, then we are sure to get a *yes* answer there. But this is because of correlations set up in the total system, and not because the first final state T' was really identical with *one* of its conditionalizations on an eigenspace of A. That conclusion can be postulated as due to a further 'acausal' transition, but it cannot be derived.

Let us sum up therefore where we are. If we are given von

Neumann's additional rule of interpretation—going beyond Born's statistical interpretation of the quantum-mechanical state—then we must agree to the Projection Postulate. This postulate fits with current quantum mechanics in a straightforward way: it says that, at the end of a measurement, some rule comes into effect which erases the distinction between the superposition φ_t and the corresponding mixture. ('Corresponding' refers here to the set of eigenvectors of the relevant observable.) I have now phrased this in a way that will call superselection rules to mind for the contemporary reader, so we should begin to pursue that idea.

7. TWO CAT PARADOXES AND THE MACRO WORLD

After superselection rules appeared in the literature, it was found that these allow a sophisticated, consistent reformulation of the Projection Postulate. That new development I shall take up in the next section. Here I shall lay the groundwork by discussing the motivation for introducing a fundamental distinction between macro and micro world. Associated with this there is a conviction that quantum mechanics as originally formulated cannot apply universally. Two ways to say that in contemporary terminology are (*a*) that the principle of superposition (i.e., for any two pure states x and y, there is a pure state $ax + by$) does not have universal validity, or (*b*) that quantum mechanics must be amended by allowing for superselection rules. We begin here with a reflection on the putatively distinctive character of the macro world.

In his reflections on the EPR paradox and elsewhere, Schroedinger pointed to 'sinister' consequences that clearly hinge on von Neumann's Projection Postulate. They also concern the jeopardy in which quantum mechanics seems to place our intuitive picture of the world we live in—the macro world. For that reason, his deliberations are a good prolegomenon to certain attempts to explain (or explain away) the 'collapse' or 'projection' as *not* a physical, acausal change, but connected with the peculiarly macroscopic character of measurement. I shall first present his famous Cat Paradox (Schroedinger 1935*b*), and then a variant in which the cat receives the body of a kitten.

Schroedinger imagined a cat in a box, in which there is an interaction formally a little like that in measurement.

A cat is placed in a steel chamber, together with the following hellish contraption (which must be protected against direct interference by the cat): In a Geiger counter there is a tiny amount of radioactive substance, so tiny that maybe within an hour one of the atoms decays, but equally probably none of them decays. If one decays then the counter triggers and via a relay activates a little hammer which breaks a container of cyanide. If one has left this entire system for an hour, then one would say that the cat is still living if no atom has decayed. The first decay would have poisoned it. The Ψ-function of the entire system would express this by containing equal parts of the living and dead cat.

The typical feature in these cases is that an indeterminacy is transferred from the atomic to the crude macroscopic level, which then can be *decided* by direct observation. This prevents us from accepting a 'blurred model' so naively as a picture of reality. By itself it is not at all unclear or contradictory. There is a difference between a blurred or poorly focused photograph and a picture of clouds or fog patches. (Schroedinger 1935*b*; trans. Jauch 1968, 185)

The cat has two possible states: alive $|b_1\rangle$ and dead $|b_2\rangle$—a shallow representation, but not incorrect. The system with which it is coupled also has two possible states: a particle *has not* passed through a certain filter and been absorbed by a detector $|a_1\rangle$, and it *has* $|a_2\rangle$. The system begins in state $|a_1\rangle \otimes |b_1\rangle$ and is isolated. The evolution of the state is described by

$$U(|a_1\rangle \otimes |b_1\rangle) = \frac{1}{\sqrt{2}}(|a_1\rangle \otimes |b_1\rangle) + \frac{1}{\sqrt{2}}(|a_2\rangle \otimes |b_2\rangle)$$

So the final state, call it Ψ, can be described as a perfectly correlated one. We see that this final state is *not* an eigenstate of the observable $I \otimes B$, which describes the condition of the cat. By von Neumann's interpretation rule, we could say:

If the state of the whole system is Ψ, then the cat is
    alive if and only if $(I \otimes B)\Psi = b_1\Psi$, and
    dead if and only if $(I \otimes B)\Psi = b_2\Psi$

but we already know here that, for the end state Ψ, neither of these is the case. Now we open the box; the look is a measurement, and surely what we see is *either* a live cat *or* a dead one. This example is disturbing, because when it comes to cats (and other macroscopic objects) we feel sure that whatever we see when we open the box was already there. Or else, the other thing we might have seen was there; it is not inconceivable that the cat was alive till we opened the box, and died at that very moment. But our sensibilities are miffed, if not outraged, by the idea that it was not true that the cat was dead and also not true that it was alive. Obviously we can put the problem more generally: if quantum mechanics is applied uncritically, it allows superpositions of any states a macroscopic object can have—but what sense can we make of this?

The following variant of the Cat Paradox also exhibits interference on the macroscopic level which appears possible if the theory applies without qualification to that level. Let us call this one the *Benign Cat Paradox*.

Let A be the observable pertaining to the cat, whose eigenvalues a_n correspond to the properties of having the body of an n-year-old. The corresponding eigenstates $|a_n\rangle$, $|a_m\rangle$ are orthogonal: obviously, no animal could have at once the body of a 1-year-old kitten and a 12-year-old cat. But we are also interested in the curious state

$$x = \frac{1}{\sqrt{2}}|a_1\rangle + \frac{1}{\sqrt{2}}|a_{12}\rangle$$

We now devise a measurement of the observable I_x. According to von Neumann, this will project the initial state, say $|a_{12}\rangle$, into either x itself or a state orthogonal to x. Now we rig up, after the first measurement apparatus, a filter which will pass state x but not states orthogonal to x. And after that we place an A-measurement apparatus. The effect of the latter will be to project state x into either $|a_1\rangle$ or $|a_{12}\rangle$ according to von Neumann. The projections in this story all have associated probability $\frac{1}{2}$. So here is what will happen: the cat, entering the process in state $|a_{12}\rangle$, has a probability $\frac{1}{2}$ of dying (because of the filter), $\frac{1}{4}$ of emerging in the same state in which it entered, and $\frac{1}{4}$ of emerging with the body of a kitten. If the cat places enough value on the possibility of rejuvenation, it might well choose to submit—as might, in a more advanced version of the experiment, a sufficiently old rational animal!

The assumptions made here are really just the same as in the more famous Cat Paradox. They are that we can ensure total isolation for this complex system; that all superpositions are allowed, for macro states as well as micro states. Another assumption involved, if not equally obvious, is that the Hermitean operators which distinguish these superpositions really represent observables. In sum, what is assumed is the unrestricted applicability of elementary quantum mechanics.

All this sounds so strange, as applied to our own familiar world, that we must wonder if quantum mechanics applies differently at the macroscopic and microscopic levels. This cannot be asserted arbitrarily. First of all, the boundary line between the two is vague: cats are macroscopic and individual electrons are not, but where is the dividing line? Secondly, the boundary is prima facie drawn in terms of human observability—but humans have no privileged status in fundamental physics. Still, it is clear that macroscopic objects, which involve many elementary particles, have a much greater level of physical complexity. Perhaps the observables whose values can be distinguished at the level of human observations are subject to restrictions that do not hold for all observables—not that it *has to be* that way, but perhaps the world *could be* that way. We turn now to some solutions of the measurement problem suggested by this line of thought.

8. MACROSCOPIC CHARACTER AND SUPERSELECTION RULES

In recent decades, von Neumann's account of measurement has in effect been improved and defended by the assertion that the apparent acausal transitions are not outside quantum mechanics but must appear as natural consequences for the correct treatment of macroscopic systems.[11] This line of thought reaches its most elegant and successful form through an appeal to superselection rules.[12] Indeed, as I shall try to show, in this form the Projection Postulate becomes not only defensible but in some sense (logically, anyway) invulnerable. However, there appear to be involved several conjectures which have at present no further substantiation, and whose implications are not entirely clear.

Attempts to make sense of measurement along these lines advance a conclusion of roughly the form:

1. When Y is a macroscopic object—such as any real measurement apparatus is—then the final state $\phi = \Sigma c_i| a_i\rangle \otimes |b_i\rangle$ is empirically indistinguishable from the mixed state $\Sigma c_i^2 I_{|a(i)\rangle \otimes |b(i)\rangle}$

Furry, who was perhaps the first to consider this possibility, pointed out quite correctly that, if all Hermitean operators represent observables, then those two states do lead to different measurement outcome predictions.[13] But the 'if', we know now, is deniable. Danieri *et al.* (1962) argued that developments in quantum thermodynamics support conclusion 1, for large systems, as an approximation. Such approaches were thoroughly criticized by Bub (1968), who made clear that results about approximation cannot remove the conceptual distinctions that engender the puzzles—as Earman was to say later (1986, 224), in such a case a miss is as good as a mile.[14] Hepp's demonstration that the two states become literally and strictly indistinguishable for the case of infinitely many degrees of freedom does not counter this objection effectively, for such a case is also at best an approximation to that of large but finite systems, which alone are actual.[15]

Let us turn now to the alternative presented by Beltrametti and Cassinelli, which appeals to a superselection rule. Indeed, conclusion 1 has exactly the form that bespeaks such a rule. For the concept of a superselection operator is the way we found in quantum mechanics of saying that certain states cannot really be superposed, whatever the mathematical formalism allows. It suffices now to insist that this is exactly the distinguishing macroscopic character of the cat's being alive or dead, or of the indicator-states of the measurement apparatus.

Let B be the pointer-reading observable for apparatus Y, with eigenspaces B_0, B_1, ... and eigenvectors $|b_0\rangle$, $|b_1\rangle$, ... We may take $|b_0\rangle$ to be the groundstate of the apparatus. Postulate now a superselection rule which entails that these eigenspaces are the coherent subspaces; in other words, B is a superselection operator.

What this means for the complex system $X + Y$, when

apparatus is coupled to system to be measured, is that $I \otimes B$ is a superselection operator for the tensor product space $H_X \otimes H_Y$. Assume that no other superselection rules are in force. Then the coherent subspaces are $H_X \otimes B_0$, which contains the initial state $x \otimes |b_0\rangle$; and $H_X \otimes B_1$, $H_X \otimes B_2$, ... If $x = \Sigma c_i |a_i\rangle$ then the final state will be $\Sigma c_i |a_i\rangle \otimes |b_i\rangle$, which does not lie in any coherent subspace. But as we know from the discussion of superselection rules, it still represents a state, namely, exactly the same state as $\Sigma c_i^2 I_{|a(i)\rangle \otimes |b(i)\rangle}$. And thus we have deduced conclusion 1.

The logical character of that conclusion, which puzzled so many writers in the past, is now in one sense beyond reproach. But we still need to ask about the conjecture of a superselection rule on which this deduction is based. There are three problems: the first noted by Beltrametti and Cassinelli themselves, the second raised in effect by R. I. G. Hughes, and the third by Leggett. All three problems are addressed in Bub's more recent solution, but there I shall point out a fourth problem that is perhaps the most fundamental of all.

The first problem is: what accounts for this postulated superselection rule? It is conjectured to be due to the macroscopic character of the measurement apparatus. If such conjectures were allowable without limit, classical behaviour could never disconfirm quantum predictions. For instance, if there had been no violations of Bell's Inequalities in, say, Aspect's experiments, we could have 'saved' the quantum-mechanical description by conjecturing superselection rules. The mere conjecture of superselection rules—though formally the very thing that substantiates suggestion 1—is too easy and universal a tool for restoring consistency. If it is to be plausible, the conjecture needs particular substantiation. Hence Beltrametti and Cassinelli very rightly ask:

whether the very mechanism of combining subsystems can cause, at least in the limit of very large numbers of subsystems, the birth of superselection rules of the compound system. ... Of course, should one have a theory predicting the appearance of superselection rules as an effect of the combination of subsystems with purely quantum behavior, one would have at hand the possibility of a deep understanding of the 'classical limit' of quantum mechanics. (Beltrametti and Cassinelli 1981*a*, 74)

In the part omitted, they note that such a prediction of the appearance of superselection rules arising from complexity could not involve any simple extrapolation by numbers. For macroscopic bodies and substances usually behave as classical systems, but there are remarkable exceptions: 'a pot of liquid helium is macroscopic but displays quantum behaviour'. The gap that needs to be filled in this account is therefore by no means trivial.

The second problem, raised in effect by Hughes (1989, sect. 9.7), is also serious. In our initial discussion of superselection rules we noted that, if the Hamiltonian H is an observable—and we were used to thinking of it as almost the paradigmatic quantum-mechanical observable—then evolution out of a coherent subspace is impossible by any process of the form described by Schroedinger's equation. In that case, there could be no evolution from the groundstate in $H_X \otimes B_0$ to any other indicator state of the apparatus! Therefore, this treatment of the measurement problem must assume that H *does not* represent an observable. The implications of this, or its consequences for the applicability of quantum mechanics to macroscopic systems, are not clear. Even if there are conditions under which a Hamiltonian can safely fail to represent an observable, measurement may not be one of those.

The third point concerns empirical fact. In connection with von Neumann's first defence, we faced the question of whether or not the Projection Postulate adds new empirical predictions, beyond those derivable from the basic theory plus the Born Rule for probability calculations. Suppose we have a more sophisticated theory of measurement, which does begin with von Neumann's interpretation rule, but derives the Projection Postulate from a superselection rule. In this version, are there new empirical predictions?

We should expect that the answer is *yes*, since superselection rules in general add empirical content. There is no doubt that, in a model with superselection rules, fewer imaginable phenomena can be embedded, because some pure states are eliminated. As we recall, a vector which does not lie in a coherent subspace represents not a pure but a mixed state, namely the same state as is also represented by a certain statistical operator which is not a projection operator. The resulting absence of 'probability interference' certainly does make a difference to predictions of measurement outcomes.

In this case, the superselection rule has not been completely specified, but we know this much about it: it places each eigenspace of some macroscopic observable inside a distinct coherent subspace. To succeed as a general account of measurement, this must happen for each macroscopic observable which can function as the 'pointer-observable' of a measurement. Finally, the class of macro observables has also not been completely specified, but we know this much about it: those observables whose values correspond to humanly directly detectable differences in humanly observable objects must be among the macro observables.

A few years ago, Anthony Leggett proposed experiments designed to test the assertion that the eigenspaces of macroscopic observables are indeed separated in this way by superselection rules (see Leggett 1980, 1986; Leggett and Garg 1985). Leggett formulated the empirical question as follows:

Is it possible, in practice, to prepare a macroscopic system ... which can be in one or two or more *macroscopically distinct* states, and then present ourselves with the choice of either measuring which of the two states it is in, or of observing the interference between the two possibilities? Further, could we then demonstrate that in the second case it *could not have been* in a definite macrostate? (Leggett 1986, 47)

Leggett observes that Bohr would probably have argued that the first question has a negative answer. But he adds that the situation has been altered dramatically by the prediction of the Josephson effect (1962), and the experiments in progress on that effect. As to the second question, Leggett shows that the answer is *yes* by a demonstration adapted from that of the Bell Inequalities.

Conceptually, we are here certainly well within the realm of possibility. Schroedinger remarked that the emission of a single photon can sink a battleship—it suffices to amplify the absorption of one photon in a photoelectric cell sufficiently to trigger a missile. So if we can strongly correlate the states of two systems, one microscopic and one macroscopic, we shall detect interference of probabilities on the macroscopic level unless there is a superselection rule which effectively turns some macro states (correlated with pure micro states) into mixtures. Nor is there a conceptual absurdity in the idea of detecting probability interference at the macroscopic level; that is exactly the message of the Bell Inequalities.

The relevant experiments are in progress (see *Proofs and illustrations*). Almost every discussant has voiced the opinion that the interference of probabilities will be detected. If that is so, and if we generalize that result to all macro-observables, then the presence of the Projection Postulate (in the derivable form it has in the superselection theory of measurement) implies false empirical predictions. Even if the predictions agree, however, we learn from Leggett that this theory of measurement does not merely interpret the basic quantum mechanics plus Born Rule, but adds empirical content.

What we had hoped for was an interpretation which would allow the interpreted theory to stand regardless of the outcome of such experiments. If no such interference of probabilities is detected, fine; then we can say that we have found that macro states are separated by superselection rules. That is the sort of discovery that 'writes theory by experimental means', and we hope for much normal progress in science of this sort. But the interpreted theory should not run ahead of the experiments and design itself so as to be incapable of accommodating a contrary experimental result. This, it seems to me, is a fundamental indictment of the Projection Postulate.

All three problems are confronted directly in a new version of the present sort of solution, recently proposed by Jeffrey Bub (1988, 1989*b*, 1989*d*). Bub proposes to model macroscopic systems as quantum-mechanical systems with an infinite number of degrees of freedom. This does not just mean that one uses a Hilbert space of countable dimensionality—we already know that in any case, by embedding, whatever can be modelled in a finite-dimensional space can also be modelled in an infinite-dimensional space. No, Bub is speaking here of a direct sum of infinite-dimensional Hilbert spaces, each of which provides a different irreducible representation of the algebra of observables that pertain to the system. But each of these spaces can also be regarded as a coherent subspace, modelling one macro state of the system, separated by a superselection rule. One simple physical model Bub discusses is an infinite one-dimensional array of spin $\frac{1}{2}$ systems. Schroedinger's cat, or even an elephant, consists of less than 10^{25} particles. Quantum-mechanical systems of the sort discussed by Bub are asymptotic idealizations.

Is this acceptable as a solution? Bub notes that the superselection rule emerges naturally here, that the Hamiltonian is not an observable but the evolution is still unitary, and that such asymptotic idealizations have very familiar uses in classical physics. He adds, *à propos* Leggett, that he has not excluded macroscopic systems with finite degrees of freedom. All this is correct, and I shall not quarrel with the radical idealization proposed here, which regards macroscopic observables as emerging in collective behaviour only in the case where idealization shades off into fiction. An empiricist point of view would not be a good basis from which to start such a quarrel!

But I have a more fundamental misgiving with all the versions of the approach discussed in this section. For I see a disparity between this sort of solution and the crucial problem to which all this is addressed—which I can state no better than in Bub's words:

The point is that we must have determinateness *somewhere* in the theoretical scheme for the probabilities [in Born's Rule] to make sense. What we want is to relate the probabilities to a generalized 'counting' (in the measure-theoretic sense) over determinate possibilities.... Measurement results are determinate.... But our most fundamental physical theory is not simply about measurements—it is about the behavior of physical systems. (Bub 1989*d*, 135)

Exactly—and surely, quantum theory, literally understood, predicts also outcomes of micro processes in the ionosphere? The only predictions are via Born's probabilities for measurement outcomes. So if an interpretation of quantum mechanics resolutely pegs measurement interactions at special macroscopic processes alone, does it not say that quantum theory makes no predictions for what happens in micro processes in the ionosphere? It is one thing to point out that all *our* practically relevant expectations concern macroscopic phenomena; it is quite another thing to interpret the theory as attaching probabilities only to those phenomena at the anthropocentrically important level. The passage just quoted reveals the central foundational problem behind all discusions overtly focused on measurement in quantum mechanics: to display this theory's significant coherent probability assignments to determinate events.

Proofs and illustrations

After a few brief historical remarks, I shall describe the experiment proposed by Leggett.

It appears that in 1980 the superselection rule solution to the problem of measurement was developed independently by Beltrametti and Cassinelli (1981*a*) and Kay-Kong Wan (1980). The suggestion that there is no superposition between macroscopic states had been in the literature for some time, in various forms (e.g. Gottfried 1966; Cartwright 1974*b*). In Chapter 6, Section 8 I stated Wan's description of quantum systems with (Abelian) superselection rules. To this he added the postulate that time evolution is unitary. Wan explicitly discusses the point that the Hamiltonian is not an observable, in his solution. He cites parallel examples elsewhere: Dirac's Hamiltonian formulation of General Relativity, and the Gupta–Bleurer formulation of quantum electrodynamics. It is not clear to what extent we can be sanguine about this with respect to the measurement interaction, however.

The Josephson effect (Josephson 1962) is one in which the temporal evolution of a macroscopic variable can be controlled by a microscopic energy, on the order of the thermal energy of an atom at room temperature. The 'Josephson device' on which Leggett concentrates is an RF SQUID ring. This is a bulk superconducting ring interrupted by a single Josephson junction. The macroscopic dynamic variable is the total flux (circulating current) in the ring. The ring has available to it two degenerate states, corresponding to the current circulating in either the clockwise or the anti-clockwise direction. The magnitude of the current is on the order of a few microamperes—it is proposed that this be counted as macroscopic, because it is a current that could certainly have been studied in a classical experimental context, before the advent of quantum theory.

If a measurement is made to detect the direction of the current at any given time, one of two results obtains: clockwise (+1), or anti-clockwise (−1). The question can now be posed: what are the probabilities of finding +1 twice for measurements made at distinct times t, t'? Let the relevant observable be denoted as $P(t)$; then with the usual idealizations the calculated expectation value is

$$(1) \quad \langle P(t)P(t') \rangle = \cos[\Delta(t' - t)]$$

where Δ is the characteristic resonance frequency of the system. This is the expectation value if we regard the total system as a quantum-mechanical system not subject to superselection rules. The formula reveals a correlation between $P(t)$ and $P(t')$; e.g., if $t' - t$ is half a cycle, then the predicated value of $P(t')$, conditional on value $+1$ for $P(t)$, equals -1 with certainty. If however we add the supposition that at an intermediate time, e.g. at quarter-cycle, the state was a mixture of the ± 1 eigenstates, we obtain a different prediction.

It is more instructive to look at the general demonstration adopted from Bell's argument than to continue with these details. One form in which Bell's Inequalities can be stated, for expectation values of observables $P(i)$, $i = 1, 2, 3, 4$ which take only value ± 1, is the following:

$$(2) \quad \langle P(1)P(2) \rangle + \langle P(2)P(3) \rangle + \langle P(3)P(4) \rangle$$
$$- \langle P(1)P(4) \rangle \leq 2$$

but if we set the difference between successive times t, $t'(t' = 2, 3, 4, t = t' - 2)$ equal to $(\pi/4)\Delta$, and hence the interval between $t = 4$ and $t = 1$ equal to $(3\pi/4)\Delta$, formula (1) implies

$$(3) \quad \langle P(1)P(2) \rangle + \langle P(2)P(3) \rangle + \langle P(3)P(4) \rangle - \langle P(1)P(4) \rangle$$
$$= 3\cos(\pi/4) - \cos(3\pi/4)$$
$$= 3(1/\sqrt{2}) - (-1/\sqrt{2}) = 4/\sqrt{2} = 2\sqrt{2} > 2$$

a violation of (2), and hence of Bell's Inequalities.

Modal Interpretation of Quantum Mechanics[1]

On the one hand, it is a truism that quantum physics describes an indeterministic world. On the other, the quantum theory of an isolated system describes its state as evolving deterministically. How can the two be reconciled?

The question can be posed more generally: how could any theory be like that? In this form I discussed it already in Chapter 5. We found three alternatives; I shall list these here, and mention the interpretations of quantum mechanics which they cover. The first is that, in certain isolated systems, the state does after all not develop deterministically. We examined this alternative, in von Neumann's version, in the preceding chapter. The second denies not the determinism but the apparent indeterminism. It says that a measurement is not a process characterizable as the evolution of an isolated system, ever: a measurement is an interaction *incompletely described*, by leaving out something or other. There are various interpretations along this line. Perhaps the most radical is the idea that quantum theory was devised to describe only situations in which an observer (or at least, the measuring environment) is involved, while leaving that part out of the description. John Wheeler noted this as a major alternative; it is certainly reminiscent of some early Copenhagen texts.[2] Equally radical, but in a different direction, is Everett's many-worlds interpretation: indeterminism is an illusion, and disappears if we also describe all the worlds there are besides our own. The third alternative is to deny neither the determinism of the total system evolution nor the indeterminism of outcomes, but to say that the two are different aspects of the total situation. Specifically, we can deny the identification of value-attributions to observables with attributions of states; the state can then develop deterministically, with only statistical constraints on changes in the values of the observables. The modal interpretation takes this third road.[3]

This separation of states and observables was always left open in our general discussion; it was ruled out only by von Neumann's interpretation. To explain the modal interpretation I shall proceed in two ways. Sections 1–5 will be kept at an intuitive level. A more rigorous presentation will be found in Sections 6–7. The reader may choose to read these in either order, depending on his or her frustration with imprecision and abstraction respectively. The discussion is limited to systems with at most two components, throughout both these basic presentations. Sections 8 and 9 take up the major question of consistency of the interpretation extended to many-body systems.

1. THE MODAL INTERPRETATION

Von Neumann's interpretation tells us how to read attributions of values to observables: they are classifications of states, just as they were in modern science before our century:

1. Observable *B* has value *b* *if and only if* a *B*-measurement is certain to have outcome *b*.

This looks agreeably classical and right. But that in itself is suspicious here, for in the classical world-picture measuring was just looking, without disturbing, as if by a disembodied intelligence. It was also assumed there that for any observable, at any time, there is a *unique* value which is *certain* to be shown by a correct measurement. All this hangs together nicely (the values are there, and will be seen if one looks), but what happens to this picture when we admit states in which measurement has uncertain outcomes? Von Neumann gave a radical answer: if the outcome of a measurement of *B* is uncertain, then *B* has no value. That answer is tenable, but then the classical look of statement 1, which it had at first sight, was deceptive. It is not classical at all, for it says that unmeasured observables have *no* value, except in the special case of certainty. So the classical look of 1, being deceptive, can't very well be cited as a reason for accepting it.

Let us now step back and look at states and observables in

the larger perspective developed in Part I. The term 'state' is familiar from classical mechanics, thermodynamics, chemistry, engineering, and so on. Beginning with this classical background, we would assume that a state is given if we know

(*a*)   the value of all observables

and that, if we knew the state, then we would know all there is to know about

(*b*)   how the system will develop if left alone and how it will react if acted upon

where such action-upon includes measurement, of course. If it is assumed that observables simply have values, there to be seen if we look, and also that the temporal evolution is determined entirely by what those values are, then there is no need to keep (*a*) and (*b*) very separate. But the times have changed. So we must be cautious, and distinguish two concepts of state, one for each of the above two roles:

Value state: fully specified by stating which observables have values and what they are

Dynamic state: fully specified by stating how the system will develop if isolated, and how if acted upon in any definite, given fashion

Note that measurement is an interaction, so the prediction of measurement outcome probabilities belongs to the role of the dynamic state. The concept of dynamic state remains the primary one.

Another way to present this same conceptual distinction was to distinquish between two sorts of propositions:

Value-attributing proposition: e.g. $\langle m, E \rangle$ says that observable *m* actually has a value in *E*

State-attributing proposition: e.g. $[m, E]$ says that the state is such that a measurement of *m* must have an outcome in *E*

The connection between these presentations is of course that the value state is what makes value-attributions true, while the dynamic state similarly determines the truth-values of state-attributing propositions. (In general, there may be other state-attributing propositions besides those of form $[m,E]$; the dynamic state may carry other sorts of information; see the discussion of mixtures in *Proofs and illustrations*.) To reject von Neumann's interpretation rule is to reject the equivalence of value-attributing and state-attributing propositions. Because of the very tight logical structure we saw in the preceding chapter, everything will look a little different once we deny this. As preliminary to a closer look, let us just see roughly and intuitively what we could now say about measurement—specifically, von Neumann measurement.

I propose the following: to suppose that the outcome of a measurement of observable A pertaining to system X is b is to suppose that the pointer-observable *has* value b (at the time marking the end of that measurement). But we do not infer that X is in an eigenstate of A. Hence the pure state $\Sigma c_i |a_i\rangle \otimes |b_i\rangle$ of the combined system $X + Y$ is not the same as, and has not become, the corresponding mixture.

However, we do think of measurement outcomes as relevant to the question what state the system was in. This suggests that $[m,E]$ implies $\langle m,E \rangle$—if the state was such that m must have a value in E, then it does actually have a value in E. More generally, *from information about what values observables did have, we can*—by looking for a statistical fit—*infer backward to the state*. In general, such inference, being statistical, will require much more than a single measurement.[4] It will be possible to conclude, in this way, that a certain source or preparation procedure produces systems in state W—or to attribute state W to the systems in an ensemble of which we have inspected a sample—but not to determine that a single given system is in this state, otherwise. Finally, if $[m,E]$ were not true—if the state were not such that m must have a value in E—then it is still possible that m does actually have a value in E anyway. This is a true *transition from the possible to the actual*.

Referring back to Chapter 8, Section 2, we can now see that this allows us to think of quantum-mechanical measurement as having indeterminism of form 1 ('indeterministic output'):

$$\text{IN,} \; W \longrightarrow W', \; \text{OUT}$$

In this schema we can now say: IN and OUT are the initial and final values of observables (value states), while W and W' are the dynamic states. The evolution of W into W' is deterministic, in accordance with Schroedinger's equation above, and without acausal jumps or collapses. But W' only tells us what is the possible and probable character of OUT (including *some* necessities, of course) and does not fully determine it.

It may be noted that IN, on this picture, plays no direct predictive role. The character of the initial values of the observables could at best be a symptom or clue to what the initial state is. The expectation and indeed character of the future is determined, to the partial extent that it is determined at all, by the dynamic state W (of the *whole* system) alone. The value state must be allowed to change unpredictably, within the limits set by the dynamic state, in order to allow for indeterminism in the phenomena.

What then is the empirical significance of actual values of observables? They do not increase predictive power if added to a description of the concurrent dynamic state. In that sense they are 'empirically superfluous'. But taken by themselves, they do have predictive value, exactly because they are symptomatic of the dynamic state.

Proofs and illustrations

Let us think a bit more about states and observables and see if there are any additional reasons for accepting von Neumann's interpretation rule. A state can always be represented by a statistical operator—if only a projection operator, projecting along a specific vector—and that is a Hermitean operator. But Hermitean operators are meant to represent observables. Does it not follow then that *being in a state* is the same as, or in some strong sense equivalent to, *a certain observable's having a certain value*? For example, to be in pure state x, isn't that just the same as for observable I_x to have value 1?

But this just assumes the interpretation rule again. To be in state x is to be in an eigenstate, corresponding to eigenvalue 1, of observable I_x. By Born, that implies being such that a measurement of I_x is *certain* to have outcome 1. But the question at issue is whether the converse implication holds. Indeed, for the mixed states we must be still more careful if we want to spell out this state-observable correspondence: to be in state W is *not* the same as for observable W to have value 1. Consider

$$W_1 = \tfrac{1}{2} I_x + \tfrac{1}{2} I_y$$
$$W_2 = \tfrac{1}{4} I_x + \tfrac{3}{4} I_y$$

where $x \perp y$. Then the yes/no observables which have value 1 with certainty in these states are the same, namely those represented by a projection operator $I_{[x,y]}$. But the two mixed states are not the same. So the correspondence between states and observables is only this: a state is uniquely specified if and only if we give the expectation values of all observables in that state—equivalently, the probabilities of all measurement outcomes. Only for the pure states does that reduce to spelling out the certainties.

This point has been noted generally as a limitation to the quantum-logical approach initiated by von Neumann. That approach, which has been very valuable in foundational research, has also been extrapolated rather far by philosophers. Great caution is needed in such extrapolation, exactly because of what we noted above. In von Neumann's and Birkhoff's early writings on this, *propositions* are identified as projection operators or equivalently as subspaces. The way this is read is, in effect, via von Neumann's interpretation rule. Thus, the proposition $[M,E]$, represented by the subspace $S = [\{x : Mx = kx$ for some $k \in E\}]$, is read as 'M has a value in E'. That proposition is *satisfied* by a state which assigns 1 to S—i.e. a state W such that $Tr(WI_S) = 1$—so that one can indeed say: the proposition $[M,E]$ is satisfied if and only if M has a value in E. (That is, the reading fits the proposition, if one accepts von Neumann's interpretation rule.) Again, all this hangs nicely together, but it must be noted that above examples W_1 and W_2 of *distinct* mixed states *satisfy exactly the same* propositions, in this sense. Thus, the family of propositions, so conceived, is not rich enough to separate distinct mixed states.

To sum up, we see an intimate connection between states and observables, but it is not intimate enough to force von Neumann's cutting of the Gordian knot. Indeed, we have found some reason against that move. States can be identified *in terms of* observables, but cannot be identified *with* them.

2. THE MODAL ACCOUNT DEVELOPED

One of the chief difficulties in the epistemology of quantum mechanics is its apparent inadequacy for describing events. The fact that there are systems which do not admit dispersion free states leads to the inevitable and irreducible probability statements concerning certain events. Such events may be the measurements associated with yes-no experiments. ... The individual occurrence of such phenomena is then completely outside the scope of the theory; only the probabilities for such events can be accounted for in our description of the state. (Jauch 1968, 173)

I will not pretend that Jauch would have endorsed the modal interpretation. Yet the above words seem to express the modal alternative exactly. They say that a *state*, which is in the scope of quantum mechanics, gives us only probabilities for actual occurrence of *events* which are outside that scope. They can't be entirely outside the scope, since the events are surely described if they are assigned probabilities; but at least they are not the same things as the states which assign them probability.

In other words, the state delimits what can and cannot occur, and how likely it is—it delimits possibility, impossibility, and probability of occurrence—but does not say what actually occurs. The transition from the possible to the actual is not a transition *of* state, but a transition *described by* the state.

We know how states are described and represented in quantum mechanics—what about the events in question? They are described by statements of form

1. Observable B pertaining to system X (actually) has value b

and the Born interpretation rule sometimes tells us the probability of this event actually occurring (this statement actually being true) at the end of an appropriate measurement. This requires, as we noted in the preceding section, that such a statement as 1 *cannot* be equated with *any* of the form

2. The system is in a state of type . . .

which has specific forms like

2(*a*). The system is in state W

2(*b*). The system is in some pure state in subspace S

2(c). The system is in some mixture of states $x_1, \ldots, x_n$

2(d). The system is in some state W such that $Tr(WI_S) = 1$

For each of 2(a)–2(d) can at best give us a probability of statement 1 being true.

But there must also be some logical connection between value-attributions (form 1) and state-attributions (form 2). What are they? Recall that, at the end of a von Neumann measurement of non-degenerate observable A on system X by apparatus Y with pointer-observable B, we have the following description of total and reduced states:

(a) X is in state $W_X = \Sigma c_i^2 I_{|a(i)\rangle}$
(b) Y is in state $W_Y = \Sigma c_i^2 I_{|b(i)\rangle}$
(c) $X + Y$ is in state $\varphi_t = \Sigma c_i |a_i\rangle \otimes |b_i\rangle$

Now the Born Rule, given our proposed modal alternative to von Neumann's interpretation rule, leads here to

(d) For some index k, pointer-observable B has value b_k; and the probability that this index k is the index i, equals c_i^2.

It is clear that (d) only establishes a link between the state at the end of an A-measurement and the statements attributing values to B. The obvious question is: could this simply be a consequence of a larger principle which connects states more generally with values of observables?

The interpretational question facing us is exactly: in general, which value-attributions are true? The response to this question can be very conservative or very liberal. Both court later puzzles. I take it that the Copenhagen interpretation—really, a roughly correlated set of attitudes expressed by members of the Copenhagen school, and not a precise interpretation—introduced great conservatism in this respect. Copenhagen scientists appeared to doubt or deny that observables even have values, unless their state forces us to say so. I shall accordingly refer to the following very cautious answer as the *Copenhagen variant* of the modal interpretation. It is the variant I prefer.

This interpretation says that, if system X has dynamic state W at t, then the state-attributions $[M, E]$ which are true are those such that $Tr(WI_E^m) = 1$. About the value-attributions, it says that they cannot be deduced from the dynamic state, but are constrained in three ways:

(i) If $[M,E]$ is true then so is the value-attribution $\langle M,E \rangle$: observable M has value in E:

(ii) All the true value-attributions could have Born probability 1 together:

(iii) The set of true value-attributions is maximal with respect to feature (ii).

Let us use the following terminology: *W makes* $[M,E]$ *true* exactly if $Tr(WI_E^M) = 1$. Then clauses (ii) and (iii) tell us that we already have a 'bookkeeping device' to identify the set of true value-attributions. Call this set S. Then (ii) tells us there must be dynamic state W' such that $\langle M,E \rangle$ is in S only if W' makes $[M,E]$ true. Adding (iii), we see that W' is pure, that $\langle M,E \rangle$ is in S if and only if W' makes $[M,E]$ true, and that W' is unique. Finally, (i) tells us that W' is, in the terminology of Chapter 6, possible relative to W.

That dynamic state W' is the bookkeeping device which identifies the true value-attributions correctly. Hence it can be used to *represent* the value state. That does not mean that value states are dynamic states, but only that each admits the same sort of mathematical representation. I will call this pure state W' the value state of X as true t, but that *does not imply* that X really has dynamic state W'. No, it has dynamic state W, as I said to begin with. We can sum all this up in a single postulate, which describes our family of models for physical situations governed by quantum mechanics:

(e) Given that system X is in state W at time t, then for all observables M pertaining to X:

    ($e1$) a state-attribution $[M,E]$ is true if and only if W makes it true;

    ($e2$) there is a certain pure state x which is possible relative to W, and the value-attribution $\langle M,E \rangle$ is true if and only if x makes $[M,E]$ true.

The first consequence we deduce from this postulate is that

(i)–(iii) hold, as required. The second consequence concerns the identification of observables. My terminology and notation 'Hermitean operator M represents observable m' allow for the possibility that two distinct observables can be represented by the same operator. If that could happen, principle (e) above would have to be read as applying to all observables represented by given operator M. But so read, it implies that, if two observables are represented by M, it is not only their measurement outcome probabilities that are always the same, but also their values. This is because of the 'if and only if' in ($e2$). Hence there is then no difference at all between these observables; they are the same. In other words, implicit in this interpretation (Copenhagen variant of the modal interpretation) is the following principle:

> *Identity of Observables*: If observables m and m' are represented by the same Hermitean operator, then $m = m'$.

This answers a question raised at the beginning of Chapter 6. It is a principle that plays a crucial role in the 'no hidden variables' theorems, a point that will be discussed further in Chapter 10. It is not a tautology, of course; indeed, it may be regarded as a peculiarly empiricist constraint upon interpretation. It does give us the advantage that we may use the same name now for an observable and its representing operator.[5]

Thirdly, we can point out that, in a certain sense, *it is as if the ignorance interpretation of mixtures were correct*. For if system X is in mixed state W, then the actual values of observables pertaining to X are exactly those it would have had if it had been in a pure state in the image space of W. But we don't know which pure state—all of them are possibilities for us, if we are told only that X is in mixed state W. We remember of course from Chapter 7 that the ignorance interpretation is not tenable, for a number of reasons, but perhaps this helps to explain its intuitive appeal. For some of what it entails for values of observables is right.

The fourth consequence is the rejection of the Classical Principle, that each observable always has one of its possible sharp values. We can derivatively attribute 'unsharp' values. To observable M there correspond a large set of observables I_E^M, as we know, having eigenvalues 1 and 0 only. Now we have the equation

$$P_W^M(E) = 1 \text{ if and only if } Tr(WI_E^M) = 1$$

Therefore we have the following result concerning state-attributing propositions:

$$[M,E] = [I_E^M,1]$$

where I abbreviate '$[\ldots,\{\ldots\}]$' by dropping the set parentheses. The Copenhagen variant of the modal interpretation entails now, because of its adoption of $(e2)$—or, equivalently, of (i)–(iii)—that, similarly,

$$\langle M,E \rangle = \langle I_E^M,1 \rangle$$

That also means that $\langle M,E \rangle$ is *not* the classical disjunction of the value attributions $\langle M,r \rangle : r \in E$. Indeed, we should note:

If $\langle I_E^M,1 \rangle$ is true if and only if Borel set $E_0 \subseteq E$, then E_0 is also the smallest Borel set such that $\langle M,E_0 \rangle$ is true

and we should say that M has *unsharp value* E_0. If M and M' are incompatible observables, which have no eigenvectors in common, and $\langle M',s \rangle$ is true, then $\langle M,r \rangle$ is not true for any value r. Yet $\langle M,R \rangle$ is still true, because it just means $\langle I_R,1 \rangle$ is true, and that is a tautology (R the set of all real numbers.) We can therefore distinguish, for value-attributing propositions, the principle of *Excluded Middle* ($\langle M,R \rangle$ is true for every observable M), which is correct, and that of *Bivalence* (either $\langle M,E \rangle$ is true or $\langle M,R - E \rangle$ is true), which is false.

A more classical looking 'anti-Copenhagen' variant would replace $(e2)$ with the assertion that for each M there is some value r such that $\langle M,r \rangle$ is true. That is logically tenable but has very curious features—features which, after acquaintance with the Copenhagen way of thought, seem much more curious than value gaps. I shall not discuss the anti-Copenhagen variant any further here, but shall return to it briefly in Chapter 10.

3. WHAT HAPPENS IN A MEASUREMENT?

There are actually two questions about measurement when it comes to interpretation. Many questions about what happens have been settled by the foundational research on which Chapter 7 reports. We have seen there that, if a physical process is to be a candidate for designation as a measurement at all, it must satisfy certain requirements. And we have also seen what implications those requirements have inside quantum theory. But it remains to ask: (*a*) what is the significance of measurement? and (*b*) what is really going on in a measurement? In the next section I shall broach the second question; here I shall stick with the first.

One recurring worry among philosophers is that the appearance of the term 'measurement' in the Born Rule bears its anthropocentric connotations essentially. That would mean that we cannot think of quantum theory as a putative autonomous description of the world in neutral physical terms and prospectively complete. In the jargon: if that were so, we could not be realists with respect to the theory, but only instrumentalists. This worry is much reinforced by 'philosophical' discussions by some of the great physicists who were involved in the development of quantum theory. I hope that the discussion in Chapter 7 has already laid this question to rest, since the requirements upon physical correlates of measurement involved no reference to us, to persons or consciousness, and not even to the macro–micro distinction.

But then, if that worrisome idea about the significance of measurement is gone, what *is* the significance? I believe it is the following. Every interpretation of quantum mechanics begins qualitatively, by speaking out on whether an observable always (sometimes, never) has sharp values, while the system is (is not) in one of its eigenstates, and so forth. But then must come a next step: assignment of probabilities. Human acts being nothing very special among the physical categories, the probabilities announced when Stern and Gerlach, or Aspect, or Compton, or Leggett designs an experiment to be actually carried out in a laboratory must be a special case. Fine; but how do we draw out consistent, self-coherent probabilities in general? Some particle in the ionosphere is capable of various spin states; is there a definite probability, given its physical past and circumstances, that the value of its spin along the (earth centre)–(North Star centre) axis equals 1? The *given* had better be handled very delicately, or else we'll give such implicitly incoherent answers as were derived from the naive ignorance interpretation.

Officially, quantum mechanics allows for only one way to assign probabilities—via Born's Rule. We can *interpret* this to extend to all processes, even on the microscopic level in the ionosphere, which meet the minimal requirements for a physical correlate of measurement. Can we go a little further? Von Neumann went a lot further. Can we go a little way along with him, exploiting for example the correlations set up in those special von Neumann, or von Neumann–Lueders, measurements? The answer is *yes*, but we must proceed as if barefooted in a field strewn with glass. This problem—the general problem of drawing a consistent general recipe for probability assignments from quantum theory—is what is *really* behind the seemingly disproportionate attention to measurement.

To spell out in detail what happens in measurement, and how the Born probabilities as interpreted in (*d*) are in accordance with the large principle (*e*), we must become a little more precise. The situation of system X at given time t is characterized according to (*e*) by two states: its *dynamic state W* and its *value state x*. What does this look like when X is a compound system? We need to characterize the situation for it and *also* for its components. So if $Z = X + Y$, we have both a dynamic state and a value state for each of Z, X, Y. I shall here discuss only the case in which Z has a pure dynamical state φ. Let us designate the mixed states assigned to X and Y, by 'reduction of the density-matrix' as $\#\varphi$ and $\varphi\#$. Then the situation is this:

(*f*) $(X + Y)$, X, and Y have dynamic states
   φ, $\#\varphi$, $\varphi\#$ respectively

(*g*) $(X + Y)$, X, and Y have as value states
   φ, x, y respectively, where x is possible relative to $\#\varphi$
   and y possible relative to $\varphi\#$.

We cannot specify what x and y are: there are a number of possibilities. We only know that they must be possible relative to $\#\varphi$ and $\varphi\#$. This means mathematically that they are vectors in the image spaces of $\#\varphi$ and $\varphi\#$.

What we would like next is a probability distribution on these possibilities. The non-unique decomposability of mixtures stands in the way of a general rule for probability assignments which is at once simple and consistent.[6] But the Born Rule is meant to assign probabilities coherently if the situation comes *at the end of a measurement*. The rule presupposes that measurement is a process which is so structured that it singles out certain observables, and tells us the probabilities for their possible values—without the inconsistencies that plague the ignorance interpretation. That means of course that the observables singled out are mutually compatible.

The major recognized classes of measurement in the literature satisfy the metacriterion which was suggested by this thought. When that metacriterion is met, reasoning of the following form:

1. Process PP was a measurement of observable A with B designated as pointer-observable, and also an A'-measurement with B' designated as pointer-observables, and . . .

2. The probability that B had value a at the end equals p_a, for $a = a_1, a_2, \ldots$; the probability that B' had value a' at the end equals $p_{a'}$, for $a' = a'_1, a'_2, \ldots$; . . .

in which statement 2 follows from 1 by Born's Rule, never leads to an incoherent probability assignment.

Well, how could it anyway? Easily enough, if we respect the functional relations among observables. That is the message of the 'no hidden variables' theorems. If the above sort of reasoning is continued with, i.e.

3. $B' = f(B)$ and $B'' = g(B)$ and . . . So the probability that B' had a value in E equals the probability that B had a value in $f^{-1}(E)$, the probability that B'' had a value in F equals the probability that B had a value in $g^{-1}(F)$, and . . .

then those theorems tell us that 1–3 will lead to incoherence (i.e. to inconsistency with the classical probability calculus) unless all those observables are mutually compatible.

Obviously, these reflections can be met with different responses. Some may object to the use of classical logic on which I relied, others may make light of the classical probability calculus. But both folklore and foundational literature honour overwhelmingly the conviction that the mere description of the outcomes of (joint) measurements does not, as such, violate classical logic or probability. In Chapters 4 and 5 we saw that, though deterministic or more generally causal underpinnings cannot be postulated, the surface state in an experimental situation is classical in the sense of logic and probability theory. Since this very same surface state clearly violated the Bell Inequalities, it can still be very non-classical in the sense that it *requires* (something like) quantum-mechanical modelling. But then, there is no reason why these very same surface states could not represent micro situations in nature just as well.

The metacriterion in question was that, if a process measures various observables on an object system jointly, then it ends with the apparatus in a mixture of joint eigenstates of the pointer-observables. It is to these, in the first instance, that the Born Rule assigns probabilities. As usual, the paradigm illustration is the von Neumann measurement of an observable with non-degenerate spectrum. We can consistently add to our previous principles:

(h)   If the situation described in (f) and (g) is at the end of an A-measurement with B designated as pointer-observable, and with $\varphi = \Sigma c_i |a_i\rangle \otimes |b_i\rangle$, then the probability is c_k^2 that $y = |b_k\rangle$

This has Born's Rule translated into our present representation of this physical situation, as corollary, since $y = |b_k\rangle$ exactly if B has value b_k. It is consistent with the preceding because x and y do, as they must, lie in the image spaces of $\#\varphi$ and $\varphi\#$. But of course, a von Neumann measurement has a special feature. It does not just allow statistical inference backward to the initial state of the object system. In addition, it effects a perfect correlation between measurement and pointer-observables. Hence we can go one step further, and add:

(h')   If the situation described in (f) and (g) is at the end of an A-measurement with B designated as pointer-observable, and with $\varphi = \Sigma c_i |a_i\rangle \otimes |b_i\rangle$, then the probability equals c_k^2 that *both* $y = |b_k\rangle$ *and* $x = |a_k\rangle$.

In the *Proofs and illustrations* I will discuss the similar principle for the more general class of von Neumann–Lueders measurements. I will also look more closely there at the obvious query about consistency.

The extension of (h) to (h') is important, because it leads immediately to the result that a pointer-reading result b_k implies with certainty that observable A had value a_k at that point:

(i) If the situation described in (g) and (h) is the end of an A-measurement with $\varphi = \Sigma c_i |a_i\rangle \otimes |b_i\rangle$, then the probability that value-attribution $\langle A, a_k \rangle$ is true, given that value-attribution $\langle B, b_k \rangle$ is true, equals 1.

The proof of this is simple. The probabilities of the combinations $(x = |a_i\rangle$ and $y = |b_i\rangle)$ sum to 1, therefore the probability of any combination $(x = |a_i\rangle$ and $y = |b_j\rangle)$ with $i \neq j$ must equal 0. Notice also, however, that a possibility which has probability 0 may still really occur. Probability 0 does not imply impossibility. The addition of (h) and (h') to our interpretation assigns probabilities; it does nothing else.

We are tempted to exclaim: *it is as if the Projection Postulate were correct*. For at the end of the measurement of A on system X, it is indeed true that A has the actual value which is the measurement outcome. But, of course, the Projection Postulate is not really correct: there has been a transition from possible to actual value, so what it entailed about values of observables is correct, but that is all. There has been no acausal state transition.

Proofs and illustrations

In general, a measurement need not effect a correlation, and the probabilities are only assigned to the pointer-observable values. In such a case, there should be no statistical inference concerning the values of observables on the object system. Even in the von Neumann case, one might worry about consistency. In Chapter 7 I took up Jon Dorling's query whether we might not have a case of dual von Neumann measurement: X measures B on Y, with pointer-observable A, while Y measures A' on X with pointer-observable B'—and A is incompatible with A'. We found that that is in fact impossible, given the general character of von Neumann measurement. This result carries over to the class of von Neumann–Lueders measurements.

There is another way in which inconsistency could result.

Imagine that, in one and the same process, Y measures A on X *and* Z measures B on X, with the measurements ending at the same time.[7] Then the inferences to the value state of X via principle (i), applied to both measurements, entail that this value state is a joint eigenstate of A and B. Will that necessarily be so?

The total system $(Z + X + Y)$ can equally be described as $X + (Y + Z)$. In the envisaged case, at the initial time, Y and Z are in their pure groundstates y and z, and end up in mixtures of their indicator states y_i and z_j. Because of the correlations, we see:

$$X + Y \text{ has final state } \Sigma c_i^2 I_{|a(i)\rangle \otimes |y(i)\rangle}$$

$$X + Z \text{ has final state } \Sigma d_j^2 I_{|b(j)\rangle \otimes |z(j)\rangle}$$

We do not know prima facie that $Y + Z$ is a measuring instrument for any observable pertaining to X. Specifically, we do not know prima facie that $Y + Z$ ends up in a mixture of states $y_i \otimes z_j$. But the same reasoning we used before removes this spectre of inconsistency. We do know from the above that

$$X \text{ has final state } T' = \sum c_i^2 I_{|a(i)\rangle} = \sum d_j^2 I_{|b(j)\rangle}$$

The coefficients were arbitrary, so let us take it that all c_i^2 are distinct. The relevant evolution operator, subject to the requirements for measurements, must act the same way, whatever the coefficients are. But in that case, the orthogonal decomposition of T' is unique, and therefore $\{|a_i\rangle\} = \{|b_j\rangle\}$. Cases of 'accidental degeneracy' do not bother us, since the probabilities are assigned *only* conditional on measurement, and that is a condition not on the particular episode, but on the type of evolution.

All of this carries over to measurement in general, *mutatis mutandis*. If the measurement is not of the von Neumann–Lueders class, then we have the analogue of principle (h) *but not* the analogue of (h'). Inferences to what the measured object system is or was like are only of a statistical sort, to its initial state. In the general Lueders case, the observable $A = \Sigma a I_a^A$ that is measured may be degenerate, and so may the pointer-observable $B = \Sigma a I_a^B$. The evolution $U: x \otimes y \longrightarrow U(x \otimes y) = \varphi$ is such however that there is a strong correlation. Using the notation '$W[A(a)]$' as short for 'the conditionalization of W on the a-eigenspace of A',

$(1a)$ $(I_a^A \otimes I_b^B)\varphi$ = the null vector if $a \neq b$

$(1b)$ $\varphi\# = \Sigma Tr(I_a^A Ix)\varphi\#[B(a)]$

$(1c)$ $\#\varphi = \Sigma Tr(I_a^A Ix)\#\varphi[A(a)]$

Here $Tr(I_a^A I_x)$ is the probability at once that A has value a on the measured system, at the end, and that pointer-observable B has value a at that time.

4. PUZZLE: HOW FAR DOES HOLISM GO?

Every interpretation of quantum mechanics has some features that bespeak the distinctively non-classical character of this theory. Adjusting 'the story' might make it sound deceptively classical, but can never remove its air of paradox altogether.

In the modal interpretation, Copenhagen variant, there are three especially puzzling features which make its world-picture truly non-classical. One we saw already; it is the possibility of observables taking unsharp values, the failure of Bivalence. Another strange possibility concerns what happens when no measurement is going on, and I shall discuss that in the next section. The one I want to discuss now is the way in which the holism of quantum states—the same holism already found in classical probability, as we saw in Chapter 3—extends also to observables and their values.

One of the virtues of our present interpretation, which I recounted in Section 2, was the principle I called Identity of Observables. It says that, if observables are represented by the same Hermitean operator, they are one and the same observable. This could be seen by looking at the model of a possible situation: since the value state is represented by means of a vector in Hilbert space, and an observable has a value only if that vector is one of its eigenvalues, there can't be any more to an observable than that. Given that we were looking at models of situations involving a single system (treated as a unit), we can restate this principle in the following way:

1. *Identity of Observables* (second version): If the probabilities for measurement outcomes for observables A and B are the same for every dynamic state of any system to which both A and B pertain, then $A = B$.

The reference to physical systems could be replaced here: whether or not an observable pertains to a given system depends solely on its (dynamic) state-space, so we could have referred to that instead. I am being very careful here for reasons that will soon be clear.

Turning now to a reflection on holism, we need to ask what statement 1 entails for compound systems. The reduced state $\#W$ assigned to compound X of system $X + Y$ in state W is determined *by definition* through the equation

$\#W$ is the state such that $Tr(A\#W) = Tr((A \otimes I)W)$ for all observables A pertaining to X.

Let H_X and H_Y be the Hilbert spaces for X and Y respectively, so $H_X \otimes H_Y$ is the Hilbert space for $(X + Y)$. The quantification 'for all A' is therefore over all the Hermitean operators on H_X, and I is the identity operator in H_Y. How does principle 1 apply to A and $A \otimes I$? We tend to think of these two observables as 'essentially' the same, since we feel it is our choice whether to model X's behaviour by itself or as part of a larger system.

But 1 does not apply to A and $A \otimes I$ at all, since they do not pertain to the same system. There is certainly a very intimate relation between them, in the Born probabilities:

The probabilities of measurement outcomes for A and for $A \otimes I$, conditional on their measurement on any system X and any system $X + Y$ respectively, are the same.

But that is not the same equivalence relationship as discussed in principle 1. The importance of this point appears in the more general context of how joint probabilities for A and B, pertaining to X and Y, are related to those for $A \otimes B$. The first thing to notice is that there is in general not a one-to-one correspondence between the outcomes:

$$(A \otimes B)(|a\rangle \otimes |b\rangle) = ab(|a\rangle \otimes |b\rangle)$$

for *ab* does not in general stand in one-to-one correspondence to the pair $\langle a,b \rangle$. Nor do the predictions of measurement outcomes taken separately relate very closely. The probabilities of outcomes a' and b', for measurements on A and B on the two components of $X + Y$ in state $\Sigma c_{ab} |a\rangle \otimes |b\rangle$, do *not* tell us what the probability of outcome $a'b'$ for a measurement of $A \otimes B$ is—not even if the correspondence $\langle a',b' \rangle \longleftrightarrow a'b'$ is unique (i.e. not even if $ab = a'b'$ only if $a = a'$ and $b = b'$). For the cross-reference in the coefficients c_{ab} determines the correlation, which cannot be read off from the two marginal probabilities. Quite independent of the details of the modal interpretation, we must therefore beware of fudging the distinctions between such observables as A,B on the one hand, and $A \otimes B$, $A \otimes I$, etc., on the other.

To see the holism of states reappear as holism of values, consider the question:

2.   Is there an observable $A \& B$, definable by the equivalence: $A \& B$ has value a_{km} if and only if A has value a_k and B has value b_m?

Here $k,m \longrightarrow a_{km}$ must at least be a unique correspondence; the probative case arises when A and B pertain to different systems X and Y, so that $A \otimes B$ pertains to $X + Y$. For the latter, we have

3.   The probability of an outcome in E for a measurement of $A \otimes B$ equals the probability of outcomes a_k and b_m such that a_{km} is in E, for joint measurement of A and of B.

Putting 1 and 3 together, we must conclude that, if there is such an observable as $A \& B$ then $A \& B = A \otimes B$. This shows that there is no such observable at all! For in the case in which $X + Y$ is in correlated pure state $\varphi = \Sigma c_i |a_i\rangle \otimes |b_i\rangle$, for example, it is *quite possible* that A and B have sharp values. But since $X + Y$ has a pure dynamic state, its value state is the same. Hence if several c_i^2 are between 0 and 1 then $A \otimes B$ has no sharp value. Question 2 must therefore be answered: in general, *no!* Of course there is no similar obstacle to the principle that $A \otimes B$ has value a_{km} if and only if $A \otimes I$ has value a_k and $I \otimes B$ has value b_m; for in our example, none of the three has a sharp value at all.

However logical this is, I fear that the reader's mind may now boggle. How could A and $A \otimes I$ have such different values? But we are not being asked to imagine a strange experimental outcome. When we translate out puzzlement into a truly empirical question, it disappears. To see this, suppose that A and $A \otimes I$ are both measured on systems X and $(X + Y)$ respectively. This entails that they are part of a large system $(X + Y + Z + W + V + \ldots)$. It may be that Z measures A on X and W measures $A \otimes I$ in $(X + Y)$, or perhaps Y itself measures A on X. The question is: will the outcomes of the measurements agree? To get a probability for that, we must assume that a further measurement is being made, by V say, on that state of $(X + Y + Z + W)$ to see if there is such agreement. My assertion is that the theory predicts that this last measurement will have outcome *yes* with probability 1.

To keep the discussion simple, suppose that X is a particle in state $x = (1\sqrt{2})(|+\rangle_A + |-\rangle_A)$, that Y is an A-measurement apparatus with groundstate y and pointer observable B:

$$|+\rangle_A \otimes y \longrightarrow |+\rangle_A \otimes |+\rangle_B$$
$$|-\rangle_A \otimes y \longrightarrow |-\rangle_A \otimes |-\rangle_B$$

and that Z is a measurement apparatus with groundstate z and pointer-observable C. The eigenvalues of C are 1 to register agreement and 0 to register disagreement:

$$|i\rangle_A \otimes |j\rangle_B \otimes z \longrightarrow |i\rangle_A \otimes |j\rangle_B \otimes |\delta_{ij}\rangle_C$$

and of course in all cases the evolution is linear. Now we see what happened:

$$
\begin{aligned}
x \otimes y \otimes z &\longrightarrow (1/\sqrt{2})[|+\rangle_A \otimes |+\rangle_B + |-\rangle_A \otimes |-\rangle_B] \otimes z \\
&= (1/\sqrt{2})[|+\rangle_A \otimes |+\rangle_B \otimes z + |-\rangle_A \otimes |-\rangle_B \otimes z] \\
&\longrightarrow (1/\sqrt{2})[|+\rangle_A \otimes |+\rangle_B \otimes |1\rangle_C \\
&\quad + |-\rangle_A \otimes |-\rangle_B \otimes |1\rangle_C] \\
&= (1/\sqrt{2})[|+\rangle_A \otimes |+\rangle_B + |-\rangle_A \otimes |-\rangle_B] \otimes |1\rangle_C
\end{aligned}
$$

in other words, the answer will be *yes*, *agreement* with probability 1.

The strange holism which allows A and $A \otimes I$ to take different values in our models therefore 'shows up' only outside measurement contexts—in other words, it has probability 0 of ever really showing up. That is why we can in practice ignore the difference between A and $A \otimes I$.

5. PUZZLE: IS THERE CHAOS BEHIND THE REGULARITIES?

Quantum mechanics gives us probabilities only for measurement outcomes, conditional on the hypothesis that a measurement is made. That is very stingy. It has also to philosophers a very empiricist or even positivist sound, but that is deceptive. At least on our present interpretation, any process at all is a measurement if it satisfies certain purely physical conditions, which make no reference to conscious observers, and need not be macroscopic. But we are still left with the question: what can happen when no measurement is being made?

The answer on our present interpretation is: *anything is possible*. Of course, 'anything' does not mean anything logically conceivable; for, according to the distinctive Copenhagen principles laid down above, no value-attributing propositions *can* be true together unless they can be certain together (in the sense of receiving Born probability 1). This is a very weak constraint on what actually happens, and it still allows things to happen in actuality which have *zero* probability of ever appearing as measurement outcomes. That is partly because zero probability is not strict impossibility, and partly because something which has Born probability 1 under *some* circumstances need not have Born probability 1 under any circumstances realized in the particular system or process you are considering. That is because, in a given particular case, the dynamic state may have correlations in it which are not to be found in all dynamic states of which the system is capable. I think we should compare this carefully with what could happen in a classical statistical theory. The probabilities it gives us are conditional on measurement; and while that is not to be construed anthropocentrically, it is *still* limited. What is more, the probabilities evolve—embodied in the quantum-mechanical dynamic states—without 'feedback' from the actual values.

Here is a classical example of evolving probability. I toss a die repeatedly; the die is fair to begin, but it is so constructed that with each toss it becomes more biased in a certain way. Let $p[n,i]$ be the probability that face 'i' comes up in trial n. Suppose $p[1,i] = \frac{1}{6}$ for $i = 1, \ldots, 6$, but

$$p[(n + 1), i] = ip[n,i]/T(n)$$

where $T(n)$ is the needed normalization factor (so that the sum of probabilities remains equal to 1). Now we have two descriptions of this world: *first*, the deterministic development of that probability distribution, and *second*, the actual series of toss outcomes. Logically speaking, the latter can be any sequence of integers between 1 and 6. But we note to ourselves that this class of possible outcome sequences has an *induced* probability measure—roughly speaking, the higher-numbered faces become continually more likely to turn up. The 'mechanics' however resides solely in the first part of the description, the temporal evolution of the probabilities.

If we study the temporal evolution of a wave function of a compound system in quantum mechanics, certain episodes qualify as the measurement of some observable on one part of the system by another part. There the Born Rule gives us probabilities for events. The total history of the system then consists of two parts: (*a*) the aforementioned, temporally evolving, quantum-mechanical state, and (*b*) the sequence of actual outcome events. The latter could logically speaking be any sequence of possible measured values of the observables in question—but, of course, we have an induced probability measure on those sequences. So far we have a parallel.

In my classical example it was clearly implied that the evolution of the probabilities was independent of the actual outcomes. What if the die had been so constructed that each time a face came up it was more likely to come up again later? Then probability-evolution and outcome-sequence would be interdependent. Should we not think analogously that in quantum physics the process of interaction designated as measurement *changes* the probabilities? What, for instance, if I am the measurement apparatus, and I decide beforehand that if I detect a red light I shall shoot the piano player? Does it not follow then that the occurrence of an actual value (and not just the total quantum-mechanical state) affects the probability of what happens next?

There is a fallacy here. The interaction affects the measured system if we think of it *by itself*, apart from the total isolated system of which it is part. But if we have before us the description of the total evolution of the state of the entire system, then the description of some internal episode *as* a measurement *adds nothing*—it is simply a classification. If we want to reflect on, for example, von Neumann's graphic 'immediate repetition' demand, we do it as follows. We ask: what about the total evolution of a system in which a certain episode qualifies as an immediate succession of two measurements of the same observable, plus a measurement which checks on whether the two outcomes agreed? What are the probabilistic predictions for outcomes of *that*? To get the answer, we look again to the wave function of a total system in which that happens, and the Born Rule allows us to deduce: we can expect the outcome *the two agreed!* with probabilistic certainty.

It is exactly at this point that our intuitions tend to declare war on our deductions. 'We don't want to know', they insist, 'what probabilities are derived for the outcome of a third measurement, if performed—we want to know purely and simply whether, if the first two measurements occur, their outcomes do agree!'

But this question has a presupposition: namely that quantum mechanics, if it is to be complete and accurate, should give us such information. In other words, it is assumed that the theory must give us more than the probabilities of measurement outcomes. And this is an assumption that does not come from physics, but from somewhere outside. If it had some independent justification of its own, it would be a desideratum which might be imposed as a criterion of adequacy for interpretation. But it does not.

This war of intuitions lies very deep, and appears nowhere so clearly as in the thought-experiments in which we imagine ourselves involved as observers and manipulators. In this area even subtle thinkers leave philosophical clarity behind, and regress to a primitive mind–body dualism or naive materialism.

In *any* supposed total description of the universe and its entire history (let us call that a Universal Story), our actions and observations must appear no different from other processes. It is that initial supposition of having a Universal Story which entails that their designation as free actions or conscious observations can only be a *classification* of certain episodes already entirely encompassed by other supposed descriptions. If I then additionally suppose myself to be somehow outside the process and capable of reporting directly on the state of the system—as if I had an 'ontological telescope'—I am supposing that there is something not encompassed in the Universal Story. But by supposition it encompasses everything! The result is incoherence.

Here is a more down-to-earth issue. If the measurement outcomes' probabilities are set by the evolving state but the outcomes do not provide a 'feedback' input to that evolution, should we not expect a very chaotic life? Imagine the following measurement (observation) outcomes in everyday life. Looking out of my window I see two cars collide; next the drivers emerge gesticulating towards each other; a crowd gathers; a police car shows up and stops at the accident. If at each moment the probabilities for possible events were given by the total state at that time, shouldn't we expect to see very often some 'unconnected series'? For example, a car stops askew in the middle of the intersection; a police car appears; a crowd gathers; and then another car, whose speed has been fluctuating noticeably, hits the first one. Stranger sequences yet can be imagined. Also, in the case of Schroedinger's Cat, the account I gave above is logically comparable with the cat being sometimes alive and sometimes dead, between the triggering of the device and the opening of the box.

This is only a more complicated version of earlier points about repeated measurement. Here we have a correlation over time; we could ask a similar question about a simultaneous correlation. In Aspect's experiment, each photon has a 50 per cent probability of passing if the filters' orientation is parallel: why should there be a correlation observed? The answer is the same to both questions: given the total state of the system, the uncorrelated outcome combinations are given a zero or negligible probability.[8]

Proofs and illustrations

There are important differences between the modal interpretation, Ballentine's ensemble interpretation, and Everett's many-worlds interpretation. The latter says that all possible measurement results are real—the world branches into many real worlds at that point, each associated with one of those possible outcomes. As a result, indeterminism is a matter of appearance only. Reality did not choose among alternatives, but embraced all of them. We are invited to classify our memory and opinion about 'what happened' as 'belonging' to one of the alternative branches.

The first difference is therefore philosophical only: the modal interpretation says that only one of the alternative outcomes is real. But in addition, the alternative possible system histories are identified differently. In the modal interpretation, if the system as a whole is isolated, and begins in a pure state, then its state remains pure, and the values of observables pertaining to the system as a whole are entirely determined by that state. At that level there is no indeterminism. However, each subsystem has a state too (found by the usual 'reduction of the density matrix'). There we see indeterminism, for the values of the observables pertaining to the subsystem correspond to one of the components of its mixed state. In this way, the modal interpretation will be like (a cleaned-up version of) the ignorance interpretation of mixtures.[9] Obviously, the holism of the total state introduces here a radical holism also for what really happens: what is really true of the system as a whole does not supervene, in any respect, on what is true of the subsystems.

In the many-worlds interpretation, the world's branching was tied to components of the total state written as a superposition. This introduced some very strange features, for superposition—unlike mixture—is nothing like the combination of alternatives that we see in possible world models. The idea of 'relative states' which Everett introduced brings some strange asymmetries into the interpretation.[10]

The 'ensemble interpretation' has often been viewed sympathetically by physicists. It too differs sharply from von Neumann's interpretation, especially in the form developed and recently defended by Leslie Ballentine (1970, 1989). According to this interpretation, the quantum-mechanical state characterizes no individual system, but rather the statistical properties of an aggregate or ensemble of similarly prepared systems. This sort of interpretation also admits of a number of variants. Ballentine's motivation certainly derives largely from the sort of dissatisfaction with von Neumann's interpretation, and specifically with the Projection Postulate, which I have also expressed.

The modal interpretation does attribute states to individual systems, but *formally* it is a sort of ensemble interpretation nevertheless. At least this is so on the standard (possible worlds) semantics for modality. The ensemble is in this case not a set of similar physical systems, but a set of ways that a given system could be. Each element is the same system configured differently, i.e. characterized by different values for the observables, but with the same dynamic state. For the dynamic state summarizes what is common to all the different ways the system could be, and does not indicate which is the actual one. However, ensemble interpretations tend to try for statistical proportions in the aggregate at all times. In the modal interpretation as elaborated in this chapter, the dynamic state is linked only to statistical distributions on the ensemble at the end of measurement interactions it could enter, and not under other conditions.

The intuitive agreement of modal and ensemble interpretations is best seen in the case of a mixed (dynamic) state W. Under this modal interpretation, the set of ways in which the system in state W could be is represented by the entire collection of alternatives $\{\langle \text{dynamic state } W, \text{value state } y \rangle : y$ in the image space of $W\}$. This corresponds to the ensemble (perhaps a virtual ensemble) in the other interpretation.

6. THE RESOURCES OF QUANTUM LOGIC[11]

In the 1960s and 1970s quantum logic became a highly developed mathematical subject. The inspiration was in part the apparent, tacit agreement to develop quantum theory itself starting with the most general notion of a theoretical model which could possibly fit any data the future might bring. The postulates of quantum logic narrowed down this class of models step by step, and eventually the representation theorems showed that the state-space was constrained to be a Hilbert space.[12] As the term 'logic' shows, the project was subject to very high standards of rigour. This gives it a special value for the project of interpretation of quantum mechanics. For the more or less axiomatic development allows us to introduce the interpretational additions also at each relevant point, with absolute clarity. In my opinion, any philosophical view on contemporary physics should eventually measure itself by this standard. Yet I also admit willingly that readers not so interested in formal logic can omit this section and the next without hindering their understanding of what follows after.

My construction will proceed in two stages. First I shall present — in one particular version — the quantum-logical approach to the structure of physical theory in general. This will continue the discussion begun on a more intuitive level in Chapter 5; if the discussion here becomes at times too uncompromisingly abstract, the reader may wish to review Section 6 of that chapter. The next section will instantiate that general format in accordance with the modal interpretation.

You cannot understand sentences of the language of physics without understanding physics. In other words, the interpretation of what we may call the language of a theory is determined by that theory. This determination occurs via the family of models which that theory offers for the purpose of modelling the physical phenomena we study. So if we want to study the logic — and hence the language — of a theory in general, we must begin with a general account of what a theory gives us. Here I propose, in a tentative spirit, a relatively shallow analysis of this subject. A theory deals with physical systems of a certain sort; it tells us what the possible states of such a system are, what observables (physical magnitudes) pertain to such a system, and what values they can have; and it specifies a relation between the states and values of those observables. Obviously, theories do much more, providing for instance dynamical laws for change of state with time. But this much will do for now.

The relation between states and observables is the subject of the first assumption I shall list. In classical physics, the state determined a precise value for each observable, and did so uniquely. In quantum mechanics, the nearest we come is Born's rules for the calculation of probabilities. Since our assumptions should cover both cases, and possible generalizations thereof, without becoming uselessly vacuous, we have the following:

First Assumption: A physical theory relates the states and observations it has specified in a way that assigns to each state x and to each observable m a classical probability measure P_x^m on the Borel sets of possible values.

To simplify the discussion, and without real loss of generality, I shall also assume that the observables are real valued, and that for each observable m there is a smallest countable set of real numbers—the spectrum of the observable, thus assumed discrete—which receives probability 1 from all these measures. The precise content of this First Assumption as it occurs in elementary quantum theory, and its interpretation, will be discussed later.

6.1. *Properties, observables, and propositions*

Properties, observables, propositions, and possible situations form a family of concepts, any of which can be taken as basic.

A physical system has some properties, and is capable of other properties. We classify systems into types or sorts by the properties which they are capable of having.

An observable (or physical magnitude) corresponds to a family of properties. That is, we may say that two properties are orthogonal if it is impossible for a system to have both at once. Then an observable corresponds to a maximal family of mutually orthogonal properties. If we index these properties as $(P_n : n \in I)$, then we can restate (X *has property* P_n) as (*the observable has value n on system X*). So either properties or observables can be taken as basic.

To take propositions as basic, we simply think of the property P as the function that takes system X into the proposition (X has P).

The notion of possible situation enters when we think of truth. Some propositions are true and others not; but if the situation had been different, those other propositions would have been true. If propositions are taken as basic, then a possible situation is a map of propositions into the value 1 (true) and 0 (not true). But in some ways the simplest course is to take possible situations as basic. Then a proposition can be identified with a set of situations (intuitively, the ones in which it is true), a property maps systems into propositions, and an observable maps values into properties. A proposition q is true in situation w exactly if w is in q, and q is impossible exactly if it is empty. The spectrum of an observable is the set of values which it does not map into the impossible property (i.e. the property that maps all situations into the impossible proposition).

Most important is not what we take as basic, but rather the facility to switch effortlessly from any one of these concepts to any other. Equally important, however, is the need to renounce certain classical intuitions which these concepts themselves do not force on us (see Feyerabend 1958; van Fraassen 1972*b*). As a simple example, consider the temperature T measured on the Kelvin scale, and the derivative quantity T^* which has integer value n exactly when T has a value in the half-open interval $[n, n + 1)$. Let us generalize the description a little:

1.   Observables m and fm are related by the fact that, if m has value r, then fm has value $f(r)$.

Because of the conceptual surroundings in which we first encountered temperature, the example will immediately mobilize the following (see Chapter 5, Section 1):

2.   *Classical Principle*: If system X is capable of having any member of the family of properties corresponding to an observable, then it must always have one of them.

In other words, if the observable m pertains to system X, then it always has a sharp value.

Using 2, we infer from 1 immediately, first, that the observable fm is uniquely determined by f and m, and second, that if fm has value r then m must have some value in $f^{-1}(r) = \{x : f(x) = r\}$. The first consequence is not denied in any interpretation of quantum mechanics I know, but the second *is* denied. So we must take care not to accept principle 2 from the outset. Neither should we deny it at once, however, for it is true of classical theories, and is also accepted in some other interpretations of quantum mechanics.

Using only the relation specified in the First Assumption, we can formulate many propositions which describe the (dynamic) state of a physical system. We could say, for instance, 'This system has a state x such that $P_x^m(E) = \frac{1}{2}$.' But quantum logic is characterized by a preoccupation with propositions in which the only probabilities are 0 and 1. Herein lie both its power and its limitations. As before, I define:

3.   A state-attributing proposition $[m, E]$ is a proposition which is true about a physical system if and only if that system is in a state x such that $P_x^m(E) = 1$.

Relative to those concerns which I have outlined so far, propositions that always have the same truth-value are not relevantly different in any way. This means that we can identify a proposition with (more accurately: represent it by) the set of states which make it true. We arrive then at the following characterization of state-attributing propositions, and of the relation among them which corresponds to valid argumentation:

4. $[m, E] = \{x : P_x^m(E) = 1\}$
5.   Propositions $[m_1, E_1]$, ..., $[m_k, E_k]$, ... together *imply* $[m, E]$ exactly if $\cap \{[m_i, E_i] : i = 1, \ldots, k, \ldots\} \subseteq [m, E]$.

There is no choice here, if the assertion that one proposition implies another is to mean that we can validly argue from the one as premiss to the other as conclusion. The family of all state-attributing propositions I shall call **P**.

What is the structure of **P**? This is the central question of quantum logic. We see in proposition 5 that **P** has a natural and significant partial order. Our First Assumption allows us in addition to apply some classical probability theory, with the following consequences:

6.   If $E \subseteq F$ then $[m, E] \subseteq [m, F]$.

7.   If index set I is countable, then
     $\cap\{[m, E_i] : i \in I\} = [m, \cap \{E_i : i \in I\}]$.

We note however that these concern only propositions relating to a single observable.

6.2. *Propositions relating different observables*

If m is any observable, let us denote as $[m]$ the class of all state-attributing propositions of the form $[m, E]$. We can introduce three partial orders on this class, which I call the proposition range of m:

8. $[m, E] \subseteq [m, F]$.
9. For all states x, *if* $P_x^m(E) = 1$ then also $P_x^m(F) = 1$.
10. For all states x, $P_x^m(E) \leq P_x^m(F)$.

By definition, 8 and 9 coincide; and of course 10 implies 9. Almost everywhere in quantum logic, we see the assumption that all three relations are the same.

Second Assumption: If $[m, E] \subseteq [m, F]$, then for all states x, $P_x^m(E) \leq P_x^m(F)$.

Within $[m]$, the states are 'strongly order-determining'. This has the consequence that we can define a complement and join, as well as a meet, within $[m]$, and that its structure is completely determined.

11. $[m, E]^\perp = [m, R - E]$
12. $\cup \{[m, E_i] : i \in I\} = [m, \cup\{E_i : i \in I\}]$ for countable index set I.

At this point it is possible to deduce:

13. $\langle [m], ^\perp, \cap, \cdot \rangle$ is a Boolean sigma-algebra.

A Boolean algebra is, of course, the algebraic notion of which a field of sets is the most familiar instance. Indeed, with our simplifying assumption of a countable spectrum, we can add to 13 that this structure is a complete lattice, and isomorphic to the Boolean sigma algebra (in fact, sigma field) of Borel subsets of its spectrum.

So far we know about the proposition ranges $[m]$ of two different observables $[m]$ and $[m']$ only that each is Boolean and that they share a common unit and zero. They overlap, and within such overlap they may 'disagree' on structure. The next assumption widens the Second Assumption about how states determine order, to the whole of **P**.

Third Assumption. If m and m' are distinct observables, and $[m, E] \subseteq [m', F]$, then $P_x^m(E) \leq P_x^{m'}(F)$ for all states.

The first consequence is that **P** is an *orthoposet*; that is,

14. The operation introduced in statement 11 determines an orthocomplement on **P**:

   (i) if $[m, E] = [m', F]$, then $[m, R - E]$

   $= [m', R - F]$

   (ii) $p^{\perp\perp} = p$

   (iii) $p \cap p^\perp = \varnothing$; $p \cup p^\perp = K$

   (iv) if $p \subseteq q$ then $q^\perp \subseteq p^\perp$

Here K and $\varnothing$ are the unit and zero element of **P**, K being of course the set of all the states and $\varnothing$ the set of 'impossible states' (for example the null vector, if we represented the states by a Hilbert space). The second consequence is that

15. The intersection of $[m]$ and $[m']$ is again a Boolean sigma algebra, and the operations of $[m]$ and $[m']$ are coincident on that intersection.

Thus we have deduced that **P** is the sort of 'pasting together' of Boolean algebras which Kalmbach and Hardegree have studied and honoured with various names; it is already *almost* what Kochen and Specker (1965*a*, 1965*b*) called a partial Boolean algebra.[13]

To provide more structure, still further assumptions need to be introduced. We define p and q to be *compatible* if there is a proposition range $[m]$ to which they both belong, and *orthogonal* if $p \subseteq q^\perp$. Common assumptions added in quantum logic are:

16. Any finite set of pairwise compatible propositions is jointly compatible.

17. Any finite set of pairwise orthogonal propositions is jointly compatible.

The first of these has the consequence that **P** is a partial Boolean algebra in the sense of Kochen and Specker. The second implies that **P** is an orthomodular orthocomplemented poset; that is, in addition to the above, whenever p and q are orthogonal their lowest upper bound exists; it may consistently be denoted $p \cup q$, for inside a single proposition range the previously defined join is that lowest upper bound.

In such a structure it is still not guaranteed that arbitrarily chosen propositions p and q have a meet (greatest lower bound). If that is so, the structure is a lattice, though generally not Boolean. Still further assumptions may be made, leading to the consequence that **P** is a lattice and, eventually, to its isomorphism to the lattice of subspaces of a separable Hilbert space (as in Piron's famous representation theorem),[14] thus establishing final contact with the subject of quantum mechanics where we began.

Every assumption introduced raises a new problem of interpretation. The mathematical content is always clear enough, but what is its physical or empirical content? For each assumption, this question is more difficult than for the one before.

7. THE MODAL INTERPRETATION, QUANTUM-LOGICALLY

In connection with the modal interpretation, what I have discussed in this chapter so far is indeed the structure of physical theory proper. But it still encompasses only the theory of the dynamic state, and leaves open entirely what exactly the probabilities are probabilities *of*. That is why I said that the precise content of the First Assumption as it occurs in quantum theory was to be discussed later. In the modal interpretation, the probabilities are of events, each describable as *an observable having a certain value*, which are parts or aspects of the value states. Section 6 was a preamble; now I shall develop the modal interpretation as a theory in the style of quantum logic. The statements P1, . . . will be postulates; D1, . . . definitions; and T1, . . . theorems.

7.1. *Value states and dynamical states*

If w is a physical situation in which system X exists, then X has both a dynamic state x and a value state λ in situation w. For the time being I shall be concentrating on one-body situations, and so I shall write, for brevity, $w = \langle x, \lambda \rangle$ or else $x = x(w)$, $\lambda = \lambda(w)$. Note that, alone, 'x' and 'λ' are variables, but '$x(\ )$' and '$\lambda(\ )$' are names of functions. The variables x, y, . . . from the end of the alphabet will range over dynamic states; λ, λ', . . . over value states.

Having renounced what Feyerabend called the Classical Principle, λ will not in general assign a value to every observable m pertaining to X. However, we might say that λ assigns m an *unsharp* value in a derivative way. Suppose that 1_E is the characteristic function of a set E of values, and consider the observable $1_E m$. If it has value 1, then it is impossible that m has a value outside E — and it is necessary that $1_{(E \cup F)} m$ has value 1 too; and so forth.

So we have two ways of treating λ. One way is to say that it is a partial function assigning values to some observables but not to others, while respecting the relationship described in 1 above between m and fm. The second way is to say that λ assigns something to each observable m, namely the least set E such that it assigns $\{1\}$ to $1_E m$. It is obviously more convenient to follow the second course; and we can then prepare ourselves for applications of probability theory by insisting that the sets assigned are Borel sets.

P1: A value state λ is a map of observables into non-empty Borel sets.

P2: If $\lambda(w)(m) = \lambda(w)(m')$ for all possible situations w, then $m = m'$.

D1: $S(m) = \{x : \lambda(w)(m) = \{x\}$ for some $w\}$.

D2: fm is the observable (if any) such that $\lambda(w)(fm) = f(\lambda(w)(m))$, where $f(X) = \{f(x) : x \in X\}$.

We call $S(m)$ the *spectrum* of m. I will limit the discussion of the theory format here to discrete point spectra, because in known theories other observables can be mathematically construed from these.

P3: $S(m)$ is a countable set of real numbers.

P4: If f is a Borel function, and m an observable, then the observable fm exists.

P5: $\lambda(w)(m) \subseteq S(m)$ for all w.

We note that, because the spectrum is countable, both $f(X)$ and $f^{-1}(X)$ are Borel sets if X is a set of possible values of m.

We can now introduce propositions as sets of possible situations, which describe the value state:

D3: $\langle m, E \rangle = \{w : \lambda(w)(m) \subseteq E\}$

$\langle m \rangle = \{\langle m, E \rangle : E \text{ a Borel set}\}$

$\mathbf{V} = \{\langle m, E \rangle : m \text{ an observable and } E \text{ a Borel set}\}$

The proposition $\langle m, E \rangle$ is thus true in situation w exactly if $\lambda(w)(m) \subseteq E$, and so exactly if $1_E m$ has value 1 in that situation. We can read $\langle m, E \rangle$ conveniently as 'm (actually) has value in E'. The set $\mathbf{V}$ of all such propositions I call the set of *value-attributions*.

The family $\langle m \rangle$ of propositions which attribute (approximate, 'unsharp') values to m, I shall call the *value range* of m. It admits the following operations:

D4: $\langle m, E \rangle^{\perp} = \langle m, R - E \rangle$

$\langle m, E \rangle \wedge \langle m, F \rangle = \langle m, E \cap F \rangle$

$\langle m, E \rangle \vee \langle m, F \rangle = \langle m, E \cup F \rangle$

$\wedge \{\langle m, E_i \rangle : i \in N\} = \langle m, \cap \{E_i : i \in N\}\rangle$

T1: $\langle m, E \rangle \subseteq \langle m, F \rangle$ iff $S(m) \cap E \subseteq S(m) \cap F$

T2: $\langle m, E \rangle \wedge \langle m, F \rangle = \langle m, E \rangle \cap \langle m, F \rangle$

That the join is well defined on $\langle m \rangle$ follows from T1, because the partial ordering of the family is isomorphic to the set inclusion ordering of the subsets of $S(m)$. That is because P5 and D1 guarantee that the correspondence of $\langle m, E \rangle$ with $S(m) \cap E$ is one-to-one, an isomorphism with the Borel subsets of $S(m)$. This fact does not depend on P3. We note that we have at present no guarantee that if $\langle m, E \rangle = \langle m', E' \rangle$ then $\langle m, R - E \rangle = \langle m', R - E' \rangle$ and that even within $\langle m \rangle$ the join of two propositions is not in general equal to their set-theoretic union. We conclude, however:

T3: $\mathbf{V}$ is the union of a family of Boolean sigma algebras $\langle m \rangle$ with common unit and zero equal to $\langle m, S(m) \rangle$ and $\langle m, \Lambda \rangle$ respectively.

The situation is very, but not entirely, classical; as before, we have the Law of Excluded Middle (every situation w belongs to $q \vee q^{\perp}$) but not the Law of Bivalence (a situation w may belong neither to q nor to $q^{\perp}$).

Our prediction of what will or would happen to a system under contemplated physical conditions is based on the dynamic state. Hence, in particular, so are our predictions about measurement outcomes. This means that the state must provide us, for each observable m, with a probability measure on the possible values of m, telling us the conditional probability $P_x^m(E)$ that a value is found in E, if m be measured on a system in dynamic state x. I will restrict the discussion here to von Neumann–Lueders measurements, in which correlations are set up. The reader will easily see, by comparison to Section 5 above, what to omit for the more general case of *any* sort of measurement.

We make several assumptions to introduce our next postulate. The first is that at the end of the measurement m will have a sharp value, and the probability that it is a given value is exactly the probability that the measurement apparatus will indicate that value. Secondly, we assume that the probability that m will have a value in E if m is measured is exactly the same as the probability that $1_E m$ will have value 1 if it is measured. Together, these assumptions imply that the derivative function

$$P_x(\langle m, E \rangle) = P_x^m(E)$$

is well defined on **V**. The third assumption is that dynamic states are experimentally distinguishable; if two such states are distinct, then there must be some observable for which they yield different predictions of measurement outcomes. Hence we can identify the state x with the function P_x:

P6:   A dynamic state x is a function from **V** into $[0, 1]$, whose restriction to each Boolean sigma algebra $\langle m \rangle$ is a probability measure.

Note that I use the term 'probability measure' only in the classical, Kolmogoroff, sense (though with the more general 'Boolean sigma algebra' replacing 'sigma field'). Also note that we have here a postulate, not a definition (similarly for P1 about value states). Merely to fix the terminology exactly, therefore, I stipulate:

P7: x and λ are a dynamic state and a value state respectively, only if there exist possible situations w and w' such that $x = x(w)$, $\lambda = \lambda(w')$.

We can now return to the traditional topic of quantum logic as introduced by Birkhoff and von Neumann: the propositions which describe the dynamic state:

D5: x is an *eigenstate* of m, with corresponding *eigenvalue* r, exactly if $x(\langle m, \{r\}\rangle) = 1$.

D6: $[m, E] = \{w : x(w)(\langle m, E\rangle) = 1\}$.

      $[m] = \{[m, E] : E$ a Borel set$\}$.

      P $= \{[m, E] : m$ an observable, E a Borel set$\}$.

P is the set of *state-attributions*, $[m]$ the *proposition range* of m, and we read $[m, E]$ briefly as 'm must have value in E', or more accurately as '$1_E m$ must have value 1'. On our present interpretation these readings are accurate, provided we take the former as concerning an unsharp value. It does not mean that there is some eigenvalue in the set E such that m has that value. On the other hand, it means more than that an outcome in E is certain if a measurement be made. If $[m, E]$ is true, then $1_E m$ does have value 1 whether any measurement be made or not. This is the way in which the interpretation goes beyond what quantum mechanics itself implies.

In Section 6 I discussed various postulates that bestow structure on **P**. Here I shall only state the major such postulate ('strong order separation') and its consequences, without further comment, though adapted to the present modal framework.

P8: $[m, E] \subseteq [m', E']$ only if, for all dynamical states x, $x(\langle m, E\rangle) \leq x(\langle m', E'\rangle)$.

D7: $[m, E]^{\perp} = [m, R - E]$
      $[m, E] \uplus [m, F] = [m, E \cup F]$.

T4: $^{\perp}$ is well defined on **P** and $\uplus$ on $[m]$.

T5: $[m, E] \cap [m, F] = [m, E \cap F]$.

T6: $[m]$ is a Boolean sigma algebra with operations $^{\perp}$, $\cap$, $\uplus$; and with unit and zero elements equal to $[m', S(m')]$, and $[m', \Lambda]$ for any m'.

T7:   The intersection $[m] \cap [m']$ is again a Boolean sigma algebra with the operations of $[m]$ and $[m']$ coinciding in this overlap.

T8: $\langle \mathbf{P}, \subseteq, \perp \rangle$ is an orthoposet (i.e. $\perp$ is an ortho-complement).

Thus $\mathbf{P}$ is an orthoposet formed by 'pasting together' a family of Boolean algebras, whole operations coincide in areas of overlap. More postulates could clearly be added, as we discussed above, so as to approach the lattice of subspaces of a Hilbert space.

7.2. *General approach to mixed states*

Some dynamic states are more informative than others: the predictions we base on them are more precise, with probabilities closer to 1 and 0. In familiar cases such as classical and quantum mechanics, the less informative cases are always *mixtures* (convex combinations) of more informative ones. What about the general case? We must first reintroduce the requisite concepts for our now very general framework. Based on the idea that giving more information amounts to ruling out more possibilities that were considered open beforehand, I introduce here the first technical use of 'possible' in the present context:

P9: If $A = \{x_i\}$ is a countable set of dynamic states and $0 < c_i \leq 1$ with $\Sigma c_i = 1$, then $\Sigma c_i x_i$ is also a dynamic state (called a *mixture* of A).

D8: x is *pure* iff $x = cy + (1 - c)z$ and $c \neq 0$ only if $x = y = z$; otherwise x is *mixed*.

D9: y is *possible relative to* x (briefly, xRy) exactly if, for all q in $\mathbf{V}$, if $x(q) = 1$ then $y(q) = 1$.

D10. w' is *possible relative to* w (briefly, wRw') exactly if, for all q in $\mathbf{V}$, if w is in q then w' is in q.

D11: x is *prime* if xRy implies that $x = y$.

Although it will not play a salient role in our discussion here, it may help to mention the idea of superposition. We can generalize D9 as follows:

y is *possible relative to the set X* of states exactly if, for all q in $\mathbf{V}$, if $x(q) = 1$ for all x in X then $y(q) = 1$.

In quantum mechanics, this relation comprises both superpositions (proper) and mixtures. (Precisely: y is in that case a mixture of pure states, each of which is a superposition of elements of X.) Some writers use 'superposition' in this general sense, and speak of 'coherent' and 'incoherent' superpositions.

In the case of quantum mechanics, as we recall, *prime* is the same as *pure* and xRy if and only if y is a state whose image space is included in that of x.[15] This result suggests at least directions of inquiry to us, and we begin by noting the following elementary consequences of our definitions:

T9: wRw' if and only if $x(w)Rx(w')$.

T10:   (*a*)   If $B \subseteq A$, x a mixture of A and y a mixture of B, then xRy.

      (*b*)   If x is a mixture of A, then $x(q) = 1$ if and only if $y(q) = 1$ for all y in A (for q in **V**).

T11:   If x and x' are mixtures of the same set A, $x(w) = x$ and $x(w') = x'$ and $q \in$ **P**, then w is in q if and only if w' is in q.

T12:   A state is prime only if it is pure.

T10 follows easily because, if x is a mixture of y and some other state, and $y(q) \neq 1$, then $x(q) \neq 1$. The immediate corollary T11 establishes that quantum-logical propositions (i.e. members of **P**) do not separate mixed states with the same components. Finally, T12 follows because if x is not pure, say $x = cy + (1 - c)z$, and $y \neq x$, then by T10 we have xRy, so x is not prime.

At this point we try to introduce minimal reasonable assumptions to secure the equivalence of *pure* and *prime*. T11 establishes that quantum-logical propositions do not separate mixed states in general; but we can postulate that they separate pure states. We can also postulate that there are enough prime states so that each dynamic state corresponds naturally at least to a set of prime states:

P10:   If x and y are pure and $x \neq y$, then there is a proposition q in **V** such that $x(q) = 1 \neq y(q)$.

P11:   For all q in **V**, $x(q) = 1$ if and only if $y(q) = 1$ for all prime states y such that xRy.

T13:   A state is pure only if it is prime.

Let us denote the set of prime states which are possible relative to a given state x as $PR(x)$. Due to P6, the state x will give 1 to some but not all value-attributing propositions, so this set $PR(x)$ will never be empty. To prove T13, suppose x is pure and let y be a prime state possible relative to x. Then by definition $PR(y)$ contains $PR(x)$. But since y is prime, $PR(y)$ contains only y. Therefore $PR(x) = \{y\} = PR(y)$. Since y is prime it is also pure, by T12. Therefore by P10, it follows that $x = y$, since they give 1 to the same value-attributing propositions. Hence x is prime.

The separation postulate P10 would of course follow from the stronger assumption that **P** is atomistic (generated by its atoms) and that pure states correspond to its atoms. This is true for the lattice of subspaces of a separable Hilbert space.

The coincidence of *pure* and *prime* does not establish that if xRy then x is a mixture of y and some (other) state. Suppose we can find a positive real number c such that $cy(q) \leq x(q)$ for all q in **V**. Then we can define a function z on **V** by

$$x = cy + (1 - c)z$$

and conceptually there is no reason why z should not be a dynamic state. But our earlier discussion in Chapter 6, Section 4, shows that the existence of such a state is connected to the dimensionality of the space.

7.3. *Relations between states and observables*

We have finally come to the point where we must choose between alternative interpretations of the probability measures in our formal theory. The guiding idea will be this: there is no 'collapse of the wave packet' in a measurement, as far as the dynamic state is concerned—yet at the end of the measurement of an observable, that observable has one of the possible values in its spectrum, with probability as given by the initial state.

Suppose we know the dynamic state at time t; can we infer anything about values of observables at that time? The obvious candidate for a principle relating the two is that, if $x(w)$ is an eigenstate of m, for eigenvalue r, then m actually does have value r. In view of P4, this will then generalize to any set E of eigenvalues. In our symbolism,

$$[m, E] \text{ implies } \langle m, E \rangle.$$

This does not follow from the P1–P11 above, but I shall assume it. It will imply that, at least in the case of an eigenstate, a measurement does reveal the value the observable had at the outset.

Secondly, how many of the value-attributions allowed by the preceding assumption are true? If we say an absolutely minimal set, we embrace what I have called von Neumann's interpretation rule. If we say absolutely maximal, then we imply that every observable always has a precise value in its spectrum, and hence the Classical Principle, which Feyerabend made explicit. The latter requires in turn that we say either that a single Hermitean operator can correspond to more than one observable, or else that the values of functionally related observables may, on individual occasions, violate those very functional relationships. (This follows from Kochen and Specker's 'no hidden variables' proof, as I shall discuss in the next chapter.) I will take a middle course and assume only that the set of propositions in **V** that are true in a given situation is *relatively maximal*: if all the value-attributions true in w are also true in w', then the converse is true as well; that is, exactly the same value-attributions are then true in both.

Finally, I shall add what I consider the distinctive Copenhagen school assumption: propositions about a system cannot be jointly true unless they can be jointly certain. In other words, the Uncertainty Principle exhibits not simply a limit to our knowledge, but a limit of what can be objectively true at the same time. All of the above is summarized in postulates P12–P14:

P12: $[m, E] \subseteq \langle m, E \rangle$

P13: If $\lambda(m) \subseteq \lambda'(m)$ for all m, then $\lambda = \lambda'$.

T14: If w is in q only if w' is in q, for all q in **V**, then also for all q in **V**, w' is in q only if w is in q.

P14: If $X \subseteq \mathbf{V}$ and w is in $\cap X$, then there is a dynamic state x such that $x(q) = 1$ for all q in X.

We can now deduce the consequences by which this theory merits at once the epithets 'modal' and 'Copenhagen'. Recalling the discussion of necessity in Chapter 1, we define modal operators on the following propositions:

D12: If q is any proposition, then
$\Box q = \{w : \text{for all } w', \text{ if } wRw' \text{ then } w' \in q\}$
$\Diamond q = \{w : \text{for some } w', wRw' \text{ and } w' \in q\}$

T15:   (*a*)   For each situation w there is a pure dynamic state y such that $x(w)Ry$, and for all q in **V**, w is in q if and only if $y(q) = 1$.

 (*b*) $[m, E] = \square \langle m, E \rangle$.

 (*c*)   Identity of observables: If $x(\langle m, E \rangle) = x(\langle m', E \rangle)$ for all dynamic states x and all Borel sets E, then $m = m'$.

To prove (*a*), let $T = \{q \in \mathbf{V} : w \in q\}$. By P14, there is a dynamic state z such that $z(q) = 1$ for all q in T. We have to show that z is possible relative to $x(w)$, that z is pure, and that also, conversely, if $z(q) = 1$ then q is in T. The relevant feature here is obviously the relative maximality depicted in P13 and T14.

Let us first take the case in which $x(w)$ is pure, which is to say prime (by T12 and T13). By P12 we see that if $x(w)(q) = 1$ then w is in q, so q is in T. So if $x(w)(q) = 1$ then also $z(q) = 1$; that is, $x(w)Rz$. But $x(w)$ is prime, so then $x(w) = z$. We have now proved the following:

Lemma: If $x(w)$ is pure, then w is in q if and only if $x(w)(q) = 1$.

Consider next the case in which $x(w)$ is not pure. It is still true by P12 that if $x(w)(q) = 1$ then q is in T, and so $x(w)Rz$, by the same reasoning as above. To bring T14 into play, we note that by P11 there is a pure state y such that zRy, and if $z(q) = 1$—hence also if q is in T—then $y(q) = 1$. By transitivity, also $x(w)Ry$. We look now at the alternative situation w' whose dynamical state $x(w')$ is that prime state y. By the lemma above, w' is in q if and only if $y(q) = 1$. By T14 and the fact that y assigns 1 to all members of T, we see that:

For all q, if q is in T then w' is in q.

Hence the antecedent of T14 holds, and we conclude that also conversely, if w' is in q, so is w. The two equivalences together now imply that w is in q if and only if pure state y assigns 1 to q.

Clearly, at this point, we can for all practical purposes represent the value state $\lambda(w)$ by that pure dynamic state y. The equation is this: $\lambda(w)(m) = E$ exactly if E is the least Borel set such that $y(\langle m,E \rangle) = 1$.

To prove (*b*) we must show that w is in $[m,E]$ if and only if all w' such that wRw' are in $\langle m,E \rangle$. But w is in $[m,E]$ if and only if $x(w)(\langle m,E \rangle) = 1$, so this amounts to

$$x(w)(q) = 1 \text{ iff, for all } w', \text{ if } wRw' \text{ then } w' \text{ is in } q.$$

By T9 (or D10), the right-hand side is equivalent to: for all w', if $x(w)Rx(w')$ then $x(w')(q) = 1$. This follows from the left-hand side by the definition of R. Conversely suppose the right-hand side is true. By P7 it follows then that, for all prime states z, if $x(w)Rz$ then $z(q) = 1$. But then also by P11 it follows that $x(w)(q) = 1$, as required.

Finally, to prove (*c*), the identity criterion for observables, note that its antecedent entails that, for all Borel sets E and all pure states y, we have $y(\langle m,E \rangle) = y(\langle m',E \rangle)$. In view of part (*a*) of the present theorem and P7, we can represent the value states by the pure dynamic states. So we conclude that, for all w, $\lambda(w)(m) = \lambda(w)(m')$, for if that equation did not hold there would be a pure state y such that $\lambda(w)(m)$ is the smallest Borel set E such that $y(\langle m,E \rangle) = 1$ but not such that $y(\langle m',E \rangle) = 1$, which is counter to hypothesis. Therefore by P2 we conclude that $m = m'$. This last consequence T15(*c*) we express as: *Statistically equivalent observables are identical*.

Returning now to our modal operators, we can ferret out the logic of value-attributions. As we have just seen, $[m, E]$ can be read as 'Necessarily, $\langle m, E \rangle$' because $[m, E] = \square \langle m, E \rangle$. However, this says only that the dynamic state assigns 1 to $\langle m, E \rangle$ if and only if the value state that accompanies any relatively possible dynamic state makes $\langle m, E \rangle$ true. Another sense of necessity is given by: it is necessary that q (in **V**) exactly if any value state that could accompany the present dynamic state makes q true.

D13:   If q is any proposition,
       $$\boxdot q = \{w : \text{for all } w', \text{ if } x(w) = x(w') \text{ then } w' \in q\}.$$
P15:   If wRw' there is a possible situation w'' such that $x(w'') = x(w)$ and $\lambda(w'') = \lambda(w')$.
T16: $[m, E] = \boxdot \langle m, E \rangle$.

To prove T16, suppose first that w is in $[m,E]$. Then by T15(*b*), w is in $\square \langle m,E \rangle$, so all w' such that wRw' are in $\langle m,E \rangle$. *A fortiori*, all w' such that $x(w) = x(w')$ are in $\langle m,E \rangle$, in view of T9. Therefore w is in $\boxdot \langle m,E \rangle$. Suppose on the other hand that w is not in $[m,E]$. Then again by T15(b) there must be some w' such that wRw' and w' is not in $\langle m,E \rangle$. That means that $\lambda(w')(m)$ is not included in E. By P15, there is now a possible situation w'' such that $x(w'') = x(w)$ and $\lambda(w'') = \lambda(w')$, and w'' is therefore not in $\langle m,E \rangle$. But then w is not in $\boxdot \langle m,E \rangle$.

While the operation $\Box$ was based on the transitive relative possibility relation R, the new operator $\boxdot$ is based on an equivalence relation. We are here not interested in saying anything about iterated or nested modal operators—where transitivity, etc., makes a difference—but we should notice that this character of $\boxdot$ establishes a correspondence between **V** and **P** (a sort of 'supervaluation' relationship).

T17: w is in $[m, E]$ iff $\{w' : x(w) = x(w')\} \subseteq \langle m, E \rangle$.

T18: $[m, E]$ implies $[m', E']$ if and only if $\langle m, E \rangle$ implies $\langle m', E' \rangle$.

Both follow at once from T16. Thus the map $[m, E] \longrightarrow \langle m,E \rangle$ is an isomorphism of the posets $\langle \mathbf{P}, \subseteq \rangle$ and $\langle \mathbf{V}, \subseteq \rangle$. Recalling definition D4 and theorems T4 and T8 concerning the orthocomplement on **P**, we infer also that:

T19: If $\langle m, E \rangle = \langle m', E' \rangle$ then $\langle m, R - E \rangle = \langle m', R - E' \rangle$.

T20: $\langle \mathbf{P}, \subseteq, {}^{\perp} \rangle$ and $\langle \mathbf{V}, \subseteq, {}^{\perp} \rangle$ are isomorphic orthoposets.

Similar inferences may be drawn from T7 concerning the intersections of Boolean algebras $\langle m \rangle$. Briefly, the logic of **V** is just that of **P**: quantum logic.

7.4. *Representation of complex systems and measurement*

So far I have concentrated on systems taken as wholes, ignoring whether they have other systems as components or, alternatively, are components of other systems. All the postulates I have laid down, it must be emphasized, concern propositions, and classes of propositions, which are about a single system treated as a unit.

The representation of complex systems is, not surprisingly, a complex subject. I have said little enough about the algebraic structure of the set of observables pertaining to one system; I shall leave undiscussed even more about the abstract relations between the algebras of observables of a system and its components.[16] But I shall introduce enough here to provide a sufficient framework for the description of measurement processes.

Relying on the preceding sections, I shall now simply identify the value state λ with the pure dynamic state y to which it corresponds by theorem T15(a). Thus we can also write $w = \langle x(w), y(w) \rangle$, where w is in $\langle m, E \rangle$ exactly if $y(w)(\langle m, E \rangle) = 1$. (Note here also that in this notation y is a variable when alone, but $y(\ )$ names a function.)

In the case of a complex system $X_3 = (X_2 + X_1)$ involved in situation w, both dynamic and value states must be ascribed to the whole system *and* to its components. The dynamic and value state of X_i in w will be denoted $x_i(w)$ and $y_i(w)$, or alternatively I write:

$$w = \langle x_3, x_2, x_1; y_3, y_2, y_1 \rangle$$

Each situation $w_i = \langle x_i, y_i \rangle$ is a possible situation of the one-body type, and must therefore satisfy all the postulates laid down so far. These constraints can now be summarized of course as: $x_i R y_i$. (In Sections 8 and 9 we shall discuss why this cannot be the end of the matter, however.)

In the case of a one-body system we had the simple principle that whatever propositions in **V** can be jointly true can be jointly certain. Something analogous, if not equally simple, must be said when we consider a class of propositions some of which are about one system, say X_3, and some about another, say X_1 or X_2. The principle I wish to impose here is that all the value-attributions true about X_3 could also, through a suitable preparation of state, be jointly certain, without making it thereby impossible that the actually true value-attributions to X_2 and X_1 are still true.

To express this more precisely, let subscripts indicate which system an observable pertains to; so, for instance, $\langle m_1, E \rangle$ says that observable m_1 has actual value in E on system X_1. Now we can look at some observable m_3 on X_3, and suppose that the value states are such that, say, m_3 has actual value r, m_2 has actual value s, and m_1 has actual value t. The principle I described in the preceding paragraph says then that the system

$X_3 = (X_2 + X_1)$ could, in some possible situations, be such that the dynamic state of X_3 is an eigenstate of m_3 corresponding to eigenvalue r, while m_2 and m_1 still have actual values s and t respectively. This consequence we could even express by means of our modal symbolism if we let $q_1 = \langle m_1, E_1 \rangle$, $q_2 = \langle m_2, E_2 \rangle$, and $q_3 = \langle m_3, E_3 \rangle$:

T21: $(q_1 \cap q_2 \cap q_3)$ *implies* $\Diamond(q_1 \cap q_2 \cap \Box \, q_3)$.

Because of the correspondence between, and our consequent present identification of value states with, pure dynamical states, we can state the general principle very simply:

P16: If w is a possible situation, then there is also a situation w' such that $x_3(w') = y_3(w)$, and $y_i(w') = y_i(w)$ for $i = 1, 2, 3$.

This implies T21 above, which of course could not be deduced beforehand.

It might be asked why P16 and its consequence T21 are not strengthened to accord more nearly with the corresponding postulate P14. Why, in fact, do we not assert:

$$(q_1 \cap q_2 \cap q_3) \text{ } implies \text{ } \Diamond\Box(q_1 \cap q_2 \cap q_3)?$$

The reason is that in quantum mechanics the complex system is capable of having certain dynamic states which are pure and yet leave open various possibilities for the value states of the components. This is crucial to the description of measurement processes with uncertain outcomes.

How are the dynamic states related to each other? This is more difficult to describe, since it concerns exactly the algebras of observables pertaining to different systems, which I shall leave largely undiscussed on the abstract level. But in quantum mechanics, the state x_3 determines the states x_2 and x_1 by reduction of the density matrix. Using that as our guide, and choosing a suggestive symbolism reminiscent of tensor products, we can introduce the following postulate:

P17: For each observable m_2 and observable m_1 pertaining to X_2 and X_1, there are observables $m_2 \otimes I$ and $I \otimes m_1$ which pertain to system $X_3 = (X_2 + X_1)$, and such that, for all possible situations w, and all Borel sets E, where $x_i(w) = x_i$:

(i) $x_3(\langle m_2 \otimes I, E \rangle) = x_2(\langle m_2, E \rangle)$

(ii) $x_3(\langle I \otimes m_1, E \rangle) = x_1(\langle m_1, E \rangle)$

The observables $(m_2 \otimes I)$ and $(I \otimes m_1)$ are uniquely determined by equations (i) and (ii), according to theorem T15(c). If we are given $x_3(w)$, then $x_2(w)$ and $x_1(w)$ are uniquely determined by (i) and (ii) since they are functions defined on the sets $\langle m_2 \rangle$ and $\langle m_1 \rangle$ consisting of the propositions $\langle m_2, E \rangle$ and $\langle m_1, E \rangle$. Hence we can introduce a functional notation:

D14: If equations (i) and (ii) hold for given dynamical states x_3, x_2, x_1 for all observables m_2 and m_1 pertaining to X_2 and X_1, and all Borel sets E, then $x_2 = \#x_3$; $x_1 = x_3\#$.

T18: If w is a possible situation involving $X_3 = (X_2 + X_1)$, then

(a) $x_2(w) = \#x_3(w)$ and $x_1(w) = x_3(w)\#$.

(b) $\#y_3(w)Ry_2(w)$ and $y_3(w)\#Ry_1(w)$.

That $\#$ is well defined, as well as T18(a), is established by the immediately preceding discussion. To see T18(b) we simply apply P16 and T18(a): Given the possible situation $w = \langle x_3, x_2, x_1; y_3, y_2, y_1 \rangle$, there must be another situation $w' = \langle x_3' = y_3, x_2', x_1'; y_3, y_2' = y_2, y_1' = y_1 \rangle$. But x_iRy_i. Since $x_3' = y_3$ is pure, it follows that $y_3' = y_3$ as well. But $x_2' = \#x_3' = \# y_3$, and $x_2'Ry_2'$; so $\#y_3Ry_2$. Similarly for $y_3\#$ and y_1.

This theorem allows a simple summary description of a situation involving a complex system:

$$w = \langle x_3, \#x_3, x_3\#; y_3, y_2, y_1 \rangle$$

where x_3Ry_3, $(\#y_3)Ry_2$, and $(y_3\#)Ry_1$.

In general, $\#x_3$ and $x_3\#$ are mixed even when x_3 is pure.

It must be strongly emphasized that, because of T15(c), the complex observable $m_2 \otimes I$ is uniquely determined by m_2 simply because its statistical distribution relative to every dynamical state is determined. This leaves open the possibility that the actual value of m_2 does not determine the actual value of $m_2 \otimes I$.

Having described complex systems in general, we can now look at measurement and ask how elementary quantum theory fits the general structure exhibited so far. The situation I shall concentrate on is the familiar ideal description of von Neumann–Lueders measurement, which requires us to look at a model of sufficient generality, namely of the interaction of two systems (object and apparatus) leading to a correlation of observables.

The evolution of the dynamic state of an isolated system is governed by a one-parameter group U of unitary transformations, with infinitesimal generator H, the Hamiltonian of the system. Suppose now that X_1 and X_2 are two systems, and the complex system $(X_2 + X_1)$ is isolated; under what conditions can we say that a measurement of observable m is performed on X_1, by means of apparatus X_2, during the time interval (t, t')? The answer I gave in Chapter 7 is that the Hamiltonian of the complex system must be of a certain sort, regardless of what else we may say about the system. In order to keep separate our abstract formulation so far from the concrete instance of extant quantum mechanics, I shall use Greek letters here for all pure quantum states.

A possible situation $w = \langle x, y \rangle$ will evolve in time, with the evolution of the dynamical state x governed by quantum theory. That is, if the system is isolated, then there is a dynamic group of evolution operators U_d such that $w(t + d) = \langle x(t + d), y(t + d) \rangle = \langle U_d x(t), y(t + d) \rangle$. Notice that this gives us no information about how the value state evolves, except for such general constraints as that $x(t) R y(t)$ for all t.

A *process* is a map $t \rightarrow w(t)$ of time into possible situations. If t_1 is the initial time of a measurement of the sort just described (with X_2 as apparatus and X_1 as object), then the situation at time t_1 is of the following sort:

(1) $w(t_1) = \langle x_3(t_1), x_2(t_1), x_1(t_1); y_3(t_1), y_2(t_1), y_1(t_1) \rangle$

with: $x_3(t_1) = y_3(t_1) = I_{\psi \otimes \varphi}$

$x_2(t_1) = y_2(t_1) = I_{\psi}$

$x_1(t_1) = y_1(t_1) = I_{\varphi}$

$\varphi = \Sigma c_i \varphi_i$

Since all three systems begin in pure states in the case we are examining (and because of P16, this is sufficiently general for discussion), the initial situation is especially simple; it is just

(2) $\qquad w(t_1) = \langle I_{\psi \otimes \varphi}, I_\psi, I_\varphi; I_{\psi \otimes \varphi}, I_\psi, I_\varphi \rangle$

During the interval d, $x_3(t_1)$ is transformed by U_d to $x_3(t_2)$ where $t_2 = t_1 + d$; the equations governing measurement allow us therefore to infer an exact description of the final situation:

(3) $\ w(t_2) = \langle x_3(t_2), x_2(t_2), x_1(t_2); y_3(t_2), y_2(t_2), y_1(t_2) \rangle$

$\qquad$ with: $\ x_3(t_2) = I_{\Sigma c(i) \psi(i) \otimes \varphi(i)}$

$\qquad\qquad\quad x_2(t_2) = \sum c_i^2 I_{\psi(i)}$

$\qquad\qquad\quad x_1(t_2) = \sum c_i^2 I_{\varphi(i)}$

$\qquad\qquad\quad y_3(t_2) = x_3(t_2)$

$\qquad\qquad\quad x_2(t_2) R y_2(t_2)$ and $x_1(t_2) R y_1(t_2)$.

We can put also the last clause in quantum-mechanical terms. Denoting as $S(x_i)$ the image space of statistical operator x_i:

(4) $\ y_2(t_2) = I_{\psi'}$, for some vector ψ' in $S(x_2(t_2))$, i.e. in the subspace spanned by $\{\psi_i : c_i \neq 0\}$.

(5) $\ y_1(t_2) = I_{\varphi'}$, for some vector φ' in $S(x_1(t_2)) =$ the subspace spanned by $\{\varphi_i : c_i \neq 0\}$.

Since these are the total constraints on the final situation, we see that the process was indeterministic, because there are many possible situations that satisfy the description of $w(t_2)$ in (3)–(5) above. They are all the same with respect to dynamic states, but not with respect to value states.

7.5. *Probabilities of measurement outcomes*

We have now arrived at the interpretation of the probabilities which are calculated from the dynamic state. Briefly, the probability that initial dynamic state $x_1(t_1)$ gives us for $\langle m, E \rangle$ is just the probability that the final value state $y_1(t_2)$ will make $\langle m, E \rangle$ true. In the present context we assume that any observable is uniquely represented by a Hermitean operator, and it will be convenient to cease distinguishing observable m

and the Hermitean operator M that represents it. We already use the terms M, M' to stand for observables pertaining to X_1 and X_2 respectively, so we can write without ambiguity:

(6) $w_{(t)}$ is in $\langle M, E \rangle$ iff $y_1(t) \, (\langle M, E \rangle) = 1$
   $w(t)$ is in $\langle M', E \rangle$ iff $y_2(t) \, (\langle M', E \rangle) = 1$

I shall label as 'PQ' special assumptions that interpret the mathematical representation in Hilbert space.

> PQ1:   The pure dynamic states y are (represented by) one-dimensional projections I_φ and the observables are (represented by) Hermitean operators $M = \Sigma r_i I_i$, with $\{I_i\}$ an orthogonal set of projection operators such that:
>
> (*a*)   the eigenvalues r_i of M are the solutions of the equations $M\varphi = r_i\varphi = r_i I_i \varphi$
>
> (*b*) $y(\langle M, E \rangle) = 1$ iff $I_E^M \varphi = \varphi$, where $y = \varphi$ and $I_E^M = \Sigma\{I_i : r_i \in E\}$

The theory was designed from the outset to equate $\langle m, E \rangle = \langle 1_E m, \{1\} \rangle$; hence we see that every proposition in **V** corresponds to a projection operator:

(7) $w(t)$ is in $\langle M, E \rangle$ iff $y_1(t) = I_\varphi$ for some vector φ such that $I_E^M \varphi = \varphi$.

We may also postulate the other half of the correspondence:[17]

> PQ2:   If P is a Hermitean operator, it is (or represents) some observable.

Looking now at an arbitrary dynamic state x, we see that it induces a map of the projection operators into $[0, 1]$ with the following properties:

 (i) $x(I) = 1$ when I is the identity.
 (ii) If P and Q are orthogonal, then $x(P + Q) = x(P) + x(Q)$.

The second property follows because if P and Q are orthogonal then we can find a Hermitean operator M with spectral decomposition $\Sigma r_i P_i$ such that $P_1 = P$, $P_2 = Q$; in which case P and Q correspond to disjoint propositions in the range $\langle M \rangle$; and our basic postulate on states, P6, then entails (ii). But now it follows by Gleason's theorem[18] that

(8)   Each dynamical state x is (representable by) a statistical operator such that $x(\langle m, E \rangle) = Tr(xI_E^M)$.

This is a large part of what was contained in Born's Rule for calculating probabilities in quantum mechanics. It follows for us here, via Gleason's theorem, from the basic pairing of observables and pure dynamic states with operators, plus our theorem T15 which established that value states stand in an exact correspondence to pure dynamic states.

But there is a second part to Born's Rule, which cannot be deduced so far. What do the calculated probabilities attach to? What are they probabilities of? The answer that they are the probabilities of measurement outcomes, conditional on the measurement having been made, must be added to the theory as an interpretative postulate. I give it here for von Neumann–Lueders measurement:

PQ3:   (Born): When M is an observable pertaining to X_1, and the process $\langle w(t) \rangle = \langle x_i(t); y_i(t) \rangle$, $i = 1,2,3$, such that $w(t_1) \longrightarrow w(t_2)$ is a measurement of M on X_1 by X_2, with pointer-observable M', then $x_1(t_1)(\langle M, E \rangle)$ is the probability that $w(t_2)$ is in $\langle M, E \rangle$, and equally the probability that it is in $\langle M', E \rangle$.

PQ3 is consistent because the measured observable is essentially unique, as we saw in our previous discussions of von Neumann–Lueders measurement. Without that, the postulate could assign probabilities to propositions attributing sharp values to non-commuting observables.

Because x restricted to the value range $\langle m \rangle$ of a single observable is a classical probability measure, we deduce:

(9)   The probabilities $x_1(t_1)(\langle M, \{r_i\} \rangle)$ that M will have sharp value r_i in final situation $w(t_2)$ sum to 1; and hence the probability is 1 that M will have some precise value at the end of the measurement.

Similarly of course for the 'pointer'-observable. We can summarize this finding, recalling the role of the value state, as follows:

(10) The probability equals 1 that $y_2(t_2)$ is one of the pure states $I_{\psi(i)}$ and $y_1(t_2)$ one of the pure states $I_{\varphi(i)}$.

What we have not yet discussed is whether the final value state of the apparatus is a certain indication of the final value state of the object. Here we can derive the following fact, simply by applying the whole preceding account to a further measurement. Note that $M' \otimes M$ has among its eigenvectors $(\psi_i \otimes \varphi_i)$:

(11) $$(M' \otimes M)\,(\psi_i \otimes \varphi_i) = M'\psi_i \otimes M\varphi_i$$
$$= r'_i\psi_i \otimes r_i\varphi_i$$
$$= (r'_i\, r_i)(\psi_i \otimes \varphi_i)$$

and we deduce, just like above:

(12) If an $M' \otimes M$ measurement be made on a system with initial dynamical state $x_3(t_2)$ (i.e. the projection along $\Sigma c_i\psi_i \otimes \varphi_i$), then the probability equals 1 that (the outcome) will be $r'_i r_i$ for one of the indices i.

Thus, if we check the whole process by looking at the apparatus and simultaneously measuring M again independently (a complex new operation performed on the complex system $X_3 = (X_2 + X_1)$ as a whole), we are certain (probability 1) to find that the measurement apparatus correctly indicated the final value of M on the object system.

The modal interpretation, Copenhagen variant, is very minimal, and I shall end with a quick look at one way in which it might be possible to extend it. Intuitive accounts of measurements often point out that the weights e_i in the orthogonal decomposition $x = \Sigma e_i I_{\varphi(i)}$ of a mixed state are non-negative and sum to 1, hence act like classical probabilities. If we can read them as genuine probabilities, is there not also some hope of *deducing* the second half, PQ3, of Born's probability rule?

It would in any case be pleasing to be able to give a realistic reading to those weights. The obvious candidate for us would be:

(∗) If the dynamic state is $x = \Sigma e_i I_{\varphi(i)}$ and $\{\varphi_i\}$ is a complete orthogonal set, then the probability equals e_i that the value state is $I_{\varphi(i)}$ . . .

Unfortunately, the non-uniqueness of the decomposition when

some of the weights e_i are equal makes that assertion logically self-contradictory if maintained in general. So we would have to add:

(∗) ... provided all the weights e_i are different.

However, we can group together those states which appear with equal weights. If e_i is positive, then there can be at most finitely many vectors such that $e_j = e_i$ (since an infinite sum of equal positive numbers would be considerably higher than $1 = \Sigma e_i$). So we have something like this:

$$x = aI_1 + aI_2 + bI_3 + bI_4 + bI_5 + cI_6 + dI_7 + \ldots$$
$$= a(I_1 + I_2) + b(I_3 + I_4 + I_5) + cI_6 + dI_7 + \ldots$$

Now $I_1 + I_2$ is again a projection operator; in this way we arrive at a decomposition of form:

$$x = \Sigma\{aI_a : a \in A\}$$
$$I_a = \Sigma\{I_{\varphi(i)} : e_i = a\}$$
$$I_aI_b = 0 \text{ if } a \neq b; \Sigma I_a = I$$

and this is unique; it does not depend on the base $\{\varphi_i\}$ initially used.

We can now clearly interpret the values e_i (that is, the values a in A) as probabilities with reference to this decomposition. I state this here as a possible new postulate, which can be added if we wish:

PQ4: If the dynamic state is $x = \Sigma\{aI_a : a \in A\}$ where the I_a are orthogonal projection operators, as above, then the probability equals $a \dim I_a$ that the value state y is such that $I_ay = y$.

Here $\dim I_a$ is the dimension of the subspace on which I_a projects, that is, the size of an orthogonal basis for that subspace. But do we derive any advantage from such an addition? If we had only PQ1, PQ2, and PQ4, we would not be able to deduce that the observable M has a precise value at the end of an M-measurement. (This would follow only in the special case in which $x_1(t_2)$ has a unique decomposition into pure states, that is, only if all the coefficients c_i in $x_1(t_1) = I_\varphi$, with $\varphi = \Sigma c_i\varphi_i$, have distinct squares.)

So we could not replace PQ3 by PQ4 without loss. On the other hand, what PQ4 adds is only information about the possible situations which is not reflected in any measurement outcomes. Though not forced by the *problématique*, PQ4 could be added as an intuitively pleasing embellishment.[19]

8. MODAL INTERPRETATION OF COMPOSITION AND REDUCTION

Let us now set the quantum-logical approach aside again, and return to the main story.

The modal interpretation can be summed up in part by saying that in salient respects it is *as if* the Projection Postulate, and the ignorance interpretation of mixtures, were true. To attribute a (dynamic) state is to assert a statistical hypothesis—that is, to assert a related cluster of probability judgements. Those probabilities must be probabilities of *something*; contrary to von Neumann, we take that *something* to be not states but events, and take an event to consist in some observable having some value.

The touchstone for the interpretation will be the quantum-theoretical treatment of compound states. That is where the probabilities and possibilities are tangled, and where the holistic character of the quantum world becomes manifest. The familiarity of certain ways of thinking about this should not blind us to the complexity of the question: is it really even consistent to say that the world is *as if* those familiar ways are correct? Many questions are disarmed when the agnostic asserts only that the world is *as if* thus or so is the case—but not the question of consistency! And we have been shielded from the full impact of this question so far, because the most complicated system we have looked at has still consisted of only two parts: measuring apparatus and measured object.

In general, we must consider an N-body system $S = X_1 + \ldots + X_N$. The order in which the components are listed has no counterpart in reality; but when we describe the states and observables, some such order creeps into our representation. The system can also be mutually divided in various ways: $X_1 + X_2 + X_3$ into a two-component system if thought of as $X_1 + (X_2 + X_3)$, or as $(X_1 + X_2) + X_3$. The system $X_1 + X_3$ is also a subsystem of $X_1 + X_2 + X_3$; all these divisions are equally good as alternative representations.

The total system S has a dynamic state; so does each subsystem. If S is in state W, and S' is a subsystem, then S' is in reduced state $\#W[S']$. This reduction was discussed at some length in the preceding chapter. It has very nice formal features, which ensure consistency of attribution. Thus, $\#W[X_1 + X_2]$ can be reduced again to give a state to X_2, namely $\#(\#W[X_1 + X_2])[X_2]$. But, as consistency requires, that is just the state $\#W[X_2]$. Of course, if one of these components is a measurement apparatus, carrying out a measurement on one of the other components, this reduction of states is part of the formalization of the so-called Projection Postulate as well.

The reduced states are in general mixtures, even if the total state W of whole system S is pure. According to the modal interpretation, the actual values of observables correspond, for a system in mixed state W', to those made certain by *some* pure state x in the image space of W'. Thus we have for each subsystem S' a dynamic state $\#W[S']$ and a value state $\lambda[S']$. If W itself was pure, let us say it is equally represented by the vector φ, we arrive then at the following picture for a three-component system:

$$\varphi;\ \lambda_{123}$$

$$\#\varphi[1, 2];\ \lambda_{12} \qquad \#\varphi[2, 3];\ \lambda_{23} \qquad \#\varphi[1, 3];\ \lambda_{13}$$

$$\#\varphi[1];\ \lambda_1 \qquad \#\varphi[2];\ \lambda_2 \qquad \#\varphi[3];\ \lambda_3$$

where I have indicated the subsystems of $S = (X_1 + X_2 + X_3)$ just by indices; e.g., $\#\varphi[1, 2]$ and λ_{12} are respectively the dynamic state $\#\varphi[X_1 + X_2]$ and the value state $\lambda[X_1 + X_2]$.

Both the ignorance interpretation of mixtures and the modal interpretation require this picture to remain consistent subject to the following requirements:

(a) $\#\varphi[S']$ is the reduction of state φ of S with respect to subsystem S'.

(b) $\lambda[S']$ is a pure state in the image space of $\#\varphi[S']$, i.e., is possible relative to the latter.

But it is clear that there is an additional requirement, which comes into play exactly when the compound has more than two components. The situation would not be as if the ignorance interpretation were true if we could not *pretend* that the dynamic state of $X_1 + X_2$ is *really* $\lambda[X_1 + X_2]$! For that interpretation said that $\#\varphi[1, 2]$ just represents incomplete knowledge. So the following alteration of part of the picture:

$$\lambda_{12}; \quad \lambda_{12}$$

$$\#\lambda_{12}[X_1]; \quad \lambda_1 \qquad \#\lambda_{12}[X_2]; \quad \lambda_2$$

must still be consistent, subject to demands (*a*) and (*b*) above. That is, λ_1 should be possible not only relative to $\#\varphi[X_1]$, or equivalently $\#(\#\varphi[12])[X_1]$, but also relative to the reduction $\#\lambda_{12}[X_1]$ of the value state λ_{12}. This extra criterion therefore amounts to

(*c*) If $S'' \subseteq S'$ then $\lambda[S'']$ is a pure state also in the image space of $\#(\lambda[S'])[S'']$.

Whenever several overlapping constraints are imposed jointly, inconsistency threatens.[20] What about here?

We can settle part of the question at once by reasoning from definitions. The consistency of (*b*) and (*c*) requires:

Theorem: If φ is in the image space of W, then the image space of $\#\varphi[S']$ is part of the image space of $\#W[S']$.

The total system, of which W and φ can be states, can here be thought of as divided into two parts, $S'' + S'$. Let the Hilbert space for this total system be $H_1 \otimes H_2$, so $\#W$ and $\#\varphi$ are states on H_2. The observable which is the projection on the image space of $\#W$ we may call A. Clearly, A has expectation 1 in $\#W$ and $I \otimes A$ has expectation 1 in W. Since φ is in the image space of W it is possible relative to W, hence $I \otimes A$ also has expectation 1 in W. Therefore A has expectation 1 in $\#\varphi$. This requires that $\#\varphi$ is really a state on the subspace on which A projects, or equivalently that all eigenvectors of $\#\varphi$ lie in that subspace, or again equivalently, that the image space of $\#\varphi$ is part of it.

But this little result, though reassuring, is not enough. For the first alteration of the picture, which had to remain consistent, is matched by another one:

$$\lambda_{23}; \quad \lambda_{23}$$

$$\#\lambda_{23}[X_2]; \quad \lambda_2 \qquad \#\lambda_{23}[X_3]; \quad \lambda_3$$

and we see λ_2 appearing in both altered pictures. To put it differently, criterion (c) demands that $\lambda[X_2]$ be possible relative to the reductions of the value states of *both* $(X_1 + X_2)$ and $(X_2 + X_3)$. Those value states were, as it were, random selections from two image spaces (of $\#\varphi[12]$ and $\#\varphi[23]$). It is not obvious, and certainly not entailed by the above theorem, that consistency allows this sort of overlap, i.e. that there will exist a state (a candidate for being $\lambda[X_2]$) which is possible relative to the two reductions of those randomly selected elements of the two image spaces.

Nevertheless, though not obvious, it is so.[21] The proof, which follows, makes precise the qualitative sense, in which mixed states are really quite classical. Reasoning about mixed states can thus be, to a very large extent, like reasoning about pure states under conditions of ignorance—even when these mixed states are attributed by reduction to interacting components of a complex system in a truly 'tangled' quantum-mechanical state.

9. CONSISTENCY OF THE DESCRIPTION OF COMPOUND SYSTEMS

Because the proof is a little intricate, I will begin with a three-body example, which will also serve to introduce the notation.[22] The total system $S = (X_a + X_b + X_c)$ has pure state $\varphi = \Sigma e_{ijk}(x_i \otimes y_j \otimes z_k)$ with all coefficients non-zero. Each subsystem is identified by an ordered set of indices, e.g. $S_{ab} = (X_a + X_b)$. The numerical indices i, j, k are of course variables ranging over positive integers, and if Σ appears without subscripts it sums over all free numerical variables. Once such a variable is replaced by a numerical constant, it is no longer summable: in $\Sigma e_{i3k}(x_i \otimes z_k)$ the sum is over i and k.

To each subsystem we assign a dynamic state by reduction in the usual way, denoted by W with subscripts. Thus:

$$W_{abc} = \text{the pure state } \varphi = \sum e_{ijk}(x_i \otimes y_j \otimes z_k)$$

$$W_{ab} = \#\varphi[X_a + X_b] = \sum |u(k)|^2 I_{u(k)}$$

$$\text{where } u(k) = \sum_{ij} e_{ijk}(x_i \otimes y_j)$$

$$W_{ac} = \#\varphi[X_a + X_c] = \sum |v(j)|^2 I_{v(j)}$$

$$\text{where } v(j) = \sum_{ik} e_{ijk}(x_i \otimes z_k)$$

$$W_a = \#\varphi[X_a] = \sum |w(j, k)|^2 I_{w(j,k)}$$

$$\text{where } w(j, k) = \sum_i e_{ijk}x_i$$

Now we shall assign a value state to each subsystem, denoted as λ with subscripts. We do this by choosing a new number each time a numerical variable needs to be instantiated. Thus:

$$\lambda_{abc} = W_{abc} = \varphi$$

$$\lambda_{ab} = u(2) = \sum e_{ij2}(x_i \otimes y_j)$$

$$\lambda_{ac} = v(3) = \sum e_{i3k}(x_i \otimes z_k)$$

$$\lambda_a = w(3,2) = \sum e_{i32}x_i$$

For brevity I have omitted the normalization factors. To obtain the unit vector representing the state λ_{ab}, divide $\Sigma e_{ij2}(x_i \otimes y_j)$ by its length which is $\Sigma|e_{ij2}|^2$; similarly for the others.

Note that, in the choice of λ_a, there was no freedom left: k had already been instantiated to 2 and j to 3. The selection of prime numbers, instead of some other set of numerical constants, and also the order in which the states are assigned by this procedure, is logically immaterial—provided that none of these vectors is *zero*, which is here guaranteed because all coefficients are assumed to be non-zero.

That the value state chosen thus is always possible relative to the dynamic state follows from the general reduction results in Chapter 7. We must now check criterion (*c*): that reduction of value states instead gives us a consistent alternative:

$$\#\lambda_{ab}[X_a] = \sum |t(j)|^2 I_{t(j)}$$

$$\text{where } t(j) = \sum_i e_{ij2} x_i$$

$$\#\lambda_{ac}[X_a] = \sum |s(k)|^2 I_{s(k)}$$

$$\text{where } s(k) = \sum e_{i3k} x_i$$

And indeed, λ_a is possible relative to both these reduced value states, because when we again instantiate k to 2 and j to 3 we find

$$s(2) = t(3) = \sum e_{i32} x_i = \lambda_a$$

All we need to do to complete the proof is generalize the procedure in attractive notation, to show why these different roads had to lead to the same result. We note also that the recipe we follow here leads to one 'possible world' for each choice of bases in which to represent the pure state φ, and of series of numerical constants used for instantiation, and of order of assignment of value states to subsystems. Thus, a single system with given dynamic total state allows a whole system of alternative possible distributions of values to observables.

To generalize all this, we need to lift two restrictions. The first is that in the expression describing φ all coefficients have been assumed to be non-zero. The second is the restriction of the number of components N to $N = 3$. To lift the first restriction is easy, for we can always write φ in such a form—an elementary point about vector geometry.[23] Next we must repeat the above for N arbitrary.

We consider therefore the general state

$$\varphi = \sum e_{i(1),\ldots,i(n)} (x^a_{i(a)} \otimes \ldots \otimes x^n_{i(n)})$$

with all coefficients again assumed to be non-zero. Let S be the total set of letter indices $a, \ldots, n$, and let T be the set of letters used as numerical indices (variables) $i(a), \ldots, i(n)$. Any integer numerical constant 1, 2, 3, . . . can replace a numerical index (instantiation). We need several more pieces of notation:

f is the formula $e_{i(a),...,i(n)}(x^a_{i(a)} \otimes \ldots \otimes x^n_{i(n)})$.

Hence, in our notation, $\varphi = \Sigma f$. If g is any formula, then

$g^{S'}$ is the result of deleting from formula g every term $x^k_{i(k)}$, for each letter index k which belongs to the subset S' of S.

$g|S'$ is the result of replacing in formula g every numerical index $i(k)$ by the integer constant $c(k)$, for each letter index k which belongs to set S'.

To illustrate this new notation, let us see how it works for our three-body example first.[24] We rewrite i, j, and k now as $i(a)$, $i(b)$, and $i(c)$. Thus the vector $v(j)$ which appeared in the description of W_{ac} should now be denoted as

$$v(b) = \sum_{\substack{i(m) \\ m \neq b}} e_{i(a)i(b)i(c)}(x^a_{i(a)} \otimes x^c_{i(c)})$$

It is formed by dropping the term $x^b_{i(b)}$ from the expression f used to describe φ as Σf, and then prefixing a summation symbol for a sum over the *other* indices, i.e. in this case $i(a)$ and $i(c)$. This is of course a prelude to the replacement of the variable term $i(b)$ with a constant term such as '3'. Now if we let T be the set containing just the symbol b, then we can write

$$v(b) = \sum_{\substack{i(m) \\ m \in S-T}} \text{(the expression } f \text{ with all those terms } x^n_{i(n)} \text{ removed for which } n \text{ is in } T)$$

To make the description more perspicuous, however, let S' be the set of all letter indices *except* b, so that $S' = S - T$ and $T = S - S'$. Then we can redescribe the same vector as

$$v(b) = \sum_{\substack{i(m) \\ m \in S'}} \text{(the expression } f \text{ with all those terms } x^n_{i(n)} \text{ removed for which } n \text{ is in } S - S')$$

$$= \sum_{S'} f^{(S-S')}$$

which shows the first bit of notation in use.

Next, when we constructed λ_{ac} we did this by choosing $v(j)$ with $j = 3$. In our present notation that means that we let the variable $i(b)$ take a certain constant value $K(b)$ uniquely associated with letter index b. So we obtain

$$\lambda_{ac} = v(K(b)) = \sum e_{i(a)K(b)i(c)}(x^a_{i(a)} \otimes x^c_{i(c)})$$

the summation being with respect to variable terms that still remain present. In this instance the set $S' = \{a, c\}$, so we can also write this as

$$\lambda(S') = \sum \text{(the result of changing expression } f \text{ by first deleting}$$
all terms $x^n_{i(n)}$ and then replacing each occurrence of the term $i(n)$ which remains by the constant term $K(n)$, for all letter indices n in $S - S'$)

$$= \sum f^{(S - S')}|(S - S')$$

The summation is with respect to all the variable terms which remain, which are just those terms $i(n)$ for which letter index n is in the set S'. The ones outside S' have now disappeared.

We can now go further and use the notation to put the General Reduction Theorem of Chapter 7 in the perspicuous form:

(A): The reduction of $\varphi = \Sigma f$ to the Hilbert space spanned by the vectors $x^c_{i(c)} \otimes \ldots \otimes x^k_{i(k)}$, with $S' = \{c, \ldots, k\}$ is the statistical operator

$$\#\varphi[S'] = \sum |u(S - S')|^2 \, I_{u(S-S')}$$

where $u(S - S') = \sum_{s'} f^{(S-S')}$

Now we make the assignment of dynamic and value states to all subsystems (represented by subsets S' of the index set S):

(B): $\quad W(S') = \#\varphi[S']$

(C): $\quad \lambda(S') = \sum f^{(S-S')}|(S - S')$

Since the pure state represented by $\lambda(S')$ is one of the summands in the formula given above to describe $\#\varphi[S']$, it follows at once that this state is in the image space of $W(S')$.

Lemma: If $S'' \subseteq S' \subseteq S$, then $\lambda(S'')$ is possible relative to $\#\lambda$ $(S')[S'']$.

To prove this, we must apply (A) to (C). This yields

(D): $\qquad \#\lambda(S')[S''] = \sum |w(S - S'')|^2 \, I_{w(S - S'')}$

$\qquad$ where $w(S - S'') = \sum_{S''} (f^{(S-S')}|(S - S'))^{(S-S'')}$

Because S'' is part of S', the only terms $x^k_{i(k)}$ that remain in $w(S - S'')$ have k in the smaller set S''. But the coefficients are still written with index variables that may be in S' though outside S''. So we can also describe $w(S - S'')$ as

$$w(S - S'') = \sum_{S''} f(S - S'')|(S - S')$$

since the removal of the terms $x^k_{i(k)}$ with k outside S' can just as well be done in one step. The only variable terms that remain here (uninstantiated, and unbound by a summation) are the terms $i(n)$ with n in $S' - S''$. Hence one of the summands in (D) is the formula

(E): $\qquad |w(S - S'')|(S' - S'')|^2 I_{w(S-S'')|(S'-S'')}$

in which all those remaining free variables are replaced by the corresponding numerical constants. But

(F): $w(S - S'')|(S' - S'') = \sum_{S''} ((f^{S-S''})|(S - S'))|(S' - S'')$

$\qquad\qquad\qquad\qquad = \sum_{S''} f^{(S-S'')}|(S - S'')$

$\qquad\qquad\qquad\qquad = \lambda(S'')$

as we were required to prove, for this shows that $\lambda(S'')$ is indeed a relatively possible state, being in the image space of the mixture.

10. INTERPRETATION AND THE VIRTUE OF TOLERANCE

When Descartes wrote (in his *Principles of Philosophy*, iv. 204) that 'touching the things which our senses do not perceive, it is sufficient to explain what they could be, though perhaps they are not as we describe them', he voiced a thoroughly empiricist sentiment. Under a given interpretation, a physical theory describes how things are—one way they might indeed be—but the story is not unique, and for an empiricist it need not be unique.

There is more than one tenable interpretation of quantum mechanics. It is true that each adds something to the theory: von Neumann surprises us with acausal transitions, and implies action at a distance behind the scenes, even if the transitions are respectably dressed as superselection rules. The modal account surprises us too, by separating observables from states, and even from each other, in unexpected ways. These additions are 'empirically superfluous'. That is a virtue if we simply want to have our questions answered in ways that do not contradict the theory's empirical adequacy, but it is a vice if empirical adequacy is all we want to take seriously. To complain about empirically superfluous, odd, or surprising additions, however, is not to deny tenability.

Quantum theory is well understood. How can I say this when there are several mutually contradictory, and perhaps equally plausible, interpretations in the field? Well, I assert it exactly because we do have interpretations that have so far stood the test of debate. Perhaps one of the above, or another such as Everett's or Kochen's which I have not discussed, will eventually prove far superior in ways we don't yet appreciate. But that is not needed, as far as I am concerned, to call quantum mechanics a well understood theory.

Could one object that I do not know what the theory says if I cannot single out the uniquely correct interpretation? In a strict sense this is true. The theory was developed with eyes fixed firmly on empirical success, in collaboration by many people. It was not a case of a Newton, who tried to delimit exactly what his theory said the world was like, and made it include the existence of Absolute Space and the hypothesis that the solar system's centre of mass is at absolute rest, before publishing. The tenability of several interpretations does mean that the theory as such was not logically complete but left various questions open. But is that so important? Newton's talk about God's sensorium was quietly ignored by later Newtonians, and the question of determinism on which they did often have an official stand was actually not nearly as well settled in the theory as they thought (cf. Earman 1986).

In my own opinion, that objection can be made but is misplaced. Suppose the theory had originally been stated with von Neumann's interpretation built in, and all its developers, in Germany as well as Copenhagen, had been unanimous. Then the theory would have been much more definite in what it said the world was like. It would also have gone considerably beyond claims about the empirical phenomena. Its empirical success would not, in my eyes, have constituted adequate grounds for belief in its truth, taken as a whole. As I see it, the result would not have been a better theory, by science's own criteria of success. The less committed way it was actually done seems all to the good.

Why then be interested in interpretation at all? If we are not interested in the metaphysical question of what the world is really like, what need is there to look into these issues? Well, we should still be interested in the question of how the world *could be* the way quantum mechanics—in its metaphysical vagueness but empirical audacity—says it is. That is the real question of understanding. *To understand a scientific theory, we need to see how the world could be the way that the theory says it is.* An interpretation tells us that. The answer is not unique, because the question 'How could the world be the way the theory says it is?' is not the sort of question to call for a unique answer. Faith in the actual truth of a good answer, so interpreted, is not required by understanding, nor does it help.

EPR: When Is a Correlation Not a Mystery?[1]

EINSTEIN, Podolsky, and Rosen presented their argument in 1935. They did not present it as a paradox, but rather as a demonstration that quantum mechanics does not, even in principle, describe all there is. The quantum-mechanical state, even if completely specified, does not encode all true (physically significant) information. This conclusion is, *in some form*, common to almost all interpretations of the theory. Thus, von Neumann would say that the state does not tell us which possible acausal transition will actually happen in a measurement, and the modal interpretation says that the state does not give full information about the values observables have or come to have. Indeed, how could indeterminism be reflected otherwise? But the EPR paper purports to establish a more specific conclusion about the values of incompatible observables. As we shall see, the fascination with the argument concerns a cluster of problems, centring on the non-classical correlations which we encountered already in the chapter on the empirical basis of quantum mechanics (Chapter 4) and have come across at many junctures since.

1. THE PAPER BY EINSTEIN, PODOLSKY, AND ROSEN

The structure of the EPR argument is both simple and clear. It is already set out in the abstract at the beginning, which I quote here in full:

In a complete theory there is an element corresponding to each element of reality. A sufficient condition for the reality of a physical quantity is the possibility of predicting it with certainty, without disturbing the system. In quantum mechanics in the case of two physical quantities described by non-commuting operators, the knowledge of one precludes the knowledge of the other. Then either (1) the description of reality given by the wave function in quantum mechanics is not complete or (2) these two quantities cannot have simultaneous reality. Consideration of the problem of making predictions concerning a system on the basis of measurements made on another system that had previously interacted with it leads to the result that if (1) is false then (2) is also false. One is thus led to conclude that the description of reality as given by a wave function is not complete. (EPR 1935, 777)

While the structure of the argument is logically clear, therefore, the same cannot be said of the terms employed in it. Such phrases as 'element of reality' are to be used lightly, and do not easily bear the weight of a searching critique. But if we can state the argument using only more familiar terms like 'value of an observable', then the problem will survive such worries.[2] Let us rephrase the second and third sentence in the abstract as follows:

(1) If the outcome of a measurement of observable A, pertaining to system X, and made without disturbing the state of X, will have value a_k with certainty, then A has value a_k (at the appropriate time, but whether or not that measurement is actually carried out).

(2) If A and B are represented by non-commuting operators which have no eigenvectors in common, then it is not possible to predict with certainty the outcomes of A and B measurements for the same time (conditional on being carried out) on the basis of any quantum-mechanical state.

The second is obvious: the Born probability cannot be 1 for outcomes a_k and b_m if A and B have no eigenvectors in common. The first is made very plausible by the 'no-disturbance' clause: if the outcome is not *made* certain by manipulating the subject, then surely it must reflect what was simply there?

To continue the argument, then, a situation must be exhibited in which we can predict with certainty for two incompatible observables. That is, we must produce a situation in which both the following are true:

(a) If A is measured, then the outcome will definitely (with probability 1) be a_k

(b) If B is measured, then the outcome will definitely (with probability 1) be b_m

where A, B are incompatible observables pertaining to the same system. And at first sight it is impossible to find such a situation, because of point (2) above.

But suppose that system X interacts with system Y and the interaction produces a correlated state (similar to that in a measurement interaction):

$$\varphi = \sum c_i |a_i\rangle \otimes |b_i\rangle$$

where $i = 1, \ldots, n$ and $|a_i\rangle$, $|b_i\rangle$ are eigenvectors of observables A and B, pertaining to X and Y respectively. In that case, if we measure B on Y—or equivalently $I \otimes B$ on $X + Y$—and also measure A on X—or equivalently $A \otimes I$ on $X + Y$—we have the following conditional certainty:

The outcomes could not be b_i, a_j with $i \neq j$; therefore, if the first measurement yields outcome b_i then the second must yield a_i.

Now this is still true if the interaction was such as to let X and Y drift very far apart from each other after the initial coupling. In such a case the B-measurement would be made in one place, and the A-measurement very far away. The distances and times could be such that even light signals cannot connect the two measurement operations. So clearly, the person who carries out the B-measurement (i) does not disturb or interact with system X, and (ii) can predict with certainty what his colleague, who measures A, is finding at that very moment, if indeed he makes the A measurement.

On the basis of (1), the B-measurer should now conclude: I have found value b_k, so observable A pertaining to X definitely does have value a_k—whether or not my colleague measures it.

So far this looks like an argument for the Projection Postulate. For it asks us to conclude something which could be explained—given von Neumann's interpretation rule—by the assertion that the B—or, better, the $I \otimes B$—measurement 'projects' φ into an $I \otimes B$ eigenstate. This eigenstate would have to be one of the $\{|a_i\rangle \otimes |b_i\rangle\}$, of course. And if that is happening instantaneously, then $X + Y$ has been put into state

$|a_k\rangle \otimes |b_k\rangle$, so no wonder the other measurer, at that very moment, finds value a_k if he 'looks'. The eventual lack of cogency in this argument, however, has already appeared in our previous discussions.

But now the EPR argument goes on: we can choose a different basis $\{|r_m\rangle \otimes |s_n\rangle\}$ for the subspace $[\{|a_i\rangle \otimes |b_j\rangle\}]$, and if all the coefficients c_i are equal, then we have an equation

$$\varphi = \sum c_i |a_i\rangle \otimes |b_i\rangle = \sum d_j |r_j\rangle \otimes |s_j\rangle$$

with all the d_j equal too ($j = 1, \ldots, n$). This point is easily illustrated, for example by rotating a basis. But consider now the observable R with eigenstates $|r_m\rangle$; it has no eigenvectors in common with A, if we chose a radically different basis. Similarly for S and B. So we have A incompatible with R, and B incompatible with S.

Now the B-measurer may reflect that, instead of measuring B, he could choose to measure S on Y. Again he has a conditional certainty:

> If we measure S and R, the outcomes could not be s_n, r_m with $m \neq n$; therefore, if the first measurement yields s_n, the second is certain to yield r_n.

So if he does measure S, and finds value s_j, he can conclude: if my colleague measures R, he is sure to find r_j. Since I can predict this with certainty, by (1) it follows that R has value r_j, whether measured or not.

2. INITIAL DEFENCE OF THE ARGUMENT

Have we now arrived at the conclusion that A and R both have definite values? No, not yet; for as Copenhagen scientists were quick to point out, the character in the story could not measure both B and S. His prediction with certainty cannot be made till he has carried out his own measurement—and then he can go through the above reasoning only for that observable which he actually measured.

The EPR paper replied to this sort of objection pre-emptively in a way that sharpens the interpretation of (1), but which also appears to constitute a rejection of the Projection Postulate. The authors use 'collapse of the wave packet', as a means of calculation, and in their phrasing suggest a real change of state. But they reject the idea that an operation on Y could affect the whole system $X + Y$—in this separated case—so as to produce any real change in X:

> We see therefore that, as a consequence of two different measurements performed upon the first system, the second system may be left in states with two different wave functions. On the other hand, since at the time of measurement the two systems no longer interact, no real change can take place in the second system in consequence of anything that may be done to the first system. This is, of course, merely a statement of what is meant by the absence of interaction between the two systems. Thus, *it is possible to assign two different wave functions . . . to the same reality*. (EPR 1935, 779)

And they add later that to suppose that what the second system X is like could be brought about by this action on Y, is unacceptable:

> This makes the reality of P and Q depend upon the process of measurement carried out on the first system, which does not disturb the second system in any way. No reasonable definition of reality could be expected to permit this. (EPR 1935, 780)

When I said before that the first part of the demonstration looked like an argument for the Projection Postulate, it was because that postulate offers an explanation for how observable A could become so predictable. Here, Einstein, Podolsky, and Rosen reject any explanation of a sort that entails a real change occurring in X, with the character of that change depending on what is done to Y, in these circumstances. The projection of φ into $|a_i\rangle \otimes |b_i\rangle$ makes X go into $|a_i\rangle$, so either that is not a physical change of state, or else they are rejecting it as impossible.

Exactly at this point, many readers suspect that Einstein, Podolsky, and Rosen have run into logical trouble of their own. But I think they could be sanguine here. The structure of the argument is that of a *reductio*, or of St Anselm's 'Even the fool sayeth in his heart . . .'. That is, the argument proceeds by calculating all probability as quantum mechanics prescribes, hence by 'reduction of the wave packet'. (This has the *formal* structure reified by the Projection Postulate, which asserts that this reduction represents a physical change, just as much as the wave function represents a physical state.) Using this calculation, we arrive at two remarkable conditional probabilities:

(3) $P(A = a_k \,|\, B = b_k$ **and** B, A are measured$) = 1$
    $P(R = r_j \,|\, S = s_j$ **and** R, S are measured$) = 1$

where '$X = x$' means 'the X-measurement has outcome x'. These conditional probabilities follow, without question, from the quantum mechanical description of the situation.

To this demonstrated conclusion the EPR paper added premisses which say that, in a situation in which these consequences are true, the following are also true:

(4) $P(A = a_k \,|\, A$ is measured$) = 1$
    $P(R = r_j \,|\, R$ is measured$) = 1$

for some (unknown) indices k and j. And finally, they add that when this is so the following are also true:

(5) A has value a_k
    R has value r_j

for these same indices k and j.

Those extra premisses which lead from (3) to (4) and (5) are somehow conveyed by the passages I have quoted.[3]

What I have said so far does not identify those premisses precisely, but it does imply that they are of the sort that make up an interpretation, that they are at odds with von Neumann's interpretation (because they forbid a change of state for X), and that they either are at odds with or at least go well beyond the modal interpretation. Therefore we are not led inevitably to Einstein, Podolsky, and Rosen's conclusion, since their premisses are not uncontroversial.

Without going further into this matter, we note that it was just the remarkable conditional probabilities (3)—the correlation in statistical predictions for separated systems—that make EPR important for everyone, regardless of interpretative agreement or disagreement. They represent the non-classical, empirically testable correlations to which I devoted Chapters 4 and 5. In the remainder of this chapter I propose to discuss the general challenge they present for any interpretation of quantum mechanics.

3. THE STEP TO EMPIRICAL TESTABILITY

The more orthodox physicists pointed out that the one observer who can calculate the probabilities for joint measurement outcomes of pairs $\langle A, B \rangle$ and $\langle R, S \rangle$ cannot after all measure both A and R. In consequence, they may well have tacitly supposed that the discussion must always remain a dispute about hidden elements of reality, with no direct experimental test possible. Quantum mechanics could then still maintain itself against classical intuitions by some 'inference' from overall success.

But this idea was mistaken. There is indeed no direct experimental test to divide opinions about what happens in measurements pertaining to *two* pairs of correlated variables. *But there is for three pairs.* This was John Bell's insight in the early 1960s. There the radically non-classical structure of quantum mechanics (QM) shines through on the macroscopic level. We have discussed this in a (logically) pre-QM context in Chapter 4. Now we should look at the exact transition from EPR to Bell. I shall describe it intuitively here and give details in *Proofs and illustrations*.

Let us begin with two pairs of observables, but introduce a more manageable notation. Suppose observables F, F' are respectively correlated with G, G' and pertain to different components in a compound system $X + Y$. Let the total state be

$$(1) \qquad \varphi = \sum a_i f_i \otimes g_i = \sum b_i f_i' \otimes g_i'$$

where the lower-case letters are used for eigenvectors. Question: can we find a violation of Bell's Inequalities in such a paradigmatic EPR situation?[4] The answer is *no*, for we can deduce (see below) that

$$(2) \qquad a_i^2 = b_i^2 \text{ and } (f_i \cdot f_i')^2 = (g_i \cdot g_i')^2$$

so the form of (1) does not really have the generality suggested by that notation.

There is however the result already noted by Schroedinger in his discussion of EPR (see further Section 6.1 below) that each correlation brings infinitely many others in tow. We need only note this point modestly as:

(3) Given (1) for distinct observables $F \neq F'$, there exists a distinct third pair of observables F'' and G'' such that

$$\varphi = \sum a_i f_i \otimes g_i = \sum b_i f_i' \otimes g_i' = \sum c_i f_i'' \otimes q_i''$$

as will be shown below. The construction will be such that F, F', and F'' are mutually incompatible. But with three pairs of such correlated observables we can make up the quantum-mechanical counter-examples to Bell's Inequalities. Thus one might say that the EPR type of situation implicitly contains the possibility of violating those inequalities. In view of (2) above, it will be a matter just of choosing the three bases $\{f_i\}$, $\{f_i'\}$, and $\{f_i''\}$ for the same component X of $X + Y$, so that the inner products are properly related. That we can do; I shall give a simple example below.

So now quantum mechanics is seen to predict, or at least to allow as possible, such violations of Bell's Inequalities as were discussed in Chapter 4. Unlike the causal theories discussed there, quantum mechanics is therefore not refuted by phenomena which exhibit such violations.

Proofs and illustrations

For definiteness, suppose first that F, F' pertain to system X, and have non-overlapping orthonormal bases $\{f_i\}$ and $\{f_i'\}$ of eigenvectors, similarly for the observables G, G' with system Y and base $\{g_i\}$ and $\{g_i'\}$. The total state φ of the system $X + Y$ could now have the following form:

(1) $$\varphi = \sum a_i f_i \otimes g_i = \sum b_i f_i' \otimes g_i'$$

The probability that a joint measurement of F and G yields the ith eigenvalue of F and the jth eigenvalue of G equals

(2) $$(\varphi \cdot f_i \otimes g_j)^2 = a_i^2 \text{ if } i = j; = 0 \text{ if } i \neq j$$

Similarly for F' and G', F and G', and so on. Let us focus on the *first* eigenvalue as example, and denote as $P(FG)$ the probability that such a joint measurement yields the 1st value of F and also the 1st value of G; similarly for $P(FG')$, $P(F'G)$, $P(F'G')$:

Lemma 1: $P(FG') = b_1^2(f_1 \cdot f_1')^2 = a_1^2(g_1 \cdot g_1')^2$

Proof:
$$(f_1 \otimes g'_1 \cdot \varphi) = (f_1 \otimes g'_1 \cdot \sum b_i f'_i \otimes g'_i)$$
$$= \sum b_i (f_1 \cdot f'_i)(g'_1 \cdot g'_i)$$
$$= b_1 (f_1 \cdot f'_1)$$

from which the first equation follows. The second follows *mutatis mutandis* if we use the first decomposition of φ instead of the second in (1):

Lemma 2: $b_1^2 = a_1^2$ and $(g_1 \cdot g'_1)^2 = (f_1 \cdot f'_1)^2$

Proof: The argument in Lemma 1 establishes similarly:

$$P(F'G) = a_1^2 (f_1 \cdot f'_1)^2 = b_1^2 (g_1 \cdot g'_1)^2$$

Hence

$$P(FG') + P(F'G) = [a_1^2 + b_1^2](f_1 \cdot f'_1)^2$$
$$= [a_1^2 + b_1^2](g_1 \cdot g'_1)^2$$

Therefore, $(f_1 \cdot f'_1)^2 = (g_1 \cdot g'_1)^2$. Combining this with Lemma 1, we conclude also that $a_1^2 = b_1^2$. Our focus on the first values in the orderings was inessential, so the equalities hold for each index i, and not just for the first.

Remark 1: If we allow F and F' or G and G' to have overlapping bases, the same proof yields the weaker result that for each i there is some index j such that $a_i^2 + b_j^2$ and $(g_j \cdot g'_j)^2 = (f_i \cdot f'_i)^2$. But by the symmetry of the situation, this will also be true for all j and corresponding i. Therefore we can re-order the bases so as to set $i = j$ and obtain the same result.

Remark 2: We can now write condition (1) in the new form:

$$(3) \qquad \varphi = \sum a_i f_i \otimes g_i = \sum \pm a_i f'_i \otimes g'_i$$

But we can eliminate the $\pm$ without loss of generality, for if the vector f_1 in basis $\{f_i\}$ is replaced by $-f_1$, the result is a new orthonormal basis anyway. All this is so even if the numbers are complex, since if $b_1^2 = a_1^2$ because $b_1 = a_1^*$, we can similarly replace one basis by another one so as to have equality there. That is so because if $a = b^*$ then there is some number r and angle x such that $a = re^{ix}$ and $b = re^{-ix}$. So then, if we replace any vector u by $u' = e^{-2ix}u$, we see that $bu = re^{-ix}(e^{2ix}u')$ $= au'$. Therefore, if equation (1) holds, then so will equation (3) for some (new) choice of bases.

Theorem: There is a distinct third pair of correlated variables if there are two pairs; i.e., condition (1) can be rewritten as

$$(4) \qquad \varphi = \sum a_i f_i \otimes g_i = \sum a_i f_i' \otimes g_1' = \sum a_i f_i'' \otimes g_i''$$

Proof: We first drop the $\pm$ sign in condition (3) as indicated. Then let U, V be the unitary operators defined by

$$Uf_i = f_i'$$

$$Vg_i = g_i'$$

Then $(U \otimes V)(\varphi) = \varphi$. So φ is an eigenvector of $(U \otimes V)$ and hence also of the unitary operator $(U \otimes V)^2$. But then

$$\varphi = \sum a_i U f_i' \otimes V g_i'$$

The sets $\{Uf_i'\}$ and $\{Vg_i'\}$ are new orthonormal bases of spaces H_1 and H_2 respectively. Therefore we have our third pair of correlated variables, unless the second step took us back to the first decomposition. But that is possible only if U^2 and V^2 act as the identity operator on the two original bases. However, a unitary operator is its own inverse only if it is itself the identity or minus the identity. (Reason: it can be written in form $\exp(iH)$, whose square is $\exp(2iH)$.) Notice also that, since the bases are distinct and related in that way, they are eigenvector bases for distinct maximal observables, which are mutually incompatible. It suffices to choose eigenvalues without degeneracy.

To align the present discussion with Chapter 4, let each of the observables only have two values ($i = 1, 2$) and label them as follows.

The observable measured is:
F if the first (left) apparatus is in setting 1
F' if it is in setting 2
F'' if it is in setting 3

and similarly for G, G', and G'' with the second (right) apparatus. What we write here as $P(F'G')$ is then the probability, designated in Chapter 4 as p_{22}, of getting the first eigenvalue ('particle goes through filter') on both the left and right side, with both apparatuses in setting 2. One member of Wigner's formulation of Bell's Inequalities will then be violated, for example the one written alternatively in the following two ways:

$$p_{12} + p_{23} \geq p_{13}$$

$$P(FG') + P(F'G'') \geq P(FG'')$$

Note that here the settings are different on the two sides; e.g., $p_{12} = P(FG')$ is the probability of getting the first eigenvalue on both sides ('particle goes through filter on both sides') with the left apparatus in setting 1 and the right apparatus in setting 2.

The probabilities (find 1st values of both observables if jointly measured) are the squared moduli of the calculated numbers, namely, as we have seen:

$$P(FG') = a_1{}^*a_1(f_1 \cdot f_1')^*(f_1 \cdot f_1')$$

$$P(F'G'') = a_1{}^*a_1(f_1' \cdot f_1'')^*(f_1' \cdot f_1'')$$

$$P(FG'') = a_1{}^*a_1(f_1 \cdot f_1'')^*(f_1 \cdot f_1'')$$

It is easy to get a counter-example to Bell's Inequalities here, if we choose angles which are not co-planar. We express all vectors in terms of the first basis:

$$f_1 = (1, 0, 0, 0)$$

$$f_1' = (0, 0, 1/\sqrt{2}, 1/\sqrt{2})$$

$$f_1'' = (1/\sqrt{2}, 1/\sqrt{2}, 0, 0)$$

To check the inequality we can drop the scalar term throughout, and we then observe

$$(f_1 \cdot f_1')^2 + (f_1' \cdot f_1'')^2 = 0 = P(FG') + P(F'G'')$$

$$(f_1 \cdot f_1'')^2 = (1/\sqrt{2})^2 = 1/2 = P(FG'')$$

Hence the inequality is violated.

4. HOW ARE CORRELATIONS EXPLAINED?

The core puzzle that remains concerning the EPR situation, even if we discount von Neumann's interpretation rule or his Projection Postulate, is this: how is it possible to have those correlations between spatially separated events? The events are measurement outcomes, and we have very surprising conditional certainties about what they are for one system, given the outcomes for the other. Of course, the general puzzle of distant correlation is not new in the history of physics (see e.g. McMullin 1989). To see what options we have, let us canvass the sorts of explanations that have traditionally been offered, and try to see what *could* in principle count as a satisfactory answer.

We speak of a (positive) correlation when we have two classes of events and a correspondence between them, and when an event in the one class is more likely to happen if its correspondent in the other class does too. The probability calculus allows us to state this in equivalent ways:

(1) $P(A|B) > P(A)$

 $P(B|A) > P(B)$

 $P(AB) > P(A)P(B)$

But these formulas leave tacit the correspondence relation that links event-types A and B. In the familiar example of smoking and lung cancer, the correlation is of a present smoking rate and a present lung condition—hence two events simultaneous in one body. There is also a correlation with past smoking rates, i.e. of non-simultaneous events; and a correlation with industrial pollution, i.e. of simultaneous but spatially separated events. The most interesting—but not the only mysterious—type is of a correspondence with *space-like separation* (simultaneous and spatially distant, in some frame of reference).

There appear, in the history of science and philosophy, six types of explanation which attempt to render correlations unmysterious. They are: chance, coincidence, co-ordination, pre-established harmony, logical identity, and common cause.

The first two were described and distinguished by Aristotle (*Physics*, ii. 4–6). Aristotelian science, like our own and unlike that of classical modern physics, eventually (at least in the Middle Ages) admitted a significant amount of indeterminism; individual events, though generally produced by some cause, could also just happen, by chance. An observed correlation could be like that. In the play *Rosencrantz and Guildenstern Are Dead*, the same coin tossed by one particular person always comes up heads. To say it is a fair coin, and the person honest, is to attribute this to chance—any sequence $A_1, \ldots, A_n$ of outcomes, after all, has the same probability $(1/2^n)$ as any other. To offer this diagnosis, however, is to assert also that the correlation is extremely unlikely to persist—it was a mere accident. The diagnosis is that there was no 'real' correlation, for one does not extrapolate the observed frequencies to a probability function of character (1).

Coincidence is not the same. When you and I meet in the market, it may be by coincidence—and this may be so even if many of our meetings are there. It is a coincidence if my going to the market is for reasons that have nothing to do with yours. But it is not chance, for our trips to the market do not just happen; we each could explain them causally. There was perhaps an initial chance alignment of circumstances; it is causally propagated, thus bringing about a correlation. Of this we have a good example in recent physical hypotheses. In E. A. Milne's cosmology, atomic clocks and astronomical clocks induced two different time metrics. But the two are related by a logarithmic function. This is a very exact correlation; it is not pure chance, and it is not because the two rates were somehow set by the same mechanism (as far as the theory goes). Here we do (if we accept the theory) predict persistence; but we attribute it to the causal propagation of an alignment in two separate parts or features of the world. The correlation is not mysterious, but only because no question remains when each part has been explained separately.

Of course, the mysteries we perceive in quantum mechanics cannot be dismissed in this fashion. So we must turn to the four remaining types.

By *co-ordination* I mean a correspondence effected by signals (in a wide sense): some energy or matter travelling from one location to another, and acting as partial producing factor for the corresponding event. The situation need not be deterministic; there can be indeterministic signalling if the signal is not certain to arrive and/or not certain to have the required effect. But the word *travel* must be taken seriously. Hence this explanation cannot work for corresponding events with space-like separation. To speak of instantaneous travel from X to Y is a mixed or incoherent metaphor, for the entity in question is implied to be simultaneously at X and at Y—in which case there is no need for travel, for it is at its destination already. I do not mean to rule out phenomena that one might be tempted to describe in this way—it is exactly how Descartes at one point described the connection between emission and absorption of a light ray—but it is still a misdescription; one should say instead that the entity has two (or more) coexisting parts, that it is spatially extended. Correlation between distant events happening to parts of the same extended entity, however, are not *ipso facto* less mysterious.

The recent experiments by Aspect and others leave no hope for co-ordination to explain the quantum mysteries. We have three types left, and the first, *pre-established harmony*, is instructive exactly because of the reasons for its proposal (by Malebranche and Leibniz). Consider the correlation between such mental events as my decision to raise my arm and the bodily rising of my arm. We cannot attribute this to chance, for we confidently predict its persistence. We do not consider it a coincidence in the sense of Aristotle, if we believe in free will; we predict that the correlation will persist even if, for the sake of experiment, I make these decisions in some new way or even randomly. Now these philosophers saw no causal mechanism that could co-ordinate these events through 'signals' from mind to body. Thus they had a phenomenon which does not fit the above sorts of explanations. To call it pre-established harmony, of course, can have only one of two functions: to postulate an Entity which has either predetermined my apparently free actions or co-ordinates the two series of events 'from outside', or else, *to admit that we have no explanation but refuse to consider the correlation mysterious nevertheless*. Taken in this sense, disdain on our part may be inappropriate, for it is an attitude that has often occurred in the history of science and is now perhaps forced upon us, if we are to maintain the completeness of physics.

Explanations through logical identity are the most beautiful and satisfying, and also the most treacherous. They were the paradigm of genuine scientific explanation for Aristotle: his favourite examples of explanation are all of the form 'phenomenon X just *is* Y' (see van Fraassen 1980*a*). In modern terms: the correlation between corresponding values of variables A and B is explained if each is definable as a function of a third variable. Then we say, A and B just *are* $f(C)$ and $g(C)$, as in 'temperature just *is* the mean kinetic energy of the molecules'. Clark Glymour (1980*b*, 1985) gave the following clear and instructive example. Ptolemy had already noted that, for the superior planets, the following correlation holds:

> Let us denote:
> n = number of solar years
> q = number of oppositions in those years
> r = number of revolutions of longitude in those years
>
> with the variables q and r being calculated for the same superior planet.
> In that case $n = q + r$ whenever n, q are whole numbers.

The explanation given by Copernicus consists in the three identifications:

> n = number of orbits of the earth around the sun during a certain period of time
> q = number of times the faster-moving earth overtakes the superior planet during that time
> r = number of orbits of the superior planet around the sun during that time

plus the strictly mathematical calculation showing that, for integral values of these variables, $n = q + r$. The correlation is therefore wholly unmysterious.

Hidden-variable models, when successful, can also take this form. A good example is Kochen and Specker's classical model for electron spin (1967), in which the observables are identified with certain random variables defined on the unit sphere. (Of course, this model cannot be extended to the whole of quantum theory.)

This sort of explanation was the paradigm of a scientific account in Aristotle's *Posterior Analytics*. In his phrasing: the explanation derives from real (as opposed to nominal, merely verbal) definitions, which state what the things really *are*. These real definitions, however, which summarize the 'essential' properties, are by post-Aristotelian lights themselves substantive hypotheses. (This is shown clearly enough by such Aristotelian examples as 'thunder is the noise of fire being quenched in the clouds', and especially by 'ice is condensed water', the one which Galileo attacked when he came to Florence.)

It should be easy to see then why I called this explanation pattern treacherous. Being convinced empirically that two terms are co-extensive, we may adopt their equation as a definition or convention. The equation is then no longer mysterious, because it follows *ex vi terminorum*. But when the procedure is carried out in that way, it sweeps under the rug the presuppositions, which were substantive. For a certain model was earlier proposed and adopted for good reasons, but its adequacy was not a priori. This is as true for Copernicus as for Aristotle. In addition, this procedure has demonstrable limits, which showed up exactly in the 'no hidden variable' theorems. To this point I shall return below.

Finally, correlation of simultaneous separated events can be explained if we can find a *common cause*, that is, if both can be traced back to some events in their common past. If two equivalent clocks, in uniform motion, are synchronized and started just at the moment when their paths cross, they remain synchronized, although no further signals are exchanged, and still without telepathy. For the two were pre-programmed at the initial point of coincidence. This concept, as we recall from Chapter 4, is a concept of causality which can be generalized so that it still makes sense in an indeterministic universe.

It seems to me that this insight of Reichenbach's, described by him in the 1920s and 1930s—and well known to Einstein, though I do not know how early—was one of great importance. For it showed exactly how, and to what extent, one could sensibly speak of a causal order in the context of indeterminism. Two points should be made to qualify this. First, Reichenbach only outlined a pattern, which is not by itself sufficient for explanation. Telekinesis at speeds slower than light would be mysterious even if it fitted the pattern! Wesley Salmon has laid down, as further necessary condition, that there be spatially continuous processes linking the correlated effects with the common cause. This requirement is, it seems to me, too strong for microphysics, and too weak elsewhere (since a wave produced in water, propagating outwards from my body but apparently produced and sustained by will power, would still be mysterious). Thus we have only necessary conditions for causal order in an indeterministic world. The second point is that these necessary conditions entail Bell's Inequalities, as we saw in Chapter 4. Therefore nature may not, and apparently does not, agree even to these necessary conditions. To this point too I shall return below.

5. ATTEMPTS AT PERFECT EXPLANATION

Pre-eminent among the above patterns of explanation are those through logical identities and through common causes. In this section I shall take up the former, which promise perfect explanations that close the subject altogether.[5] Recall that such an explanation accounts for a correlation between two things, in effect, by asserting that they are really but two aspects of the same thing. In the case of quantities A and B, this means that they are definable as two functions, $A = f(C)$ and $B = g(C)$, of a third. What classical features do such functional relationships retain in the quantum-theoretical environment, and how far can they get us?

5.1. *Functional relationships everywhere*

The first point to be appreciated about the quantum-mechanical formalism is that it is, as it were, *made* for representing correlations in an indeterministic world. We can see this already in the representation of a single, non-compound system, as follows.

Let A and B be two observables that have certain eigenvectors in common, call them $x_1, \ldots, x_n$, with corresponding eigenvalues: $Ax_i = a_i x_i$ and $Bx_i = b_i x_i$. Consider now solely the Hilbert (sub-)space H_0 spanned by these vectors. Any vector therein is a superposition $y = \Sigma c_i x_i$, and we can predict with certainty that, if observables A and B are both measured on a system in this state, the outcome will be (a_k, b_k) for some k. We have a *conditional certainty*: if a measurement of A yields value a_k, then a measurement of B is certain to yield the corresponding value b_k (if all the a_i are distinct; or else, a value in the set $\{b_i : a_i = a_k\}$).

The supposition here was really that A and B (restricted to subspace H_0) are compatible, that is, jointly measurable. In general, we would expect this not to be an onerous requirement, and to imply no logical connection between them, merely a certain irrelevance, a mutual non-interference. But *sub specie* the quantum-mechanical formalism, the world is much more tightly organized and internally connected than our general intuitions suggest. For we have at once an explanation by way of *logical identity* for the above correlation: there exists a maximal observable C on the space H, and functions f and g such that $A = f(C)$ and $B = g(C)$. A measurement of A or B is *ipso facto* a partial and/or indirect measurement of C—so no wonder we find correlations!

I am trying to make the best possible case for this sort of explanation of the quantum-mechanical mysteries. We can certainly take the motivation still a step or two farther. This logical identification is not embarrassed by composition or spatial separation. To see this, we look at compound systems.

For any two observables A pertaining to system X, with eigenvectors $\{|a_i\rangle\}$ and B pertaining to system Y with $\{|b_j\rangle\}$, we can define a state $\varphi = \Sigma c_i |a_i\rangle \otimes |b_i\rangle$ for $(X + Y)$ in which we have not just conditional probability, but *conditional certainty* of the value found in a B-measurement, given the result of an A-measurement. But this is exactly like the first case I considered. Thinking about the complex system $(X + Y)$, the observables measured are $A \otimes I$ and $I \otimes B$, which are both simple functions of a single observable $A \otimes B$. Thus A and B measurements are, despite appearances, partial and/or indirect measurements of a single observable. No wonder that correlations can be found in suitably chosen states!

Finally, we should add to this von Neumann's observation about compound states in general, which we noted in Chapter 7. If φ is any two-body state at all, then it has *some* bi-orthogonal decomposition ('canonical decomposition') $\varphi = \Sigma c_i(x_i \otimes y_i)$ such that the reduced states $\#\varphi[1]$ and $\#\varphi[2]$ are diagonal in the bases $\{x_i\}$ and $\{y_i\}$. If A and B are the observables such that $Ax_i = e_i x_i$ and $By_j = d_j y_j$ (with $e_i \neq e_k$ and $d_j \neq d_m$ if $i \neq k$ and $j \neq m$), then the quantities A and B are thus perfectly correlated in the two components. So the situation of perfect correlation is not rare or unusual—the mathematics of the theory entails that *every* situation displays some perfect correlation if looked at in a certain way.

This pattern of logical identity explanation of correlations, is very appealing. It was perhaps taken further in the so-called quantum-logical interpretation, and I think that it lay also at the heart of Bohr's 'holistic' reply to EPR.[6] But now we must look back on the chain of thought we have just gone through, and consider it more critically.

Of the logical identity explanation pattern, I said that it was in principle the most satisfying, but also the most treacherous: it tends to insinuate more than it can establish. If measurements of A and B are partial and/or indirect measurements of the same observable C, should we really at once conclude that it is no wonder if their outcomes are correlated? This looks so innocuous. But let us begin this way: certainly it is no wonder, provided C already, prior to measurement, has a certain value, and the measurements merely reveal that value. Then the correlations—if it is merely a matter of looking twice, three times, ... at the same thing, through different glasses, as it were—are indeed no wonder.

But having said this, we at once remember that Bohr's reply came in a context in which this way of regarding observables and measurement *had already been given up*. The measured observable was thought to take on its value in the context of a measurement situation. Now when the two parts of the complex system are far apart, and apparently not in communication, and are made subject to very different sorts of measurement (which could not have been jointly imposed on either alone)—then what?

To this rhetorical question, we must add here that a single measurement operation on part Y is a partial and/or indirect measurement of *many*, mutually incompatible observables defined on the complex system $(X + Y)$. Somehow, when the measurement is completed by a second operation carried out on the distant part X, the system must know what answer to give. That this further undermines the proffered explanation is strongly suggested by Schroedinger (see below). The assertion of identity of what is measured, though initially satisfying, is now surrounded by doubts about how far it can carry us. In fact, these doubts are implicitly borne out in the limitative theorems concerning hidden variables.

5.2. *The limits: 'no hidden variable' proofs*

The history of the 'no hidden variable' proofs can be seen as the investigation of these doubts and as uncovering the limits of this sort of explanation of the mysterious correlations.

This history began with von Neumann's proof in 1932. I do not wish to discuss this in detail, but we can get to von Neumann's conclusion via an illustration which Schroedinger gave in 1935 (in his famous 'Cat Paradox' paper) which makes the same point about equally well. To see the point, we must first carefully consider the picture painted by the above reflections on identification of observables. It is especially crucial in this context to insist on the conceptual distinction between observables and the operators which represent them. Let us start simplistically. We assume that each Hermitean operator M (on a certain Hilbert space) represents a unique observable m (pertaining to a certain system). A little later we shall have need for a somewhat less simplistic outlook.

If operator N has eigenvectors, each of which corresponds to a distinct eigenvalue, which form a basis of the space, then N is a maximal Hermitean operator. The following fact is very useful: if F is the family of these maximal operators, and M is any given Hermitean with discrete spectrum, then there exists a member N of F and a function f_M such that $M = f_M(N)$. This function corresponds to a simple relation between the eigenvalues: if x is an eigenvector of N corresponding to eigenvalue b, then x is also an eigenvector of M corresponding to eigenvalue $f_M(b)$. In other words,

$$Nx = bx \text{ implies } Mx = f_M(b)x$$

This determines M uniquely, because it assigns M a basis of eigenvectors with corresponding eigenvalues.

Recalling the correspondence of operator M to observable m, we can pose von Neumann's question this way:

When $M = f_M(N)$ is it also true (or tenable) that $m = f_M(n)$?

For the answer to be *yes*, we should be able to say that whenever observable n has value b then m has value $f_M(b)$— and of course, that when m has value c, then n must have some value in the set $f_M^{-1}(c)$.

At first sight it seems that the answer must be *yes*. But quantum mechanics itself only tells about probabilities of measured values. Thus in the above case it tells us that, for measurements made in state x, the probability of an M-measurement yielding outcome c equals the probability of an N-measurement yielding one in set $f_M^{-1}(c)$. Now the temptation is to think that there must be a reason for this connection among probabilities, and that the only sort of reason there can be must entail a *yes* answer for that question. That is, we are tempted or invited to think that relations among probabilities must come from patterns of what must happen in any individual case. ⸜

The temptation can also be described in a different way. Suppose that M is a non-maximal and N a maximal observable, and that I make an M-measurement. That is too crude to count as a proper N-measurement. But if I had made the latter, a certain value b would have been found. Therefore the M-measurement must have outcome $f_M(b)$. Put in this form, the word 'temptation' will seem very apt, for now we have made a connection with the discussions in Chapter 5 about this sort of reasoning, involving 'counterfactual definiteness'.

And the fallacy is very easy to spot here. For since M does not determine N, there will also be some other maximal operator N' such that $M = g_M(N')$. Obviously N and N' are incompatible, and it is not possible to measure them jointly. If we had measured N' instead of N, then a certain value c would have been found (reasoning as we did for N above). Hence the M-measurement must have outcome $g_M(c)$. Putting the two bits of reasoning together, we would have to conclude that $f_M(b) = g_M(c)$. What this means is that we have now concluded the complex, nested counterfactual conditional: if we measured N and found value b, then if we had measured N' instead we would have found a value in $g_M^{-1}(f_M(b))$.

If it seems that I am needlessly balking at counterfactual conclusions here — owing to some empiricist prejudice — recall from Chapter 5 that the assumption of counterfactual definiteness here at work leads to a proof that Bell's Inequalities cannot be violated. But it is perhaps easier to see how counter-intuitive the consequences are if we make up a simpler example. Suppose the eigenstates of N are the orthonormal triple x, y, z corresponding to eigenvalues 1, 2, 3. Let M have this triple as an eigenbasis as well, but corresponding to eigenvalues 1, 1, 3. For our other maximal observable choose N' with orthonormal eigenbasis x', y', z, and corresponding eigenvalues 1, 2, 3 as well. The subspace on which M takes value 1 is thus $[x, y] = [x', y']$. The above reasoning then looks like this:

Suppose we measure N and find value 3. Therefore (!?) if we had measured M with a coarser apparatus that registers only 3 and not-3 for N (equivalently, 1 and 3 for M), we would also have found value 3. So (!?) if we had measured N' instead, we would have found value 3 as well.

Is this good reasoning? Test it in imagination on the case in which the observed system has initial state $w = (1/\sqrt{2})(x + z)$. Then quantum mechanics tells us, for each of the three measurements in question, that it has outcome 3 with probability $\frac{1}{2}$, if performed. This is statistical information, saying nothing about the individual case, and does not sanction the above reasoning at all. Indeed, that passage suggests nothing so much as the fallacy: we got outcome 3, so the system must have been in eigenstate z.

Perhaps it will be countered that the nested counterfactual conditional to which I objected might be true, even though quantum mechanics does not imply it (on e.g. my own or von Neumann's interpretation) and although it looks as if it was derived from a fallacy. It is hard to evaluate this, and I think useless to try unless the conclusion in question turns out to be implied on *some* interesting interpretation. But I also think that the general connection between counterfactual definiteness and Bell's Inequalities speaks strongly against it.

Schroedinger's example of how such functionality breaks down for the single case—and has therefore been mistakenly extrapolated from functional relationships in the statistics—is quite amusing. Quantization is a curious thing, and refuses to let this story, of an intricately organized world of logically connected observables, be carried to its 'natural' conclusion. Schroedinger discusses the EPR example in which two systems are characterized each by a single coordinate q and Q respectively, with the canonically conjugate momenta p and P respectively. The state is so entangled that $q = Q$ and $p = -P$, yielding the conditional certainties that measurement of q or p will establish at once the outcome of a measurement of Q or P respectively. So, he says, it looks as if to ask for values of Q or P, when q or p has already been measured, is like asking a student for the right answer to a question—and this student is so clever that he or she always knows the right answer. But, and this is the important point here, it was not sufficient for this student to have hit on only two correct answers to begin with, and then to calculate the answer to any other question we ask. For suppose we ask for the value of

$$P^2 + a^2 Q^2$$

with a an arbitrary constant. In the EPR example, the student will always give the right answer. But here, quantum mechanics tells us that the answer must have form

$$ah, 3ah, 5ah, 7ah, \ldots$$

Now suppose he had memorized the right answer for the questions about P and Q, say $P = 4$ and $Q = 7$; and suppose he now calculates $(4^2 + a^2 7^2)$ and finds the answer nah for a certain odd integer n. He will have been extremely lucky, and he cannot be that lucky all the time. For if this works for particular constant a, it is guaranteed not to work for infinitely many other numbers that a could have been. For example, if we change a to $a + d$, and take d small enough, the difference between $(4^2 + a^2 7^2)$ and $(4^2 + (a + d)^2 7^2)$ will be less than $2ah$, so he will come up with a coefficient between the two odd integers n and $n + 2$—and that will be wrong.

This illustrates elegantly what von Neumann's proof established: that we cannot associate with every Hermitean operator B a sharp value $V(B)$ and have even the relationship

$$V(cB + dC) = cV(B) + dV(C)$$

Thus, if the observables have sharp values, they are not functionally related in the same way as the operators that represent them!

Now it will be noticed that in Schroedinger's example, as has also often been pointed out about von Neumann's proof, essential use is made of incompatible observables (represented by non-commuting operators). Could the mystery be localized here?

Subsequent work by Jauch, Gleason, Kochen and Specker, and Bell, among others, established that this is not so. (See the discussion of the 'no hidden variables' implications of Gleason's theorem and Kochen and Specker's result above, in Chapter 6.) Assignment by hidden variables of sharp values to all observables, plus the requirement that functional relations among the operators reflect those among *compatible* observables, already leads to inconsistency. This cannot be surprising once we see that compatibility is not transitive. Indeed, it seems almost obvious, given our reflection above that it is possible to have two non-commuting maximal observables N and N', and functions f and g such that a given operator M is $f(N)$ and is also $g(N')$. Thus the functionality requirement for compatible observables entails indirect restrictions on the values assigned to incompatible observables as well. The exact import of this result, obtained in different ways in the famous papers of Kochen and Specker and of Bell, is clearly presented in Fine and Teller (1978), and the further equivalence to postulation of hidden joint probability distributions is in Fine (1982b, 1982c).

Thus presented, the impression has sometimes been gained that the results obtained in this subject by Kochen and Specker on the one hand and by Bell on the other are the same. I think that ignores the distinct crucial insights to be found in both papers. In the remainder of this section I shall describe what I see as the 'abstract' achievement here, which shows the exact limits to any attempt to solve quantum mysteries of correlation through logical identities. Here I think the Kochen–Specker paper contains exactly the crucial points. In the next section I shall take up the import of the distinctive contribution by Bell.

Recall that I introduced the notation: operator M represents observable m. That presumes that a Hermitean operator represents a single observable only. But quantum mechanics itself only uses the operators to represent the statistics of measurement outcomes. The following is conceptually possible: two distinct observables m and m' have the same set of possible values, and in each quantum-mechanical state the probability of a given outcome for an m-measurement is the same as for an m'-measurement. Obviously, m and m' would then have to be represented by the same Hermitean operator M. In what sense could m and m' still be distinct? Well, we can tenably add to the above that m and m' always each have a definite value (so an actual situation is only incompletely represented by a quantum-mechanical state) and their values are not all the same. The statistical predictions could still coincide. That this is indeed a tenable hidden-variable model is established by a theorem due to Gudder (1968*a*).

Indeed, this 'de-occamization' (as Michael Redhead calls it) need not be performed everywhere: the model allows that maximal Hermitean operators do each correspond to a unique observable. The little bits of reasoning we went through above, and which got us into trouble, assumed that a non-maximal operator M could be described as a function f_M of one maximal operator and as a function g_M of another, and could *also* still be assumed to represent a unique observable. That was (logically speaking) the trouble.

What is missing in the description of this de-occamized hidden-variable model is the assumption that identity of measurement statistics implies identity *tout court* for the observables. Without that, the 'no hidden variable' proofs do not work. This is made clear by Kochen and Specker when they explicitly assume

(K–S): If two observables pertaining to a system have the same expectation value in every quantum-mechanical state (are, that is to say, empirically equivalent, according to quantum mechanics), then they are identical

which is equivalent to the assertion that if $M = M'$ then $m = m'$ in the above notation.

Measurement procedures are identified in terms of the Hermitean operator; that is why it is sometimes said that to deny (K–S) is to say that the same observable has different values depending on how it is measured—e.g. with the apparatus rotated along one of its axes of symmetry. That is a mis-statement, for it confuses two diagnoses. If we have a 'de-occamized' hidden variable model, and it agrees in its empirical predictions with quantum mechanics, thus violating (K–S), such a rotation of the apparatus *is* the measurement of a different observable. Although $f(N) = g(N')$, we have two distinct observables, identified by the couples f and N, g and N'—they are distinct though empirically equivalent.[7]

So now we have found the exact limits. As a purely logical or metaphysical exercise, we can indeed take the logical identity explanation very far. A look at Gudder's proof will show that we could have done this almost no matter what quantum theory could have been like. But here the metaphysical explanation shades off into a merely verbal one—made possible because, as far as logical consistency is concerned, almost any thesis can be superimposed on the facts. Once we impose a fairly minimal constraint—formulated as (K–S) above—on the interpretation of quantum mechanics, the putative explanation of the statistical correlation among measured observables becomes hollow. For then, to maintain logical consistency itself, we must deny functional relationships among the individual values that could have explained those found in the joint probability distributions (of the compatible observables) to which we addressed ourselves to begin with!

6. SINISTER CONSEQUENCES AND SPOOKY ACTION AT A DISTANCE

If we discount as despair the idea of a pre-established harmony, there are only two patterns of explanation that might prove equal to the mysterious correlations predicted by quantum mechanics. Pursuit of logical identity having yielded only illusory gains, we turn to the last, that of causal explanation—where the correlated effects are traced back to a real or hypothetical common cause. Perhaps because this is the form of every would-be scientific response to claims of telepathy or psychokinesis, it is often thought of as the explanation-type that

eliminates appearance of action at a distance. As a first objective in this section, I will therefore discuss how one might get an
impression of action at a distance in EPR correlations—and
whether that impression is justified. As second objective, I will
try to understand the limits to causal explanation of such
correlations.

6.1. *No empirically verifiable action at a distance*[8]

In Schroedinger's reply to—or perhaps I should say, elaboration
of—the EPR paradox in his 1935–6 papers on relations between
separated systems, he purports to demonstrate a 'sinister' consequence. This consequence, which Einstein called in 1947, in a
letter to Born, 'spooky action at a distance', is exactly that we
can affect what happens at a distance without intervening causal
chains.

An exposition of this argument will certainly show graphically
why there has been an impression of (instantaneous) action at a
distance in quantum mechanics of the very sort that Einstein
was thought to have banished in relativity theory. But the
exposition will also reveal—as I shall argue—that the arguments
crucially involve what I called von Neumann's interpretation
rule (equivalently, his Projection Postulate), and that the action-
at-a-distance exists only on one interpretation of the phenomena. On other interpretations it does not happen; and
indeed, any attempt to verify it empirically along Schroedinger's
lines is ruled out by quantum mechanics itself.

Schroedinger begins by noting what he takes to be not '*one*
but rather *the* characteristic trait of quantum mechanics, the one
that enforces its entire departure from classical lines of thought'
(1935*a*, 555). It is that if two systems, initially in pure states,
enter into interaction, their joint state generally develops in such
a way that they can no longer individually be ascribed pure
states at all. All knowledge of the component systems taken
individually could be lost, and only through measurements can
we arrive at a new attribution of a pure state to either
component. But then, knowledge of the initial states and the
interaction allows us to *infer* without further measurement a
pure state for the other one.

The 'sinister importance' of this fact lies, Schroedinger says, in its being involved also in the quantum-mechanical description of measurement. But, as we have seen in earlier chapters, the move *from* the fact that at the end of measurement a certain value is recorded *to* the ascription of a corresponding eigenstate is not part of the theory but part of an interpretation. The inference just pointed to by Schroedinger is valid on the standard interpretation, but not on all interpretations. Let us continue however with Schroedinger's argument. Schroedinger elaborates the sinister consequences as follows. Imagine a complex system $(X + Y)$ which has become entangled in the above way, and is now as a whole characterized by the state φ in the tensor Hilbert space $H_X \otimes H_Y$. Imagine also that we should like to steer component system Y into some pure state (or more generally, some subspace of H_Y), but we have no direct access to it because it is by now far away from X, where we are located. Indeterminism being ineliminable, we cannot steer Y into that desired (sort of) state with an absolute guarantee. But, and this is what Schroedinger proves here, we can design measurements to be performed on X which will have that effect on Y, generally with some positive probability.

To see, in rough outline, how the proof proceeds, suppose that y is a possible pure state of system Y, and belongs to the image space of statistical operator W'. It follows then that there is some non-degenerate observable measurable on X, let us say A, with eigenvectors $|a_i\rangle$ such that

$$(1) \qquad \varphi = \sum c_i |a_i\rangle \otimes x_i$$

and $y = x_m$ for some value m, with $c_m \neq 0$. (The set $\{x_i\}$ will in general not be orthogonal.) A measurement of A on X, i.e. of $A \otimes I$ on the combined system, sends $(X \times Y)$ via the Projection Postulate, into state $|a_m\rangle \otimes x_m$—and hence Y into state x_m—with non-zero probability. This non-causal change in what Y is like, over which the experimenter appears to have non-negligible control, happens instantaneously at the end of the measurement performed on X, on this analysis.

We have now arrived at a troublesome conclusion which appears to go some way beyond those drawn in the EPR paper. The conditional certainties were difficult enough to interpret. But here, analysing exactly the same situation, we have arrived at a consequence which Einstein was quite right to describe as 'spooky action at a distance'. The question arises immediately whether it could be eliminated from the interpretation of quantum mechanics. It would sound decidedly hollow if at this point I simply repeated my claim that other interpretations (for example the modal interpretation) are tenable and do not license Schroedinger's inference. For has he not just demonstrated an empirically testable consequence, namely that we can effectively (that is, with calculable probability) steer a distant component of a two-body system into almost any state we like?

The answer is *no*, Schroedinger has not demonstrated an empirically verifiable possibility. He gave only half of the argument that would be needed to show that.

What Schroedinger neglected to ask was: what would have been the probability of finding system Y in the desired state if we had not carried out any measurement on X first? The answer is: *it would have been just the same*. Mind you, we must interpret the question and answer purely empirically: if we measure the observable I_y—the projection along vector $y = x_m$—what is the probability of finding the value 1, on the supposition that we have first measured the observable A on X? And the answer is: exactly the same as without that supposition. (See *Proofs and illustrations*.)

To sum up, then: the expectation of $I \otimes B$ must be the same for both experimenters, since it involves no reference to measurements on X—but each calculates that this equals his own expectation for measurements of B on Y, on the supposition that he measures his own choice of A or A' on X. The fallacy all too easily committed is to confuse the latter supposition with some assumption about the information actually obtained from the chosen measurement on X, as opposed to information obtained from the choice itself.

The action at a distance described by Schroedinger thus takes place, according to the von Neumann interpretation of what happens in measurement, as a real collapse of the wave packet —a real counterpart to the mathematical operation used in calculating probabilities. Under this description, the wilful experimenter indeed makes the desired result happen, with non-negligible probability. *But there is no empirical manifestation of this*; if we make measurements on the systems for which he has those desires, we cannot tell from our findings whether he is engaged in his nefarious activity or not.

As far as that is concerned, his efforts designed to steer Y into desired state x_m are just magic—sticking pins into a voodoo doll. There is, in the EPR situation, no empirically verifiable action at a distance.

Proofs and illustrations. To prove the assertion that Schroedinger's sinister experiments would have no empirically discernible effect on the distant system, consider the most interesting, i.e. doubly correlated, EPR state:

$$(1) \qquad \varphi = \sum c_i |a_i\rangle \otimes |b_i\rangle = \sum d_i |a_i'\rangle \otimes |b_i'\rangle$$

where the first orthogonal expansion is in terms of observables A and B, and the second is in terms of A' incompatible with A, and B' incompatible with B. Now there is a unitary transformation which connects the alternative bases; let

$$(2) \qquad |b_i'\rangle = \sum_j e_{ji} |b_j\rangle$$

We note that the probability of getting value b_k in a measurement of B on the second component equals c_k^2, whether or not A is measured first, because of the perfect correlation of A and B.

As first step we prove

Lemma: $\qquad c_k^* c_k = \sum_i (d_i e_{ki})^* (d_i e_{ki})$

The proof is by the consistency of two ways of calculating the probability of the answer (1) if $I \otimes B$ is measured. From the first equation in (1), we see that the answer must be c_k^2, using our notational conversion for complex numbers. From the second equation and (2), we get

$$(I \otimes I_{|b(k)\rangle}) \varphi = \sum_i d_i |a_i'\rangle \otimes e_{ki} |b_k\rangle$$

whose squared length is $\sum_i (d_i e_{ki})^2$. Therefore the two numbers are the same.

Now we imagine someone saying to himself: if I were to

measure A' first, and then B, what is the probability of getting b_k in the second measurement? He then says to himself: I must get one of the combinations a_i' followed by b_k, if that is what happens. But each such combination has as probability the squared length of

$$(I \otimes I_{|b(k))})(I_{|a'(i)\rangle} \otimes I)\varphi = (I \otimes I_{|b(k))})(d_i|a_i'\rangle \otimes \sum e_{mi}|b_m\rangle)$$

$$= d_i|a_i'\rangle \otimes e_{ki}|b_k\rangle$$

which is $(d_i e_{ki})^2$. Adding up all those possible combinations, we arrive at total probability $\Sigma_i (d_i e_{ki})^2$—which, by the above lemma, is just the probability of finding b_k after a preceding measurement of A.

6.2. *No communication by 'Bell telephones'*

What I showed in the preceding section is a fact that has appeared in the literature in a number of ways. The apparent instantaneous action at a distance was also an apparent violation of relativity theory, and that was shown to be illusory.[9] In a more general way, exploiting violations of Bell's Inequalities for communication was shown to be impossible almost as soon as it was mooted.[10] The point, put very briefly, is that, since observables pertaining to two regions with space-like separation commute, the expectation for measurements in one such region is independent of the decision as to what measurements to make in the other.

Herbert (1981) proposed a way to circumvent this equality of expectation values, by exploiting not a series but a single measurement. His idea was refuted by Dieks (1982); it will be instructive to see why the ruse could not work.

We consider again system $X + Y$ described in the preceding subsection, a case of double correlation without degeneracy. But this time, as soon as the measurement—whatever it be—on X has been carried out, Y enters a multiplier (think of a laser) which emits a whole burst of particles of the same type as Y and in the same state. We have two experimenters with access to X; one would like to measure A and the other would prefer to measure A'.

Now, the first experimenter argues, once I measure A, I will predict with certainty an eigenvalue b_k for a measurement of B on Y. So I will predict that, if measurements of B are made on samples from that burst, they will all yield the same value— namely, b_k.

Meanwhile, the second experimenter argues: once I make my measurement of A', I will predict with certainty an outcome b'_m for a B' measurement on Y. So I predict that, if the whole burst is subjected to B' measurement, that value b'_m occurs with relative frequency 1. But on that basis I calculate that, if instead B is measured, value b_i will be found with a certain frequency $|e_{im}|^2$ which is not 1 for any index i. So I predict that, if I measure A', then a B-measurement on samples from the burst will not all give the same result.

If both are right, then making B-measurements on the burst into which Y has been 'multiplied' will reveal whether A or A' was measured on X. Thus the choice will be communicated—by what Einstein might have called 'spooky multiplication at a distance'.

What Dieks points out is that, although quantum mechanics may allow for multipliers, it cannot allow for multipliers of *this* sort. The reason is the linearity of evolution operators. In the above reasoning the Projection Postulate was again assumed, but we can leave it uncontested here. Suppose that I measure A, find value a_k, and the other component Y goes into eigenstate $|b_k\rangle$. Now it enters the multiplier, presently in its groundstate, and out comes a burst of N particles just like Y, and all of them in state $|b_k\rangle$. We describe the process that has just happened in a 'black box' form, but can still say this. At the initial moment we have [multiplier] + Y in state

$$|m_0\rangle \otimes |b_k\rangle$$

and there is an evolution of this system into a system [multiplier] $\times$ [N-particle system] in state

$$|m_1\rangle \otimes [|b_k\rangle \otimes \ldots \otimes |b_k\rangle]$$

Since the multiplier was itself presumably a compound system with some structure, this supposition presents no difficulty in principle. That is, the description $|m_0\rangle \otimes |b_k\rangle$ may just be a very incomplete or shallow description of an $(N + 1)$-body system, and may really stand therefore for a vector in an $(N + 1)$-dimensional space. What is important is that the transition must be governed by a unitary and hence linear evolution operator.

Suppose then that I measure A' instead, find value a'_n, and Y goes into eigenstate $|b'_n\rangle$. The initial state of [multiplier] $+ Y$ is now

$$|m_0\rangle \otimes |b'_n\rangle = |m_0\rangle \otimes \sum_k e_{nk}|b_k\rangle$$

$$= \sum_k e_{nk}|m_0\rangle \otimes |b_k\rangle$$

By linearity that must now evolve into the exactly similar superposition of the final states we encountered before:

$$\sum_k e_{nk}(|m_1\rangle \otimes |b_k\rangle \otimes .. \otimes |b_k\rangle)$$

Suppose that now we make a measurement to see how many of the particles in the burst are in state $|b_1\rangle$ and how many in $|b_2\rangle$. For example, what is the probability that we will find half in the first state and half in the second?

The answer is *zero*. The reason is that any state of, for example, form $|m_1\rangle \otimes |b_1\rangle \otimes |b_1\rangle \otimes \ldots \otimes |b_2\rangle \otimes |b_2\rangle$ is orthogonal to *every* state of form $|m_1\rangle \otimes |b_k\rangle \otimes \ldots \otimes |b_k\rangle$. It is the latter sort of state which we may detect with probability $|e_{nk}|^2$. Thus, anything we may detect with non-zero probability will bear out the expectation of finding the same eigenvalue of B throughout the burst.

Thus we see that, even with the assumption of the Projection Postulate uncontested, the designed device cannot exist.[11]

6.3. *The empirical content of Bell's Inequalities*

It is certainly true that violation of Bell's Inequalities is testable, and was ruled out in all the models that classical physics made available to us. But we must be very clear on how exactly quantum mechanics enriches our modelling so as to allow for it. All statistical predictions for a quantum-mechanical model which pertain solely to a set of mutually compatible observables can be duplicated in a classical (Common Cause) model. (It may be of course that such a model would leave us empirically up a tree, because there might be no empirical trace of the postulated common cause.) That is the exact sense in which, classically, instantaneous action at a distance was not inconceivable.

In the statistical correlations in measurement results for a set of observables not all mutually compatible, a terrible new beauty is born. We shall remain mystified unless we carefully distinguish two possibilities:

(a) violations of the probability calculus itself;
(b) violations of expectations calculable on the basis of any Common Cause model whatever.

We could get from (b)—which we now fully believe ourselves to have obtained in actual experiments—to (a) only on the basis of an assumption that reduces all the new riches of quantum mechanics to nought. That is the assumption that the conditional probabilities of measurement outcomes for incompatible observables—conditional on their being measured—can just be thought of as unconditional probabilities, or probabilities all of which are conditional on the same single complex physical situation.

The mechanics of mystification are quite clear. If we imagine that the conditional probabilities

p_{ij} = probability ($\langle 1,1 \rangle|$ observable L_i and observable R_j are measured)

are equal to unconditional probabilities,

$$p'_{ij} = \text{probability } (f(L,i,1) \& f(R,j,1))$$

then, no matter how the propositional matrix $f(-,-,-)$ is constructed, we can deduce from the probability calculus that Bell's Inequalities hold (given the apparently harmless assumptions that $f(L,i,1)$ is disjoint from $f(L,i,0)$ etc.).

The fact is, there is logically nothing harmless about such a reduction from conditional to unconditional probability. Probability theory *never* implied that this is always possible. Logicians who investigated the matter could only find the most *outré* reduction along such lines, and never so as to preserve those 'harmless assumptions' in general.

The whole content of Bell's Inequalities, as far as the empirical facts are concerned, lies in their conflict with (*b*) above. But that is quite enough.

7. THE END OF THE CAUSAL ORDER?

We have now come to the end of the line; we have exhausted all the traditional ways to explain correlations, and seem to have found no satisfactory way to remove the EPR mystery. *But is it a mystery*? The word 'mystery' is not merely descriptive—to call something a mystery is not so much a statement as a demand, a demand for explanation. Demands need not always be met. The Aristotelian question about the law of inertia—'But what keeps a body moving if there are no forces impressed on it?'—was not answered but discarded by the seventeenth century. It appears to be very hard to treat such demands ruthlessly, for even Newtonians would still speak of a *vis inertiae*, paying lip service to that old demand for explanation.

Each scientific age comes with its own philosophical propaganda. The Enlightenment told us that the new physics would give a complete and deterministic explanation of all empirical phenomena *and* that it would be unscientific not to demand that sort of theory. When the propaganda gets into trouble, such an aim becomes more easily attainable if standards are lowered, if some why-questions are discarded—and that is exactly what happened. For Newton had still regarded the correlations described by his law of universal gravitation as demanding explanation.[12] But a century later, the law itself was accepted as the explanation!

The force of attraction mutually exerted by two bodies of mass m and m' respectively, separated by distance r, equals $F = Gmm'/r^2$. Read in the context of the Second Law of Motion, $F = ma$, we see that, if the second body's position is made to change ever so slightly, there is an immediate effect on the velocity of the first. What explanation could be given for this? Not co-ordination through travelling influences or signals, for no travel time is available. It is true that this correlation is not *formally* as mysterious as the EPR kind. Formally speaking, it admits of a deterministic variant of the Common Cause explanation. Suppose the bodies with masses m and m' are like earth and moon, and the disturbance is that a small but exceedingly fast travelling meteor hits the earth. The dislocation of the 'earth' is at once accompanied by a disturbance in the motion of the 'moon'. But perhaps we can treat the three bodies together as an isolated system; given the initial positions, velocities, and masses a million years ago, we can calculate this very disturbance in the motion of the 'moon' today.

However, without any mechanism for telling the mutually separated 'earth', 'moon', and 'meteor' at that initial time where the others are and what they are doing, we still have here a coincidence on a grand scale. Even if the bodies were close together a million years ago, and could have 'perceived' each other's states, there appears to be no mechanism by which they can 'remember' this information. The seventeenth-century scientists liked the image of synchronized clocks running at equal rates though separately. But here we have something like clocks that run in synchrony, but have no clockworks inside! Thus, among our patterns of explanations only pre-established harmony—to the determinist, the name for despair—fits gravitational 'interaction'.

The difference between this and the EPR correlation is that in the Newtonian case the structure of the law taken by itself presents no formal obstacle to reductive explanation. It would have been consistent to postulate some explanation of the sort Newton wanted, though actual history of science did not proceed in that way. Newton's disciples in the next century discarded the demand for explanation in the same way as his predecessors had discarded the demand for an explanation of the law of inertia. The law, as now, described what was to be taken as normal.

We should not disdain these changes in attitude as metaphysical cowardice. Certainly the new generation became blind to the conceptual difficulties that disturbed the old—but there are close and necessary connections between blindness and insight. What the actual development of scientific ideas denies here is merely what certain philosophers and philosophically minded scientists assert, namely, that strict demands for explanation act as regulative ideals for science. That conception makes physics continuous with metaphysics, and is characteristic of the so-called 'realist' traditions in philosophy. There is also an alternative tradition, exemplified by the fourteenth-century nominalists (whose William of Ockham issued the first explicit rejection of the demand for reasons for inertial motion), the British empiricists, the French positivists and conventionalists, and the Vienna and Berlin Circles. This alternative empiricist, anti-metaphysical tradition (not exactly monolithic in outlook in other ways, and replete with less admirable traits as well) equates scientific explanation with the provision of relevant descriptive information.

The intricacies of the universe, and those of the quantum-mechanical world-picture, are astonishing. But attempts to explain the correlations and conditional certainties there can be about separated systems, by interpretations which 'explain' them through action at a distance 'behind the phenomena', simply add mystery to mystery. They are like the talk about *vis inertiae* to 'explain' the First Law of Motion (or like the short-lived idea of 'Pauli forces' to explain the Exclusion Principle). To understand quantum mechanics means to understand how the world could possibly be the way quantum mechanics says it is. This search for understanding would be not aided but hindered by insistence that every regularity must have a reason.

11

The Problem of Identical Particles

ONE of the earlier and also most persistent problems in philosophy is that of the One and the Many. In natural philosophy that problem emerges especially as the question: does a compound or aggregate have characteristics which do not simply follow from or supervene on the properties of its parts? Is an aggregate a 'mere' aggregate? Are there principles which are trivial or vacuous as far as non-compound systems are concerned (if there be any!) but take on force when compounds are considered?

We have already encountered a holism in quantum theory at a number of junctures. Specifically, the state of $X + Y$ determines but is not determined by the reduced states for its parts X and Y. In Chapter 3 we saw that this sort of holism characterizes even classical probability functions; it becomes non-trivial in quantum theory because there probability is 'irreducible'. But even when all that is understood, there is something more.

The something more is the principle of Permutation Invariance for compounds with distinct but 'identical' parts. The sense of 'identical' needs to be made precise; two conjectures suggest themselves and must be investigated. The first is that Permutation Invariance accounts for the well-known but puzzling departures from classical statistics of aggregates. The second is that, when 'identical' is properly understood, it entails Permutation Invariance tautologically. In this chapter and the next, I shall argue that both conjectures are false, though each has a core of truth.[1]

1. ELEMENTARY PARTICLES: AGGREGATE BEHAVIOUR

An atom ... possesses two kinds of symmetry properties:
(1) the laws governing it are spherically symmetric, i.e.

> invariant under an arbitrary rotation about [its centre]; (2)
> it is invariant under permutation of its . . . electrons.
>
> Herman Weyl (1929)

An elementary particle is characterized first of all by certain constant features, which serve to classify it. Mass is such a constant: baryons have large mass, mesons intermediate, and leptons small. Other constants, such as charge, subdivide these classes. In terminology that philosophers dislike, physicists have often referred to particles characterized by all the same constants as 'identical'. In that sense, two particles can be identical, and yet be in different states of motion—so the identity is not strict numerical identity, or even strict qualitative identity. But it is also possible that two identical particles are in the same state. Then they are certainly qualitatively the same, in all the respects representable in quantum-mechanical models—yet still numerically distinct. If that is so, the particles are 'indistinguishable' in a sense going beyond that of 'identical' as used above. At this point, philosophical puzzles went beyond terminology. As we shall see, physicists too became uneasy, and began to speak of a 'loss of identity'.

Let us describe first the two ways in which non-classical aggregate behaviour appeared. (For more historical details see also Dieks 1990.)

1.1. *Bose, Einstein, and the boson*

The introduction of Bose–Einstein statistics in Bose (1924) was the last step in a historical development directly concerned with electromagnetic radiation and statistical–mechanical analogies. If a certain amount of light, say, is introduced into an evacuated enclosure with perfectly reflecting walls, we have a situation in some ways similar to an enclosed body of gas. Specifically, the 'radiation gas' exerts a pressure on the walls and work must be expended to decrease the volume. If now a piece of matter is introduced, capable of emitting radiation in every frequency, then emission and absorption will happen until their two rates are equal, and remain so: an equilibrium is reached. The intensity of light of a given frequency in the enclosure is a function solely of that frequency and the temperature of that enclosure. The description of that function was the subject of Stefan's, Wien's, Rayleigh's, and finally Planck's laws of radiation. While Stefan's law is based on experimental results, and was accepted as a partial constraint on the required function, Wien's and Rayleigh's were based respectively on a thermodynamical argument and a deduction from the classical laws of electromagnetism (both using additional assumptions). These latter two turned out to be erroneous on the whole although approximately correct in certain limits. At this point Planck introduced his 'quantum theory', and was able to deduce his empirically satisfactory radiation law. But the deduction was based partly on classical assumptions and partly on assumptions incompatible with classical physics—not a theoretically satisfactory situation.

Einstein's treatment of the photoelectrical effect, in which corpuscular properties were attributed to energy quanta (radiation of frequency v consisting of photons having energy hv and momentum hv/c), made the statistical mechanical view more than a mere analogy. The pressure which the radiation exerts on the walls can now be attributed to the impact of the photons, exactly the same mechanism as for an ordinary gas. If Boltzmann's classical statistical mechanics is now applied to the distribution of numbers of photons over the various energy levels (corresponding to intensities of radiation over various frequencies) for an equilibrium situation, we obtain Wien's law. But Planck's law should result! The innovation introduced by Bose was in effect a non-classical assumption of equiprobability. The classical assumption would be that each arrangement of individual particles, classed together when they have the same energy level, is equiprobable. As we now construe Bose's work, the identity of the particles is ignored, and each possible assignment of occupation numbers to the different energy levels is equiprobable. This is analogous to the assertion that, if two coins are tossed, the possibilities of 2 heads, 1 head, and 0 heads are equiprobable. The move was *ad hoc*: but it led to Planck's law.

Bose himself may or may not have appreciated the radical step he had taken. Einstein did, and explained it in articles published immediately afterward (1924–5). Hence we refer to the new statistics as Bose–Einstein (BE) statistics, contrasting it

with the 'normal' or classical Maxwell–Boltzmann (MB) statistics. To illustrate their differences, imagine them applied to the
tossing of two 'identical' coins mentally labelled as a and b:

Heads	Tails	MB	BE
a, b		1/4	1/3
a	b	1/4	1/3
b	a	1/4	
	a, b	1/4	1/3

If we don't want to use labels, even mental or imaginary, we
would describe the different cases and their probabilities as
follows:

Cases	MB	BE
2 heads	1/4	1/3
1 head, 1 tail	1/2	1/3
2 tails	1/4	1/3

At this point we suspect a poor motive that might explain why
some writers spoke of 'loss of identity'. That poor motive is the
Indifference Principle, whose fortunes and demise we already
discussed in Chapter 3. For it would say: all distinct possible
cases must be given equal probabilities. Thus if BE (rather than
MB) is correct, then the second table (and not the first) divides
the possibilities correctly. Hence the use of labels must introduce an unreal distinction. But that use presupposed only
that distinct labels be assigned to distinct particles. Hence this
distinctness (or its opposite, identity) corresponds to no real
division. At this point it is very puzzling how we can still speak
of two, rather than one, particle at all. I will leave this issue for
now; since the Indifference Principle had already been generally
rejected by this time, this bit of reasoning looks anyway too
spurious to be of help.[2]

1.2. *Pauli and the Exclusion Principle*

Within a year, it became clear that Bose's treatment could not
apply to aggregates of identical particles in general. For in 1925
Pauli introduced his Exclusion Principle for electrons in an
atom. Pauli wrote in that article:

In an atom there cannot be two or more equivalent electrons for which the values of all four quantum numbers coincide. If an electron exists in an atom for which all of these numbers have definite values, then this state is 'occupied'. (Pauli 1925, 766)

The assumption is that the four quantum numbers suffice for a complete description of the state, and that an 'occupied' state cannot be entered by another electron.

The assumption of completeness here may have a 'metaphysical' air, but it is involved in the very application to atomic structure for which the Exclusion Principle was introduced. To show this, let us look at the application of this principle in the theory of atomic structure, and its reconstruction of the periodic table of chemical elements. For the structure of the hydrogen atom, three quantum numbers n, l, m, were already introduced, which together determine the hydrogen atom wave functions. The *principal* number n determines the total energy E_n, and the number of nodes (radial and angular) which is $n - 1$. This number n can take any positive integral value. The *azimuthal* number l is the number of angular nodes; it is thus less than or equal to $n - 1$, and can otherwise take any non-negative integral value. It determines the square of the angular momentum; the *magnetic* quantum number m determines one component of angular momentum, which equals $m\hbar$. This quantum number takes the values 0, ± 2, ... , $\pm l$. Considering only one electron, when $n = 1$ (electron in the lowest orbit), l and m are obviously constrained to be 0. Hence if these three numbers told us all, and the Exclusion Principle applied, there could be only one electron in the lowest orbit. But this is not so. In 1925 Goudsmit and Uhlenbeck introduced a fourth property to characterize the atomic electron, the *spin*, quantum number s being associated with total spin ($\frac{1}{2}$ for all electrons) and the quantum number m_s associated with one component thereof ($m_s = \pm\frac{1}{2}$). Having thus a new parameter with two possible values, we have at least two possible states for the case $n = 1$.

If we now assume the description to be complete, there are exactly two possible states available for $n = 1$ and, if the Exclusion Principle is then applied, a maximum number of two electrons in the first orbit. This gets us as far as the model of the helium atom, with a nucleus of charge $+2e$ and two orbital electrons. Application to the three-atom lithium atom entails that the third electron cannot also have the lowest orbital state. For the second energy level ($n = 2$), there are four orbital states: as we have already seen, l can then have value 0 and m value 0, or l can have 1 and $m = 0$, ± 1. Adding that each of these states can be further distinguished by m_s, it follows that there are at least eight possible states available. Assuming completeness of this description, there are exactly eight states, and applying the Exclusion Principle, we conclude that there can be at most eight electrons in the second orbit. And so forth. So the completeness assumption, used above in the suggested deduction of the Exclusion Principle, is also essential to its primary application in atomic theory, and is not extraneous to the scientific context.

If we now ask about the statistics applicable to an aggregate of identical particles, of a type for which the Exclusion Principle holds, we must get a different answer from before. To use the analogy of coins again, two heads cannot occur (unless there are relevant properties not describable in terms of *heads* and *tails*). Let us use the alternate analogy of two dice: if the Exclusion Principle holds, there are only $6 \times 5 = 30$ and not $6^2 = 36$ possible combinations. That the face values of the dice add up to 11 (combination of 6 and 5), Maxwell–Boltzman and the ordinary gambler give probability 1/18, while the Exclusion Principle statistics—called Fermi–Dirac (FD) after its creators—gives it 1/15. Bose–Einstein statistics gives the same possibility the value 1/21. In any such example we have BE < MB < FD, because, intuitively speaking, FD favours combinations of distinct numbers, and BE favours 'doubles'. So also the possibility that the face-values add up to 12 (combination of 6 and 6) receives 0 from FD, 1/36 from MB, and 1/21 from BE. Thus for 'doubles' we have FD < MB < BE. Once more, we face a puzzle of statistical correlation.

That bit of cogitation about identity and indistinguishability received support, of a sort, from the general quantum-theoretical treatment of aggregates. This emerged at the hands of Dirac and of Heisenberg in 1926, and was properly generalized in 1926/7 by Wigner. Identity and indistinguishability were given the precise form of a symmetry principle. I will explain this in more detail below.

2. PERMUTATION INVARIANCE AND THE DICHOTOMY PRINCIPLE

Given two systems X and Y which are capable of the same states, we represent the composite system $X + Y$ by means of a tensor product. Let H be the Hilbert space of individual pure states; then a pure state of $X + Y$ is presumably represented by a vector in $H \otimes H$, and, in general, any state by a statistical operator on $H \otimes H$. We recall also that, in the presence of superselection rules, this representation of states is no longer unique.

Any given vector in $H \otimes H$ takes the form $\varphi = \Sigma c_{ij} x_i \otimes y_j$ where $\{x_i\}$ and $\{y_j\}$ are bases for H. Intuitively, we describe this as a superposition of states $x_i \otimes y_j$ in which X has state x_i and Y has state y_j. What about the permuted situation, in which X has y_j and Y has x_i? Mathematically, vector $y \otimes x \neq x \otimes y$. Similarly, the general state φ has permutation

$$\varphi' = \sum c_{ij}(y_j \otimes x_i) \neq \varphi \text{ in general.}$$

Here we have two particles only, X and Y, and the group of permutations has only one non-trivial element and one trivial one (the identity). In the case of N particles there are $N!$ permutations, forming a larger group. Every permutation s can be represented by a unitary operator P_s which transforms the states in the obvious way, illustrated here by the change of φ into φ'.

Does the mathematical change in the vector by such a permutation lead to the representation of a new state? Or is a permutation like our previous example of rotation through 360 degrees? For identical particles X and Y, the assertion is that, indeed, there is no 'real' difference in states represented by permuted vectors such as $x \otimes y$ and $y \otimes x$.

Permutation Invariance: If φ is the state of a composite system whose components are identical particles, then the expectation value of any observable A is the same for all permutations of φ.

The Permutation Invariance Requirement requires a limitation either on the states (perhaps not all vectors represent pure states that can really be found in nature) or else on the observables (perhaps not all Hermitean operators represent measurable physical quantities). And indeed, these two sorts of limitations must go hand in hand. If the vector φ represents no state that can really occur, what of the superpositions of φ with other vectors? If we say they don't either, then the projection operator I_φ would at best represent the *yes/no* observable whose measurement must always have a negative (null) outcome. The link of the phenomena with this Hermitean would be no different from that with the projection along the null vector. On the other hand, if not all Hermitean operators represent observables, then some distinct vectors appear to represent states which are empirically indistinguishable.

The thesis of Permutation Invariance has therefore two aspects. If we regard ourselves as already knowing what the observables are, we can try to state it in the following form:

Permutation Invariance 1: If some observable A does not have the same expectation value in φ and its permutation φ', then φ does not represent a (physically possible) pure state of an aggregate of identical particles.

In this formulation the requirement has bite if we add for example that all Hermitean operators, or a certain subclass thereof, represent (real, measurable) observables.

If instead we regard ourselves as knowing already what states there are, we can attempt to formulate Permutation Invariance as follows:

Permutation Invariance 2: If some state φ is such that Hermitean operator A does not have the same expectation value in state φ as in its permutation φ', then A does not represent a (real, measurable) observable.

This formulation acquires bite if for example we add that all vectors in H^N represent states (even if not uniquely).

The two formulations may or may not have the same amount of bite. Whether or not they do, and how well they succeed in capturing the intuitively stated principle, is to be seen.

Some vectors are affected only inessentially by permutation, in that they remain parallel to themselves. In the case $N = 2$, there is only one non-trivial permutation, which replaces $x \otimes y$

by $y \otimes x$; extended by linearity, it replaces $\Sigma c_{ij}(x_i \otimes y_j)$ by $\Sigma c_{ij}(y_j \otimes x_i)$. If φ is the original vector, let φ' be its replacement by this non-trivial permutation. Then we have two special cases:

$$\varphi' = \varphi : \varphi \text{ is a } symmetric \text{ vector}$$

$$\varphi' = -\varphi : \varphi \text{ is an } anti\text{-}symmetric \text{ vector}$$

In both cases, the expectation value of any Hermitean operator A remains the same, i.e. $(\varphi \cdot A\varphi) = (\varphi' \cdot A\varphi')$. Hence we are sure that such a vector φ can be admitted as representing a state which satisfies Permutation Invariance 1. For the two-particle case, these form two subspaces:

Symmetric states: $\qquad \varphi = \Sigma c_{ij}(x_i \otimes y_j + y_j \otimes x_i)$

Anti-symmetric states: $\qquad \psi = \Sigma c_{ij}(x_i \otimes y_j - y_j \otimes x_i)$

The definition can be extended to the general case: φ in H^N is *symmetric* if every two-particle permutation leaves it the same, and *anti-symmetric* if every such permutation turns it into $-\varphi$. Note the place of 'every': we are not including e.g. the case in which some such permutation turns φ into $-\varphi$ and some other one turns it into itself! Having noted this, we can postulate the *Dichotomy Postulate*:

Dichotomy: Every type of particle is such that its aggregate can take only symmetric states (*boson*), or else such that its aggregates can take only anti-symmetric states (*fermion*).

I do not maintain as self-evident that this is the best choice of definitions, nor that the postulate is an inalienable part of quantum mechanics. It will look familiar enough to many readers, but is nevertheless debatable.

Below we will see that the given definitions do indeed imply the applicability of Fermi–Dirac statistics to fermions, and of Bose–Einstein statistics to bosons. The postulate therefore 'explains' or 'accounts for', or at least fits the dichotomy of, statistical behaviour found in nature. It also accounts in some sense (to be discussed) for the Exclusion Principle which holds for fermions.

We must however sharply distinguish the Dichotomy Postulate I have just described from the preceding weaker postulate of Permutation Invariance. Looking back, you will see that I asserted the following: we find that in the two-particle case two sorts of states automatically satisfy Permutation Invariance for all Hermitean operators. The postulate of Dichotomy, presented next, was a proposed generalization. For we must reflect that (*a*) the general *N*-particle case is not a simple extrapolation of the two-particle case, and (*b*) Permutation Invariance concerns observables, and it is possible that not all Hermitean operators represent observables.

Proofs and illustrations: Spin and Statistics

Pauli (1940) proved an important theorem for relativistic quantum mechanics, which was sometimes taken as establishing Dichotomy. This was his famous Spin and Statistics theorem. All known particles were (and are) bosons or fermions. In addition, all known bosons have integral spin and all known fermions, half-integral spin. This is too beautiful a connection to be a coincidence, one would think; and indeed it is not. Pauli proved that in the relativistic formulation, (*a*) the field operators of particles with integral spin cannot obey the fermion commutation relationship, and (*b*) the field operators of those with half-integral spin cannot obey the boson commutation relationship.

Logically this does not lead to Dichotomy at all: if particles with integral spin cannot be fermions, it does not follow they are bosons.[3] Indeed, in the 1950s Green (1953) and Volkov (1959, 1960) showed that relativistic quantum mechanics admits field operators which are of neither type. These were exploited in the literature on 'parastatistics' in the next decade, in connection with conjectures about quarks. Parastatistics are statistics of neither the Bose nor the Fermi type, and it is therefore clear that Pauli's theorem had not established Dichotomy even for relativistic quantum mechanics.

3. THE EXCLUSION PRINCIPLE

Before continuing our discussion of Dichotomy, let us pause to see what happened to Pauli's Exclusion Principle. In 1925

Wolfgang Pauli formulated his principle for orbital electrons in the atom:

> In an atom there cannot be two or more equivalent electrons for which the values of all four quantum numbers coincide. If an electron exists in an atom for which all of these numbers have definite values, then this state is 'occupied'. (Pauli 1925, 965)

In general terms, if states are specified completely, one would say then that two fermions of the same type cannot be in the same state. But this, though intuitively satisfying, is not accurate in relation to the general treatment. Suppose an aggregate of two fermions, $X + Y$, is in the anti-symmetric state:

$$(1) \qquad \varphi = \frac{1}{\sqrt{2}} [(x \otimes y) - (y \otimes x)]$$

It is indeed true that $x = y$ is excluded, for that would make this the zero vector. But in what sense do the two particles 'occupy' different states? Could we say that one is in state x and the other in state y? That does not sit well at all with our previous discussion of composition and reduction. The two reduced states are the same:

$$(2) \qquad \#\varphi[1] = \#\varphi[2] = \tfrac{1}{2}I_x + \tfrac{1}{2}I_y$$

So Reduction assigns the same state to both. If we amend the slogan to 'fermions cannot occupy the same pure state', we do indeed rule out $x \otimes x$ as a fermion state. But we do not thereby rule out the boson state

$$(3) \qquad \tfrac{1}{3}[(x \otimes x \otimes y) + (x \otimes y \otimes x) + (y \otimes x \otimes x)]$$

whose reductions are all three equal to

$$(4) \qquad \tfrac{2}{3}I_x + \tfrac{1}{3}I_y$$

so it is not true that two particles occupy the same pure state in this boson case, either.

To capture the distinction between fermions and bosons which the literature keeps alluding to in terms of exclusion, define an *exclusive product state* to be any tensor product state $x_1 \otimes \ldots \otimes x_n$ in which $x_i = x_j$ only if $i = j$. Then we have the following:

QM Exclusion Principle: All anti-symmetric states are super-positions of exclusive product states only; not all symmetric states are.

This distinguishes the anti-symmetric and symmetric subspaces from each other, but not the individual states.

It may be that the ignorance interpretation—which functions often enough as a good *pons asinorum*, whatever its philosophical difficulties—has helped to keep the slogan formulation of Exclusion alive. There is something intuitively feasible about a picture of identical fermions all occupying mutually orthogonal states, only we don't know (or it makes no sense to ask) which particle is in which state. The precise truth behind this feeling lies perhaps in the following corollary:

QM Exclusion Corollary: If $X_1 + \ldots + X_N$ is in anti-symmetric state φ, then the reduced states $\#\varphi[X_J]$ are all the same, and have an image space of at least N dimensions.

In other words, each particle is in the same reduced state, which is a mixture of at least N mutually orthogonal pure states. I shall prove these results in *Proofs and illustrations*; their significance for interpretation will be discussed below.

Proofs and illustrations

Let us here prove the QM Exclusion Principle for the anti-symmetric states. To say that φ is anti-symmetric means that a permutation of two particles turns it into $-\varphi$. A permutation of three particles, leaving none in its own place, is a product of two-particle permutations (two 'exchanges'), so turns φ into $-(-\varphi) = \varphi$ again. (Thus, if we write S_{ij} for the permutation of the ith and jth element in a sequence, then $S_{12}S_{23}$ $(\langle a, b, c\rangle) = S_{12}(\langle a, c, b\rangle) = \langle c, a, b\rangle$. We can accordingly designate $S_{12}S_{23}$ as S_{321} which is read 'cyclically': replace 3rd by 2nd, 2nd by 1st, 1st by 3rd. (In this notation, then, $S_{321} = S_{213} = S_{132}$.) All permutations are products of exchanges. We call a permutation *odd* (*even*) if it is the product of an odd (even) number of exchanges. The *signature* $s(P)$ of a permutation is $+1$ if P is even and -1 if P is odd.

Now suppose φ is an anti-symmetric vector in H^N, the pure state of an aggregate $X_1 + \ldots + X_N$. Then let B be a base for

H, so the set $\{y_1 \otimes \ldots \otimes y_N\}$, with all y_i in B, is a base for H^N. Obviously many elements of this product basis are permutations of other elements. Suppose P is an exchange, that ψ is a product basis vector, and $\varphi = a\psi + bP\psi + \Sigma d_k\xi_k$, where the ξ_k are the remaining base vectors. We now apply P to φ, to obtain

$$\begin{aligned} P\varphi &= aP\psi + bPP\psi + \sum d_k P\xi_k \\ &= aP\psi + b\psi + \sum d_k P\xi_k \\ &= -\varphi \end{aligned}$$

Since permutation is one-to-one, the vectors $P\xi$ are base vectors distinct from, and hence orthogonal to, ψ and $P\psi$. So unless $a = -b$, $P\varphi$ cannot be $-\varphi$. We conclude that $a = -b$.

Imagine now that ψ had a duplication in it; specifically that $P = S_{ij}$, where the ith and jth component of the tensor product ψ are the same. Then $P\psi = \psi$. But then

$$\begin{aligned} \varphi &= a\psi - aP\psi + \sum d_k\xi_k \\ &= a\psi - a\psi + \sum d_k\xi_k \\ &= \sum d_k\xi_k \end{aligned}$$

i.e., ψ does not appear in the expansion of φ.

This proves the QM Exclusion Principle. It also establishes that φ takes the very specific form

$$(1) \qquad \varphi = \sum c_R(s(P_1)P_1\psi^R + \ldots + s(P_{N!})P_{N!}\psi^R)$$

where the collection $\{\psi^R\}$ is an enumeration of *exclusive* base vectors, one from each permutation equivalence class, and P_1, $\ldots$, $P_{N!}$ form the group of permutations for N particles. The vectors

$$(2) \qquad \varphi^R = \sum_P s(P)P\psi^R$$

span the subspace of anti-symmetric vectors, of course.

By an exactly similar argument, any symmetric vector ψ takes the form:

$$(3) \qquad \psi = \sum c_R(P_1\xi^R + \ldots + P_{N!}\xi^R)$$

with the signature omitted, since any permutation must leave ψ the same as it was. Of course, the exclusion argument cannot be given in this case. So here the ξ^R are an enumeration of basis vectors, one from each permutation equivalence class, but with duplications inside them allowed.

We can now also add the QM Exclusion Corollary. Having proved that φ takes form (1), and that it is the state of an N-particle system $X_1 + \ldots + X_N$, we can ask for a reduction, to assign a (mixed) state to each particle. Each reduction $\#\varphi[X_j]$ will be the same for $j = 1, \ldots, N$, because of the permutation invariance of the anti-symmetric vector. What is noteworthy is that its image space must have dimension $\geq N$. If we could accept the ignorance interpretation of mixtures, we could then say: each particle is in a *distinct* pure state, with a certain probability.

I shall not advocate the ignorance interpretation, but I do want to prove that corollary.

We can rewrite (1) as

$$(4) \qquad \varphi = \sum a_r \varphi^r = \sum a_r(x_1^r \otimes \ldots \otimes x_N^r)$$

where $\{\varphi^r\}$ is an enumeration of the exclusive product states, and $a_s = s(P)a_r$ whenever $\varphi^s = P\varphi^r$. But that we can in turn rewrite as

$$(5) \qquad \varphi = \sum_s \psi^s \otimes \sum_t (a_r x_N^t : x_1^t \otimes \ldots \otimes x_{N-1}^t = \psi^s\}$$

where $\{\psi^s\}$ is an enumeration of the $(N-1)$-ary exclusive product vectors. By General Reduction, the vectors

$$(6) \qquad w(s) = \sum \{a_r x_N^t : x_1^t \otimes \ldots \otimes x_{N-1}^t = \psi^s\}$$

span the image space of $\#\varphi[X_N]$. Without loss of generality, let the enumeration be such that $a_1 \neq 0$ and

$$\varphi^1 = \psi^1 \otimes x_1 = (x_2 \otimes \ldots \otimes x_N) \otimes x_1$$

$$\varphi^2 = \psi^1 \otimes x_2 = (x_3 \otimes \ldots \otimes x_N) \otimes x_2$$

$$\varphi^N = \psi^N \otimes x_N = (x_1 \otimes \ldots \otimes x_{N-1}) \otimes x_N$$

where we clearly have $a_r = \pm a_1$ for $r = 1, \ldots, N$. Looking now at definition (6), we see that $w(1) = a_1 x_1 + by$, where y must be a sum of vectors in base B which are distinct from and hence orthogonal to $x_1, \ldots, x_N$. Moreover, both components of $w(1)$ are orthogonal to $x_2, \ldots, x_N$; hence $w(1)$ is too. This observation generalizes easily to $w(2), \ldots, w(N)$. So we give the proof about dimensionality as follows:

1. $w(2) \not\perp x_2$ and $w(1) \perp x_2$.
2. Hence $w(2) \notin [w(1)]$.
3. Hence $[w(1), w(2)]$ has dimension 2.
4. Suppose that for all $s \le r < N$, $[w(1), \ldots, w(s)]$ has dimension s.
5. $w(r + 1) \not\perp x_{r+1}$ but $w(1), \ldots, w(r) \perp x_{r+1}$.
6. Hence $w(r + 1) \notin [w(1), \ldots, w(r)]$.
7. Hence $[w(1), \ldots, w(r), w(r + 1)]$ has dimension $r + 1$.

By mathematical induction, and so $[w(1), \ldots, w(N)]$ has dimension N. But it is included in the image space of $\#\varphi[X_N]$, which therefore has dimension $\ge N$.

4. BLOKHINTSEV'S PROOF OF THE FERMION–BOSON DICHOTOMY

The assertion that all elementary particles are either fermions or bosons means that only symmetric and anti-symmetric states are allowed for aggregates of identical particles. In the case of a two-particle aggregate, this can be deduced from Permutation Invariance, as we shall see. Indeed, in this section we take up a proof, which establishes this for an N-particle aggregate, for any finite number N. But this proof is based on an assumption which becomes questionable exactly when we note that Permutation Invariance can be viewed as a superselection rule. With that assumption questioned, the proof will remain valid for the case of two particles, but not for $N > 2$.

The proof, due to Blokhintsev (1964b, 399 ff.), purports to demonstrate that there can be only fermions and bosons. It is not the only proof of its sort, but it is a very clear one.[4]

Let us temporarily submit:

Assumption: Every pure state is represented by a vector, and if a vector x represents a pure state then the projection operator I_x represents an observable.

This assumption gives us a handle on what states and observables there are and, again, looks relatively familiar. The proof to follow shows that under this assumption Permutation Invariance entails the Dichotomy postulate.

It follows at once from the Assumption that, if the expectation values of all observables are the same in pure states x and y, then $x = ky$ for some scalar k. For one of these observables is I_x, and the supposition entails therefore that it has expectation value 1 in y.

Next, consider state $\varphi(q_1, \ldots, q_n)$ of an assembly of N particles, with these labels or coordinates q_i representing the particles. Let S_{ij} be the linear operator which represents the permutation of the ith and jth particles. Hence $S_{12}\varphi(q_1, \ldots, q_n) = \varphi(q_2, q_1, \ldots, q_n)$ and so forth. According to the Permutation Invariance Requirement, the expectation value of any observable must be the same for φ and for $S_{12}\varphi$, so, as we just saw, these two vectors must be constant multiples of each other. Therefore, if $S_{12}\varphi = k\varphi$, then

$$S_{12}S_{12}\varphi = S_{12}(k\varphi) = k(S_{12}\varphi) = k^2\varphi$$

but since S_{ij} is its own inverse, that means that $\varphi = k^2\varphi$, so $k = \pm 1$.

In this deduction I assumed that the group of permutations of particles is represented by a group of unitary operators on the Hilbert space, in just the way we examined already in Chapter 7. (When we look at the example $N = 2$ below, this will be quite obvious.) The permutation of the 1st and 2nd component, let us call it S_{12}, changes $x_1 \otimes x_2 \otimes x_3 \otimes \ldots \otimes x_N$ into $x_2 \otimes x_1 \otimes x_3 \otimes \ldots \otimes x_N$. By linearity it extends to other vectors: $S_{12}\Sigma\varphi_i = \Sigma S_{12}\varphi_i$. I designate the unitary operator by the same name S_{ij} as the permutation it represents.

So now each permutation operator S_{ij} classifies the states allowed by Permutation Invariance into two sorts for each index couple (i, j):

φ is symmetric for (i, j) if $S_{ij}\varphi = \varphi$.

ψ is anti-symmetric for (i, j) if $S_{ij}\psi = -\psi$.

The fundamental dichotomy to be proved is clearly that a vector is (anti-)symmetric either for all couples (i, j) or for none.

Theorem: If φ satisfies the Permutation Invariance requirement, then there exists a number $k = \pm 1$ such that $S_{ij}\varphi = k\varphi$ for all couples (i, j).

This result is certainly impressive, but let us go back to the beginning of our argument, and look closely at the assumptions. The crucial step was the assertion that $S_{ij}\varphi = k\varphi$ for some k. This was supposed to hold for any state φ which satisfies the Permutation Invariance requirement. And we could derive this from the following assertion:

For any observable A, the expectation value of A in φ is the same as in $S_{ij}\varphi$, if φ is a pure state allowed by Permutation Invariance,

only by means of an auxiliary premiss about the existence of some observables, such as that in our listed Assumption:

If φ is a pure state allowed by the Permutation Invariance requirement, then I_φ represents an observable;

for I_φ has expectation 1 in any state ψ if and only if $\psi = k\varphi$ for some scalar k. It is possible to replace the Assumption by something more sophisticated.[5] But as we shall see in the next section, the step can be denied, whatever assumption it relies on.

Notice what has happened here. The intuitively stated Permutation Invariance principle speaks of observables. If every Hermitean operator represents an observable, then every vector represents a pure state uniquely characterized by the observables of which it is an eigenstate — and two vectors represent the same such state only if they are parallel. Hence in that case the requirement becomes: each permutation is a symmetry of each possible state, and each vector representing a pure state is an invariant of every permutation operator. Under those conditions, Dichotomy can be deduced.

There are other proofs of this same sort. One result that significantly improves on Blokhintsev is due to another Russian physicist, I. G. Kaplan (1976). When an n-body system of one sort is in pure state φ, we can assign a state $\#\varphi[i]$ to its ith component, $i = 1, \ldots, N$, by 'reduction of the density matrix'. Intuitively, $\#\varphi[i]$ is the state in which any observable which 'really' pertains only to the ith component has the same expectation value, whether we consider the state φ of the whole or the reduced state $\#\varphi[i]$ of the component. This reduced state $\#\varphi[i]$ is generally mixed rather than pure.

Kaplan assumes that, if the system is an aggregate of identical particles, then $\#\varphi[i]$ is the same for all $i = 1, \ldots, N$. From this he deduces Dichotomy. He does not purport to deduce his own assumption from what I have called Permutation Invariance 1, as indeed he could not. That thesis will tell us only that all observables pertaining to a single particle must have the same expectation value in each $\#\varphi[i]$. Indeed, this reduced state is represented by a statistical operator; let us call it W_i for perspicuity. Then if $N = 2$, for example, he requires $W_1 = W_2$. But Permutation Invariance 1 entails only that, if A is an observable we can measure, then it will have the same expectation value for both. If we now want to deduce that $W_1 = W_2$, we need the postulate that, whatever these reduced states are, there are always sufficient observables to 'separate' them. That is, for any two states W_i, W_j there is some observable A such that $Tr(AW_i) \neq Tr(AW_j)$ if $W_i \neq W_j$. I doubt that any concrete assumption weaker than Blokhintsev's about what observables there are will secure this separation. But the point stands in any case: a strong, independent assumption about the class of observables is needed, to get us from Permutation Invariance 1 to Dichotomy.

The same point can be made for the proof given by Sarry (1979). As Permutation Invariance principle, Sarry assumes that the permutation operators commute with all observables. Then he assumes that a certain Hermitean A represents an observable. He offers the following justification:

As a rule this question [whether an operator represents an observable] turns out to be very difficult, but a simple criterion is sometimes helpful ...: '$\hat{O}$ is an observable if it satisfies an algebraic equation of a polynomial type.' (Sarry 1979, 680)

The operator A in question satisfies $A^3 - |\lambda|^2 A = 0$, and so meets the criterion. For this criterion, however, Sarry cites Dirac's text, which dates well before the understanding of superselection rules that puts the question into the limelight for us. The criterion itself could be listed as a (rather strong) assumption concerning what observables there are. Hence the same point emerges very clearly: Permutation Invariance implies Dichotomy only via assumptions concerning what observables there are.

Proofs and illustration

We need to prove the theorem, and also to spell out exactly what a vector needs to be like in order to be (anti-)symmetric.

Lemma 1: If $S_{ab}\varphi = S_{bc}\varphi = \varphi$, then $S_{ac}\varphi$ cannot be $-\varphi$ (triangle equality).

To prove this, let us use a 'double labelling' picture. Let $\varphi = \varphi(q_1, \ldots, q_n)$ and a, b, c be three 'places' between 1 and N inclusive. Suppose these places are occupied by q_i, q_j, q_k; depict it as follows:

$$(1) \qquad \varphi = \varphi\left(\begin{array}{ccc} \ldots a \ldots b \ldots c \ldots \\ \ldots q_i \ldots q_j \ldots q_k \ldots \end{array}\right)$$

We now look at the effect of the permutations S_{ab} (exchange coordinates that used to occupy places a and b) and so forth:

$$(2) \qquad S_{ac}\varphi = \varphi\left(\begin{array}{ccc} \ldots a \ldots b \ldots c \ldots \\ q_k \ldots q_j \ldots q_i \ldots \end{array}\right)$$

Assume, for reductio, that $S_{ab}\varphi = S_{bc}\varphi = \varphi$ and $S_{ac}\varphi = -\varphi$. Then we have

$$(3) \qquad S_{bc}S_{ac}\varphi = S_{bc} - \varphi = -S_{bc}\varphi = -\varphi \qquad \text{by hypothesis}$$

$$(4) \qquad S_{bc}S_{ac}\varphi = S_{bc}\varphi\left(\begin{array}{ccc} \ldots a \ldots b \ldots c \ldots \\ \ldots q_k \ldots q_j \ldots q_i \ldots \end{array}\right) \qquad \text{by (2)}$$

$$= \varphi\left(\begin{array}{ccc} \ldots a \ldots b \ldots c \ldots \\ \ldots q_k \ldots q_i \ldots q_j \ldots \end{array}\right)$$

$$(5) \qquad S_{ab}S_{bc}S_{ac}\varphi = S_{ab}(-\varphi) = -\varphi \qquad \text{by (3) and hypothesis}$$

$$(6) \qquad S_{ab}S_{bc}S_{ac}\varphi = S_{ab}\varphi\left(\begin{array}{ccc} \ldots a \ldots b \ldots c \ldots \\ \ldots q_k \ldots q_i \ldots q_j \ldots \end{array}\right) \qquad \text{by (4)}$$

$$= \varphi \begin{pmatrix} \dots a \dots b \dots c \dots \\ \dots q_i \dots q_k \dots q_j \dots \end{pmatrix}$$

(7) $S_{bc}S_{ab}S_{bc}S_{ac}\varphi = S_{bc}(-\varphi) = -\varphi$ by (5) and hypothesis

(8) $S_{bc}S_{ab}S_{bc}S_{ac}\varphi = S_{bc}\varphi \begin{pmatrix} \dots a \dots b \dots c \dots \\ \dots q_i \dots q_k \dots q_j \dots \end{pmatrix}$

$$= \varphi \begin{pmatrix} \dots a \dots b \dots c \dots \\ \dots q_i \dots q_j \dots q_k \dots \end{pmatrix}$$

$$= \varphi \quad \text{by (1)}$$

But now we have a contradiction between (7) and (8), thus refuting the hypothesis and hence proving the lemma.

Lemma 2: If $S_{ab}\varphi = S_{bc}\varphi = -\varphi$, then $S_{ac}\varphi$ cannot be φ.

This is proved exactly like Lemma 1, *mutatis mutandis*.

The theorem is now proved by *reductio ad absurdum*, namely, of the hypothesis that $S_{ab}\varphi = \varphi$ and $S_{cd}\varphi = -\varphi$.

First, consider S_{bc} and S_{ac}, and let us show that $S_{bc}\varphi = S_{ac}\varphi = \varphi$. If $S_{bc}\varphi = \varphi$, then by Lemma 1, $S_{ac}\varphi = \varphi$ as well. If $S_{bc}\varphi = -\varphi$ we have two subcases:

(i) $S_{bc}\varphi = -\varphi$, $S_{ac}\varphi = \varphi$, $S_{ab}\varphi = \varphi$. But $S_{ab} = S_{ba}$, so by Lemma 1, $S_{bc}\varphi = \varphi$ *contra* supposition.

(ii) $S_{bc}\varphi = -\varphi$, $S_{ac}\varphi = -\varphi$, $S_{ab}\varphi = \varphi$. But $S_{bc} = S_{cb}$, so $S_{ac}\varphi = S_{cb}\varphi = -\varphi$ and thus by Lemma 2 we have $S_{ab}\varphi = -\varphi$, *contra* supposition.

Second, we consider S_{ac}, S_{ad}. By an argument exactly like the preceding and with reference to the hypothesis that $S_{cd}\varphi = -\varphi$, we deduce that $S_{ac}\varphi = S_{ad}\varphi = -\varphi$ as well.

But now we have two arguments, one of which leads to $S_{ac}\varphi = \varphi$ and the other to $S_{ac}\varphi = -\varphi$, thus refuting our supposition. This establishes the theorem.

To finish this section, we need to be sure exactly which vectors are symmetric for all permutations (i, j) and which are anti-symmetric. The definition we have is enough to single them out abstractly, but do they really have to look the way we say? For concreteness, I shall just discuss this for the case $N = 2$ explicitly.

Our Hilbert space here is the tensor product $H^2 = H \otimes H$, with basis $\{x \otimes y : x$ in basis B_1 for H and y in basis B_2 for $H\}$. A permutation of $x \otimes y$ is $y \otimes x$, and this is definitely not $\pm(x \otimes y)$. Hence this does not represent a state allowed by Permutation Invariance. However,

$$S_{12}[(x \otimes y) - (y \otimes x)] = [(y \otimes x) - (x \otimes y)]$$
$$= - [(x \otimes y) - (y \otimes x)]$$
$$S_{12}[(x \otimes y) + (y \otimes x)] = [(y \otimes x) + (x \otimes y)]$$
$$= [(x \otimes y) + (y \otimes x)]$$

So these are examples of anti-symmetric and symmetric vectors. Consider the subspaces

H^2_+, spanned by $\{(x \otimes y) + (y \otimes x) : x \in B_1, y \in B_2\} = B_+$

H^2_-, spanned by $\{(x \otimes y) - (y \otimes x) : x \in B_1, y \in B_2\} = B_-$

We prove first that all vectors in H^2_- are anti-symmetric:

$$S_{12}\left[\sum a_{ij}\varphi_{ij}\right] = \sum a_{ij}S_{12}(\varphi_{ij}) = \sum a_{ij}(-\varphi_{ij}) = - \sum a_{ij}\varphi_{ij}$$

where $\{\varphi_{ij}\} = B_-$. Similarly, all the vectors in H^2_+ are symmetric. Next, $H^2_- \otimes H^2_+$ is the whole space, for every basis vector $x \otimes y$ can be rewritten

$$\tfrac{1}{2}[[(x \otimes y) + (y \otimes x)] + [(x \otimes y) - (y \otimes x)]]$$

so superpositions of members of basis $B_1 \otimes B_2$ can be rewritten as superpositions of members of B_+ and B_-.

Finally, could any such superposition be symmetric for all permutations or anti-symmetric for all? Well, in this simple case we only have to look at permutation S_{12}:

$$S_{12}\left(\sum a_{ij}\varphi^+_{ij} + \sum b_{ij}\varphi^-_{ij}\right) = \sum a_{ij}\varphi^+_{ij} - \sum b_{ij}\varphi^-_{ij}$$

It is clear that in this case the operation does not multiply by ±1 unless all the a_{ij}, or else all the b_{ij}, are zero. Thus we have found that these two sets of vectors, the symmetric ones and the anti-symmetric ones, are indeed subspaces.

5. PERMUTATIONS AS SUPERSELECTION OPERATORS

In the preceding section we began by ignoring the problem of what observables there are, and focusing on which states are allowed by permutation invariance. Since the two questions are not independent, as we noted, it turned out that the derivation of the statistical dichotomy involved a special assumption about observables. Pure states were assumed to be represented by vectors, and projections along those vectors were assumed to represent observables.

When we discussed symmetries before, and found the notion of superselection rule, we also found that not every Hermitean operator can just automatically be assumed to represent an observable. After their introduction, therefore, we do not pretend to have any simple recipe for what represents what. Shortly after their introduction, this was the assessment:

The preceding discussion reflects our knowledge on the limitations of the measurability of operators. This is an incomplete knowledge, and it is generally believed that the measurability of *most* operators is open to question. It is, in fact, not very likely that the limitations on measurability can be all formulated in terms of superselection operators with which all measurable operators commute. However, it is also believed that no incorrect conclusion will be arrived at by assuming the measurability of all self-adjoint operators which commute with all superselection operators. (Houtappel *et al.* 1965, 616)

At this point therefore it becomes advantageous for us to think of Permutation Invariance in the second formulation: an operator can represent an observable only if its expectation value is not affected by permutation.

In Permutation Invariance 2 the Permutation Invariance requirement reads: there are no observables whose measurement could distinguish between a state and its permutation. We can restate this in terms of the unitary representation of the permutation group:

Permutation Invariance 2: if Hermitean operator A has a different expectation in states φ and $S_{ab}\varphi$, then A does not represent an observable.

Now this entails that, if A is (represents) an observable, then A commutes with S_{ab}. This operator S_{ab} is a unitary operator, so

Permutation Invariance 2 entails, for any φ and for any Hermitean operator A representing an observable,

$$(\varphi \cdot A\varphi) = (S_{ab}\varphi \cdot AS_{ab}\varphi)$$

Because S_{ab} is unitary and its own inverse, this can be rewritten as

$$(\varphi \cdot A\varphi) = (\varphi \cdot S_{ab}AS_{ab}\varphi)$$

If we now assume that the set of states is rich enough to separate any two observables, this entails that $A = S_{ab}AS_{ab}$ or, equivalently, since S_{ab} is its own inverse,

$$S_{ab}A = AS_{ab}$$

Thus we have a new formulation of Permutation Invariance:

Permutation Invariance 3: Each permutation operator commutes with all observables.

But that means, by definition, that the permutation operators are superselection operators. Hence there are no genuine pure states which are superpositions of vectors taken from distinct eigenspaces of such a permutation operator. (And if the Hamiltonian is an observable, we recall that this means that a state cannot evolve out of one such subspace into another.)

This observation brings us a source of new insights into permutational symmetry. Looking back to the beginning of Blokhintsev's proof, we note that, if the number of particles $N = 2$, then the permutation operator S_{12} has two eigenvalues, $+1$ and -1. Knowing now that S_{12} is a superselection operator, we can immediately infer the following corollaries. Let H^+ and H^- be the subspaces made up respectively by the symmetric and anti-symmetric vectors in H^2. Then it follows that:

1. A vector in H^+ or in H^- may represent a pure state, but a non-trivial superposition $a\varphi + b\psi$ with φ in H^+ and ψ in H^- does not represent a pure state, but only a mixture of φ and ψ.

2. If the Hamiltonian H is an observable, then no state in H^+ can evolve into one in H^-; nor vice versa.

Thus, for $N = 2$, Dichotomy is clearly established. Moreover, the slogan *for every symmetry a conservation law* is cashed out in corollary 2, which is a consequence derived previously in our general discussion of superselection rules.

But since the permutations are now themselves supposed to be non-trivial superselection operators, we conclude that membership in the family of observables is definitely not something to be taken for granted in this context. Hence we cannot follow Blokhintsev's way of generalizing this demonstration.

Looking already at the case $N = 3$, we find that the permutations do not commute:

$$S_{12}S_{23}(x \otimes y \otimes z) = S_{12}(x \otimes z \otimes y) = (z \otimes x \otimes y)$$
$$S_{23}S_{12}(x \otimes y \otimes z) = S_{23}(y \otimes x \otimes z) = (y \otimes z \otimes x)$$

This non-commuting group of superselection operators must be treated more delicately. What can be made of the assertion that the superselection operators 'split' the total Hilbert space into a family of 'coherent' subspaces—superposition of states being genuinely possible within, but not between, these coherent subspaces?

We begin by finding maximal sets of mutually commuting superselection operators, constructed from the permutation group. In any group G, two elements P and Q are called *conjugate* if they are related by an equation

$$P = SQS^{-1}$$

for some other element S of the group. This conjugacy is an equivalence relation, and the group is thus partitioned into a set of equivalence classes $[P] = \{SPS^{-1} : S \text{ in } G\}$, called *conjugacy classes*.

For each such conjugacy class C, define the 'character' XC of C to be the operator $\Sigma\{P : P \text{ in } C\}$. If we are dealing with the group of permutations of N particles—or its unitary representation—these sums are of course all finite.

Now these operators XC commute with all the permutation operators, and with each other. They form a set which is maximal in this respect. In addition, all its members commute with all observables. Thus we have here a maximal set of mutually commuting superselection operators. They have all eigenvectors in common, and we can define their eigenspaces:

$H_i =$ is the class of vectors φ such that

$$X[S_{12}]\varphi = k_i^{12}\varphi, \ldots, X[S_{mn}]\varphi = k_i^{mn}\varphi, \ldots$$

These are the coherent subspaces allowed by this set of super-selection rules, similar to H^+ and H^- in the case $N = 2$. The fundamental dichotomy is now clearly not a corollary, because there will in general be more than two such coherent subspaces. The number of operators $X[P]$ is the number of conjugacy classes of the N-particle permutation group—a number that goes up very quickly with N—and this is also the number of irreducible representations of that group. The symmetric and anti-symmetric subspaces correspond to the ('one-dimensional') representations, but from a general point of view these are just a special case.

The conclusion is therefore this: if we do not begin with such a special assumption as was used e.g. by Blokhintsev, then the Permutation Invariance requirement does not by itself lead to the fundamental dichotomy of symmetry types.

In the *Proofs and illustrations* we shall look at the case of $N = 3$ particles and exhibit a two-dimensional subspace which is invariant under all permutations. It is quite possible then that this subspace also cannot be 'empirically subdivided', i.e. that for all observables the expectation value is the same for all vectors in that subspace. This must be allowed as a possibility if only some, and not all, Hermitean operators correspond to observables. And if it is so, then this subspace is at best a 'generalized ray': its vectors all represent the same state. Thus, the negative criticism of Blokhintsev's proof is backed more positively by an illustration which is a clear counter-example, if anyone were to suggest that his or any such proof could succeed on the basis of Permutation Invariance alone.

Where do we stand now? The dichotomy 'found in nature' can be explained by the permutation symmetry, if we are given some extra assumption. And our reflection on superselection rules, which illuminate this symmetry, undermines exactly the previously uncontroversial status of that assumption. This is why, in the 1960s, scientists could contemplate denying the dichotomy for quarks, and talk of intermediate 'para-statistics'. The idea appears to have been dropped; but conceptual possibilities never fade away. Of course, as things presently stand,

Dichotomy appears to be correct, and to that extent those assumptions (and the strong version, Permutation Invariance 1) are plausible. They are just not inherently more plausible than Dichotomy itself, nor incontrovertible.[6]

Proofs and illustrations

We need to show that the 'character' operators $X[P]$ commute with each other and with all the permutation operators. And then we are to display, for the case of $N = 3$ particles, a two-dimensional subspace invariant under all permutations, and yet not representing either an anti-symmetric (fermion) or a symmetric (boson) type of state.

Let **S** be the permutation group for N objects, and let **U** be its representation, a group of unitary operators. The relation of conjugacy, $\approx$, is defined on **S**, and on **U**, in the usual way: $A \approx B$ exactly if the group also contains some X such that $A = XBX^{-1}$. This is an equivalence relation, and thus partitions the group into equivalence classes $[A] = \{B : B \approx A\} = \{XAX^{-1} : X \text{ in the group}\}$. When A is in **U**, we define Σ_A to be the sum $\Sigma[A]$, which is finite because the number N of objects is finite.

Lemma for the group **U**:

(a) $AXBX^{-1} = YBY^{-1}A$ for some Y
(b) $XBX^{-1}A = AYBY^{-1}$ for some Y

For (a), let $Y = AX$ so $Y^{-1} = X^{-1}A^{-1}$ and we get $YBY^{-1}A = AXBX^{-1}A^{-1}A = AXBX^{-1}$ as required. For (b), let $Y = A^{-1}X$ and we get $AYBY^{-1} = AA^{-1}XBX^{-1}A = XBX^{-1}A$ also as required.

Now define $A[B] = \{AX : X \in [B]\}$ and $[B]A = \{XA : X \in [B]\}$. Then we have of course $A[B] = \{AXBX^{-1} : X \in \mathbf{U}\}$ and $[B]A = \{XBX^1A : X \in \mathbf{U}\}$. Therefore the lemma shows that $A[B] = [B]A$. But it is also clear that $A\Sigma_B = \Sigma(A[B])$ and $\Sigma_B A = \Sigma([B]A)$, so we see that $A\Sigma_B = \Sigma_B A$. Thus all the character-operators Σ_A for A in **U** do indeed commute with all of **U**. That they also commute with each other follows *a fortiori*:

$$\sum_B\sum_A = \sum\left\{\sum_B X : X \in [A]\right\} = \sum\left(\sum_B[A]\right) = \sum\left([A]\sum_B\right)$$

$$= \sum\left\{X\sum_B : X \in [A]\right\} = \sum_A\sum_B$$

We note that this proof assumes only that **S** is a finite group, and thus holds for any finite unitary group representation.

Now we turn to the example of the conceptual possibility of a generalized ray which represents a state which is clearly not of the boson or fermion type, and yet satisfies Permutation Invariance. The smallest number of particles for which this can be done is three.[7]

As in the section on Blokhintsev, I shall use 'double labelling', but since $N = 3$ the top is always abc, and after a few lines I shall drop it. The permutations of three particles are $3! = 6$. The first is the identity, and the second is:

$$S_{ab}\varphi\begin{pmatrix} a & b & c \\ 1 & 2 & 3 \end{pmatrix} = \varphi\begin{pmatrix} a & b & c \\ 2 & 1 & 3 \end{pmatrix}$$

the others are S_{ac}, S_{bc}, S_{abc}, S_{acb} where I use the cyclic notation: S_{abc} replaces a by b, b by c, and c by a; thus

$$S_{abc}\varphi\begin{pmatrix} a & b & c \\ 1 & 2 & 3 \end{pmatrix} = \varphi\begin{pmatrix} a & b & c \\ 2 & 3 & 1 \end{pmatrix}$$

The last two permutations are products of the others: $S_{abc} = S_{ab}S_{ac}$ and $S_{acb} = S_{ac}S_{ab}$, and they are each other's inverse while S_{ab}, S_{ac}, S_{bc} are each their own inverse.

The example $S_{ab}S_{ac} = S_{abc}$ shows how to read Table 11.1. Starting with vector $\varphi = \varphi(\begin{smallmatrix} abc \\ 123 \end{smallmatrix})$, I will now for brevity just denote it as (123) and $S_{ab}\varphi(\begin{smallmatrix} abc \\ 123 \end{smallmatrix}) = \varphi(\begin{smallmatrix} abc \\ 213 \end{smallmatrix})$ I will just denote (213) and so forth. Let

$$\varphi_1 = \varphi + S_{ab}\varphi - S_{ac}\varphi - S_{acb}\varphi$$

$$= (123) + (213) - (321) - (312)$$

$$\varphi_2 = 2S_{bc}\varphi + 2S_{abc}\varphi - \varphi - S_{ab}\varphi - S_{ac}\varphi - S_{acb}\varphi$$

$$= 2(132) + 2(231) - (123) - (213) - (321) - (312)$$

and let H_q be the subspace which they span.

TABLE 11.1   *Product table for the N = 3 permutation group*

	I	S_{ab}	S_{ac}	S_{bc}	S_{acb}	S_{abc}
I	I	S_{ab}	S_{ac}	S_{bc}	S_{acb}	S_{abc}
S_{ab}	S_{ab}	I	S_{abc}	S_{acb}	S_{bc}	S_{ac}
S_{ac}	S_{ac}	S_{acb}	I	S_{abc}	S_{ab}	S_{bc}
S_{bc}	S_{bc}	S_{abc}	S_{acb}	I	S_{ac}	S_{ab}
$S_{ab}S_{ac} = S_{abc}$	S_{abc}	S_{bc}	S_{ab}	S_{ac}	I	S_{acb}
$S_{ac}S_{ab} = S_{acb}$	S_{acb}	S_{ac}	S_{bc}	S_{ab}	S_{abc}	I

To show that this space is two-dimensional, we have to see
that $(\varphi_1 \cdot \varphi_2) = 0$. To see this, note first of all that φ itself can
be taken as a vector $x \otimes y \otimes z$ in H^3, with x, y, z all
orthogonal to each other. Then $(123) \perp (213)$, etc. Also, we can
take x, y, z and have φ as of unit length, so $(\varphi \cdot \varphi) = 1$ and
similarly for (213), etc. Thus we keep only the terms $(123)^2$,
$(213)^2, \ldots = 1$, the others being zero, and we find

$$\varphi_1 \cdot \varphi_2 = -(123)^2 - (213)^2 + (321)^2 + (312)^2$$
$$= -1 - 1 + 1 + 1 = 0$$

To show next that this subspace is not reduced by Permutation
Invariance, we must show that, if any vector is in it, so are all
its permutations. Such a vector looks like this:

$$\varphi_3 = a\varphi_1 + b\varphi_2$$

and if S is a permutation, $S\varphi_3 = aS\varphi_1 + bS\varphi_2$; hence it suffices
to show that all permutations of φ_1, φ_2 are in this same
subspace. However, since S_{abc} and S_{acb} are products of the
others, we do not need to check them. So we must calculate six
equations of form $S\varphi = \varphi'$. This will also show very graphically
that we are not dealing with either symmetric or anti-symmetric
vectors here. Thus, if these vectors themselves (as opposed to
this subspace) represented a pure state, they would violate
Permutation Invariance. The example is not a counter-example
to Blokhintsev's proof, but shows instead exactly how his special
Assumption was substantial and crucial to the argument. Using
the product table to verify steps, we see that:

$$\varphi_1 = \varphi + S_{ab}\varphi - S_{ac}\varphi - S_{acb}\varphi$$

$$\varphi_2 = 2S_{bc}\varphi + 2S_{abc}\varphi - \varphi - S_{ab}\varphi - S_{ac}\varphi - S_{acb}\varphi$$

$$\varphi_1 - \varphi_2 = 2\varphi + 2S_{ab}\varphi - 2S_{bc}\varphi - 2S_{abc}\varphi$$

$$= 2(\varphi + S_{ab}\varphi - S_{bc}\varphi - S_{abc}\varphi)$$

Some permutations are as follows (I won't check all six):

$$S_{ab}\varphi_1 = S_{ab}\varphi + \varphi - S_{abc}\varphi - S_{bc}\varphi = \tfrac{1}{2}(\varphi_1 - \varphi_2)$$

$$S_{ac}\varphi_1 = S_{ac}\varphi + S_{ac}S_{ab}\varphi - \varphi - S_{ac}S_{acb}\varphi$$

$$= -\varphi - S_{ab}\varphi + S_{ac}\varphi + S_{acb}\varphi = -\varphi_1$$

Similarly,

$$S_{bc}\varphi_1 = (\tfrac{1}{2})(\varphi_1 + \varphi_2)$$

$$S_{ab}\varphi_2 = (-\tfrac{1}{2})(\varphi_2 + 3\varphi_1)$$

$$S_{ac}\varphi_2 = \varphi_2$$

$$S_{bc}\varphi_2 = (\tfrac{1}{2})(3\varphi_1 - \varphi_2)$$

Thus, the subspace *as a whole* is invariant under permutations, even though the individual vectors in it are not.

6. QUANTUM STATISTICAL MECHANICS

In the two-particle case, Permutation Invariance leaves us only two sorts of states, symmetric and anti-symmetric. Let us accept without question for now the extrapolation of this dichotomy to larger aggregates of fermions (capable only of anti-symmetric states) and of bosons (having only symmetric states). How does this lead us to Fermi–Dirac and Bose–Einstein statistics?

First, we should state the basic topic of statistical mechanics in its new quantum setting. Suppose there is an aggregate of particles, sufficiently diffuse so that we can ignore energy exchange between them. Suppose also that we know what sort of particles they are, and perhaps such global features of the aggregate as number, total mass, and total energy—but nothing beyond this about the individual states of the particles. What predictions should be made for outcomes of measurement on this aggregate?

The problem situation is one of ignorance, and we propose to model it as follows. We describe the possible pure states for the aggregate, and then choose an 'informationless' mixture thereof.[8] That is, we assign a mixed state which gives equal weight to all possible pure states. Then we answer the question of measurement predictions by calculating them from this mixture.

I shall do this now in some detail for a very finite case—two particles and a two-valued observable—of distinguishable particles, and also of bosons. The results for N particles of any sort I shall describe more abstractly and without proof, but the nature of the extrapolation will be clear.

Let our observable A of interest have two eigenstates: $Ax = x$ and $Ay = 0$, so its eigenvalues are 1 ('heads') and 0 ('tails'). We are interested to begin with in the very small Hilbert space $H = [x, y]$. But we wish to model two particles, so we form the tensor produce H^2 spanned by the product states:

$$\alpha = x \otimes x \qquad \beta = y \otimes y$$
$$\gamma = x \otimes y \qquad \delta = y \otimes x$$

This will do for two particles of different types, distinguishable by constant features such as different mass. Suppose we wish to model two bosons, so that all allowable states are symmetric, with γ, δ ruled out. Then we look at a subspace H^+ of H^2, spanned by

$$\alpha = x \otimes x \qquad\qquad \beta = y \otimes y$$
$$\eta = (x \otimes y) + (y \otimes x) \text{ or} \quad \xi = \eta/|\eta|$$

We note that $\{\alpha, \beta, \eta\}$ is a base, and $\{\alpha, \beta, \xi\}$ is an orthonormal base. To see this we must check for orthogonality. For example,

$$\alpha \cdot \eta = (x \otimes x) \cdot (x \otimes y) + (x \otimes x) \cdot (y \otimes x)$$
$$= (x \cdot y) + (x \cdot y)$$
$$= 0$$

which equals zero because $x \perp y$ (utilizing also the fact that x and y are unit vectors).

So H^+ is a three-dimensional subspace of the four-dimensional space H^2. We should check also that H^+ would have been just the same if we had begun with a different basis for H. For let $\{z, w\}$ be another base of H, for example

$$z = x + 2y \qquad w = 2x - y$$

which are orthogonal because

$$(z \cdot w) = (x + 2y) \cdot 2x - (x + 2y) \cdot y$$
$$= (x \cdot 2x) + (2y \cdot 2x) - (x \cdot y) - (y \cdot 2y)^*$$
$$= 2 + - 0 - 2 = 0$$

Then we find a new basis of symmetric vectors

$$\mu = z \otimes z \qquad v = w \otimes w$$
$$\sigma = z \otimes w + w \otimes z$$

which are orthogonal for similar reasons as before. But this basis is included in H^+, and therefore (because the dimension is 3) must be just another basis for it. For μ, v, and σ can be expressed as superpositions of α, β, η because the tensor product is bilinear:

$$\mu = z \otimes z = (x + 2y) \otimes (x + 2y)$$
$$= (x + 2y) \otimes x + (x + 2y) \otimes 2y$$
$$= (x \otimes x) + 2(y \otimes x) + 2(x \otimes y) + 4(y \otimes y)$$
$$= \alpha + 2\eta + 4\beta$$

$$\sigma = (z \otimes w) + (w \otimes z)$$
$$= [(x + 2y) \otimes (2x - y)] + [(2x - y) \otimes (x + 2y)]$$
$$= 2(x \otimes x) - (x \otimes y) + 4(y \otimes x) - 2(y \otimes y)$$
$$\quad + 2(x \otimes x) + 4(x \otimes y) - (y \otimes x) - 2(y \otimes y)$$
$$= 2\alpha - \eta + 4\eta - 2\beta + 2\alpha - 2\beta$$
$$= 4\alpha + 3\eta - 4\beta$$

and similarly for v.

Having the needed Hilbert spaces now at hand, we turn to the informationless mixture. Let us first see what it is for our smallest space H:

$$W = \tfrac{1}{2}(I_x + I_y) = \tfrac{1}{2}I$$

This is an equal mixture of the pure states represented in the basis $\{x, y\}$. It is exactly the same mixture if we choose another base, because the result can only be $\tfrac{1}{2}$ times the identity operator on the space. If the space is N-dimensional, it is of course $(1/N)$ times the identity. So now we know what our required mixed states are for H^2 and H^+:

$$W^{(2)} = \tfrac{1}{4}(I_\alpha + I_\beta + I_\gamma + I_\delta) = \tfrac{1}{4}I^2$$

$$W^+ = \tfrac{1}{3}(I_\alpha + I_\beta + I_\eta) = \tfrac{1}{3}I^+$$

where I^2 and I^+ are the identity operators on H^2 and H^+ respectively.

What is the probability of a measurement result in this sort of state? Let us concentrate on an observable B which has only 1 and 0 as eigenvalues, so that the probability of outcome value 1 equals the expectation value of B:

$$\text{Expectation value of } B \text{ in } W = Tr(\tfrac{1}{2}IB)$$

$$= \tfrac{1}{2}Tr(B)$$

Similarly, in $W^{(2)}$ it is $\tfrac{1}{4}Tr(B)$ and in W^+ it is $\tfrac{1}{3}Tr(B)$, if B is a projection operator on H^2. (The trace itself is also calculated with reference to the appropriate space.)

So far so good. Recall that our basic observable was A, with eigenvalues 1 ('heads') and 0 ('tails') in space H. Let us choose three new 'counting observables' which, as it were, count the number of 'heads':

$$B0 = (I_y \otimes I_y) \qquad \text{'no heads'}$$

$$B1 = (I_x \otimes I_y) + (I_y \otimes I_x) \qquad \text{'one head'}$$

$$B2 = (I_x \otimes I_x) \qquad \text{'two heads'}$$

Here a measurement of $B2$ on any state in H^2 has as expectation value exactly that of a combined measurement of $A \otimes I$ and $I \otimes A$, i.e. of A on each of the individual particles. (The reason is that x and y are the two eigenvectors of A in H.) Thus '$B2$ has value 1' corresponds to the intuitive 'two heads'; and so on.

The probability that a measurement of Bi in $W^{(2)}$ has value 1 equals $\tfrac{1}{4}Tr(Bi)$. But in H^2 each of $B2$ and $B0$ have trace 1 and

$B1$ has trace 2. For example:

$$Tr(B1) = (\alpha \cdot B1\alpha) + (\beta \cdot B1\beta) + (\gamma \cdot B1\gamma) + (\delta \cdot B1\delta)$$

$$= (\alpha \cdot 0) + (\beta \cdot 0) + (\gamma \cdot \gamma) + (\delta \cdot \delta)$$

$$= 2$$

because, for example, $B1\gamma = (I_x \otimes I_y + I_y \otimes I_x)(x \otimes y) = (I_x \otimes I_y)(x \otimes y) + (I_y \otimes I_x)(x \otimes y) = (I_x x) \otimes (I_y y) + (I_y x) \otimes (I_x y) = (x \otimes y) + 0$. So we have the results:

$$Pr(2 \text{ heads}|W^{(2)}) = (1/4)$$

$$Pr(1 \text{ head}|W^{(2)}) = (1/4)2 = 1/2$$

$$Pr(0 \text{ heads}|W^{(2)}) = (1/4)$$

Note that these are exactly the probabilities you would give in a fair toss of two fair coins, the paradigm of Maxwell–Boltzmann statistics.

Now we do the same calculation for W^+. Here we have a probability equal to $(1/3)$ the trace, and we find that all three traces are 1. For example:

$$Tr(B1) = (\alpha \cdot B1\alpha) + (\beta \cdot B1\beta) + (\eta \cdot B1\eta)$$

$$= (\alpha \cdot 0) + (\beta \cdot 0) + (\eta \cdot \eta) = 1$$

for $\quad B1\xi = (I_x \otimes I_y + I_y \otimes I_x)(x \otimes y + y \otimes x) = (I_x x) \otimes (I_y y) + 0 + 0 + (I_y y) \otimes (I_x x) = x \otimes y + y \otimes x$. So here we have

$$Pr(2 \text{ heads}|W^+) = 1/3$$

$$Pr(1 \text{ head }|W^+) = 1/3$$

$$Pr(0 \text{ heads}|W^+) = 1/3$$

and this equiprobability of distinct occupation numbers is indeed the paradigm example of Bose statistics.

In this way we get statistical behaviour which is quite extraordinary, because it displays such strong correlations in the behaviour of separated particles. We can depict this situation in the abstract by imagining coins (characterized only by whether they land heads up (characteristic H) or tails (not-H). The classical, Maxwell–Boltzmann, statistics prescribes a probability measure p^{mb} for two such coins in which there is zero correlation. The Bose statistics applied to this case gives us a probability measure p^{be} showing positive correlation. The Exclusion Principle requires the Fermi–Dirac probability distribution p^{fd}, which shows negative correlation tabulated:

	H	not-H	p^{mb}	p^{fd}	p^{be}
Case 1	2	0	1/4	0	1/3
Case 2	1	1	1/2	1	1/3
Case 3	0	2	1/4	0	1/3

This table shows the probabilities of 2 heads, 1 head and 1 tail, 2 tails. If we refer to the particles by distinct names a, b, the second case must be re-described as two cases:

	H	not-H	p^{mb}	p^{fd}	p^{be}
Case 1	a, b		1/4	0	1/3
Case 2(i)	a	b	1/4	1/2	1/6
Case 2(ii)	b	a	1/4	1/2	1/6
Case 3		a, b	1/4	0	1/3

Thus, whether Maxwell–Boltzmann or Bose–Einstein gives equal probabilities to equipossible cases depends on whether the first or second table reflects the 'real' equipossible partition. In terms of the second table, we can calculate conditional probabilities of having the *same* characteristic:

$$p^{mb}(Ha|Hb) = p^{mb}(Ha) = 1/2 - \text{no correlation}$$

$$p^{fd}(Ha|Hb) = 0 < p^{fd}(Ha) = 1/2 - \text{negative correlation}$$

$$p^{be}(Ha|Hb) = (1/3) \div (1/3 + 1/6)$$

$$= 2/3 > p^{be}(Ha) = 1/2 - \text{positive correlation.}$$

A more graphic illustration appears if we consider two particles a, b and a partition into four characteristics ('cells') F_1, F_2, F_3, F_4. Table 11.2 presents a tabulation. Here I calculate conditional probabilities of having *different* characteristics (being in *different* cells)—hence the correlations are reversed. As done above, if we introduce labels a and b, we must replace each of the first six cases by two subcases each. We have a picture then in which p^{fd}, p^{be}, p^{mb} each assign 1/4 to the event F_1a (a in cell F_1). The conditional probabilities calculated then are not the same:

$$p^{fd}(F_1a|F_3b) = 1/3$$

positive correlation ('repulsion') for different cells

$$p^{be}(F_1a|F_3b) = 1/5$$

negative correlation ('attraction') for different cells

$$p^{be}(F_1a|F_1b) = 2/5$$

positive correlation for same cell occupancy

$$p^{mb}(F_1a|F_3b) = 1/4$$

no correlation

TABLE 11.2

p^{fd}	p^{be}	p^{mb}	F_1	F_2	F_3	F_4
1/6	1/10	2/16	1	1		
1/6	1/10	2/16	1		1	
1/6	1/10	2/16	1			1
1/6	1/10	2/16		1		1
1/6	1/10	2/16		1		1
1/6	1/10	2/16			1	1
0	1/10	1/16	2			
0	1/10	1/16		2		
0	1/10	1/16			2	
0	1/10	1/16				2

It looks therefore as if, in the behaviour of indistinguishable particles, we have exactly those sorts of distant correlations which give such trouble to traditional causal models.[9]

7. CLASSICAL 'RECONSTRUCTION' VIA CARNAP AND VIA DE FINETTI'S THEOREM

If classically conceived particles could in principle exhibit Bose–Einstein or Fermi–Dirac statistical behaviour, then the subject might perhaps harbour no conceptual mysteries after all. And they can; but I shall argue that the classical 'reconstructions' are so partial that they do not remove any conceptual puzzles related to Permutation Invariance in quantum theory.

If we look at the probabilities displayed above, which pertain to the counting observables $B0$, $B1$, $B2$, we do not see anything very non-classical. It appears that we can describe systematic differences between the different cases (distinguishable particles; bosons, fermions) in terms of three different classical probability functions.

That appearance is deceptive, for it leaves out (*a*) the peculiarities of predictions made on the basis of any single state, as opposed to the uniform mixture W, and (*b*) the peculiarities even W has for predictions about non-commuting observables. The above family $B0$, $B1$, $B2$ all commute with each other, and that is a special case. Still, the above table is so striking that it will be worth our while to ask: in general, what statistics does W entail for a partition of 'cells' (joint values of commuting observables)?

Since this question, so restricted, coincides with a 'classical' question, we can actually find the answer in the literature on probability and foundations of statistics. As we shall also see, various authors have already related that literature to quantum statistics, though sometimes with misguided suggestions derived for the interpretation of quantum mechanics. I shall discuss two different sorts of classical 'reconstruction' of quantum statistics, one via Carnap's logical theory of probability and the other via De Finetti's Representation Theorem.

Carnap (1950) gave a very straightforward description of simple probability measures. He assumes the cells to be described by conjunctions of simple sentences and their negations. A *state-description* for him is a conjunction of this sort:

$$Fa \ \& \ \sim Ga \ \& \ \sim Fb \ \& \ Gb \ \& \ Fc \ \& \ Gc$$

where &, $\sim$ stand for 'and', 'not'. The constants a, b are here to be thought of as labels of particles. Some sentences are logically equivalent to permutation of individual constants in others: let us call them *isomorphic*:

$$Fa \ \& \sim Fb \text{ is isomorphic to } \sim Fa \ \& \ Fb$$

for example. A *structure-description* is the set of all state-descriptions isomorphic to a given one. A *symmetric* probability function always gives the same value to isomorphic sentences. Some symmetric probability functions are defined as follows:

1. m^+ gives the same value to all state-descriptions with the same individual constants.
2. m^* is symmetric and gives the same value to all structure-descriptions with the same constants.
3. m^F gives *zero* to each state-description in which two individual constants are indiscernible, and the same value to all other state-descriptions.

By 'indiscernible' I mean here that permutation of these constants just produces the same sentence, except for some syntactic rearrangement. (Thus, upon permutation of a and b, $Fa \ \& \ Fb \ \& \sim Fc$ becomes $Fb \ \& \ Fa \ \& \sim Fc$, which differs only by the syntactic order of the conjuncts.)

Two remarks are at once in order. The first is that the set of all state-descriptions in given constants a, b, c, ..., k is a logically exhaustive and disjoint set—so the probabilities of its members must add up to 1. Hence 1, 2, 3 define their functions uniquely. The second is that m^F is obviously m^+ conditionalized on some premiss that amounts to: *no two named individuals are indiscernible*. But m^F is exactly the same conditionalization of m^* (see *Proofs and illustrations*). Thus, both m^+ and m^* become m^F in the presence of a sort of Exclusion Principle, which makes m^F very natural indeed.

It is clear that m^+, m^*, and m^F are the classical counterparts of Maxwell–Boltzmann, Bose–Einstein, and Fermi–Dirac statistics. Also, definitions 1–3 show us exactly how to reconstruct tables such as we had above, in this way:

		m^+	m^*	m^F
(i)	$Fa \ \& \ Fb$	1/4	1/3	0
(ii)	$\sim Fa \ \& \ Fb$	1/4	1/6	1/2
(iii)	$Fa \ \& \sim Fb$	1/4	1/6	1/2
(iv)	$\sim Fa \ \& \sim Fb$	1/4	1/3	0

because the structure descriptions are {(i)}, {(ii), (iii)}, and {(iv)}. These structure descriptions correspond to the cells of a partition, as discussed above: $C1$ to (i), $C2$ to [(ii) or (iii)] and $C3$ to (iv).

Costantini and his colleagues have given us a general classification of a family of probability functions to which these belong, in terms of the correlations we see in joint behaviour in the quantum statistics (Costantini *et al.* 1982, 1983); see also Costantini (1979, 1987). Suppose for example that I have two particles a, b and three cells. Each of the statistics gives the same value to any one particle being in any given cell:

$$m^+ \ (b \text{ in } C2) = m^* \ (b \text{ in } C2) = m^F \ (b \text{ in } C2) = 1/3$$

and similarly for a, $C1$, $C3$. But if a is in $C1$, the probabilities may change:

$$m^+ \ (b \text{ in } C2 | a \text{ in } C1) = 1/3$$

$$m^* \ (b \text{ in } C2 | a \text{ in } C1) = 1/4$$

$$m^F \ (b \text{ in } C2 | a \text{ in } C1) = 1/2$$

So for m^+ it stayed the same (no correlation, independence), while for m^* it went down (negative correlation) and for m^F it went up (positive correlation). We can easily explain this. For m^F I argue: if a is in $C1$, then b cannot be there, so it must be in $C2$ or $C3$, and of these, each is equally likely. In the case of m^* I have to see that some structure-descriptions have only one state-description in them (e.g. a, b both in $C1$), while some have two (a in $C1$, b in $C2$, or a in $C2$, b in $C1$) but still get the same probability.

The above numbers are characteristic of those probability functions, in Costantini's classification, where they appear as $(1/K)e(m)$, with

$$e(m^+) = 1$$

$$e(m^*) = K/(K + 1)$$

$$e(m^F) = K/(K - 1)$$

being called the *relevance quotient*. In our example the number of cells $K = 3$, so the three numbers are $(1/3)$, $(1/3)(3/4) = (1/4)$, and $(1/3)(3/2) = 1/2$. In the case of m^F, it is crucial that there be no more than K particles, and also at least two cells, for it to make sense at all. Thus, in the case of two cells and three particles we can calculate only the first two numbers, as $(1/2)$ and $(1/3)$ respectively, which shows the same independence for m^+ and negative correlation for m^*.

These statistical correlations can be expressed intuitively as follows. Bosons tend to aggregate in the same cells, fermions in different cells, while distinguishable particles show no tendency either way. This colourful language about the statistics should not be taken too piously. As we shall see in the next section, the m^* correlations have a simple classical ignorance model as well. These probability functions, because they represent only the joint probabilities for any given set of *commuting* observables in quantum-statistical mechanics, may look odd, but they do not nearly exhaust the latter's non-classical character.

We turn now to a second approach, which essentially utilizes De Finetti's Representation Theorem (see above, Chapter 3, Section 4).

There are actually simple and familiar examples where correlation in our probabilities results not from objective lack of independence, but from our ignorance. Suppose I have two coins, one fair and one a magician's coin which comes up 'heads' 90% of the time. You don't know which I have in my hand, and so your probability that I will get 'heads' if I toss that one equals $(50\% + 90\%)/2 = 70\%$. If I ask you, after it has come up 'heads' the first time, what the probability is of doing it again, both the following answers are correct:

- (*a*) The (objective) probability is the same as it was for the first, because it is the same coin and the tosses are independent.
- (*b*) My probability for 'heads' next is no longer 70% but 75.7%.

This is because I now have some reason to think that you are using the magician's coin. So in my personal probability, which is an average of two objective chances, successive tosses are not independent.

In 1983, J. Tersoff and D. Bayer published 'Quantum Statistics for Distinguishable Particles', with the following abstract:

Quantum statistics can be reconciled with such classical ideas as distinguishable particles. Bose–Einstein and Fermi–Dirac statistics are derived for distinguishable particles by making an assumption which is different from the traditional assumption but equally reasonable. (Tersoff and Bayer 1983, 553)

The main result they proved had actually appeared earlier in the philosophical literature. It appears as a corollary to De Finetti's Representation Theorem in Richard Jeffrey's discussion in 1965, which relates that theorem to Carnap's programme (Jeffrey 1983, 199–202).

Alexander Bach's 'On Wave Properties of Identical Particles' (1984) noted that Tersoff and Bayer's result is a corollary to De Finetti's theorem.[10] Bach uses the theorem to reconstruct the apparent indistinguishability of particles in terms of De Finetti's purely classical notion of exchangeability. Bach (quite rightly) maintains, contrary to Tersoff and Bayes, that the result does not amount to a reconciliation of quantum statistics with classical concepts.

Let us here concentrate on Bose–Einstein statistics for N particles and two cells ('heads' and 'tails'). The measure is the same as Carnap's measure m^* and can be characterized uniquely as the probability function P such that

BE1: P is *exchangeable*, that is, invariant under permutations of individuals. Thus $P(H(1) \& \ldots \& H(m))$ depends only on m and is the same regardless of which particles are designated as $1, \ldots, m$

BE2: $P(H(m)|H(1) \& \ldots \& H(m-1)) = m/(m+1)$ for all numbers $m \leq N$.

De Finetti's theorem applies to exchangeable probability measures generally. It allows us to reconstruct any such measure as a (subjective) mixture of (objective) chance functions. These chance functions are probability measures of the following sort:

For $w \in [0, 1]$, the chance function c_w is the probability function which assigns probability w^k to each conjunction $H(1) \& \ldots \& H(k)$, for any k particles $1, \ldots, k$.

Thus c_w treats each particle the same way, and treats them as independent of each other. For example, $c_w(H(1)) = w = c_w(H(1) \& H(2)) + c_w(H(1) \& {\sim}H(2)) = w^2 + c_w(H(1) \& {\sim}H(2))$, so we deduce that this last term equals $w - w^2$; and so forth. If we now symbolize our probability that the number w (objective probability of 'heads') is no greater than x as $F(x)$, we can represent our personal probability function as:

$$P(X) = \int_0^1 c_w(X)\mathrm{d}F(w)$$

for any proposition X concerning the toss outcomes (see Jeffrey 1983, 199). Indeed, by De Finetti's theorem, this is the form of *any* exchangeable function on that domain. The greatest possible ignorance, or agnosticism, is perhaps one in which all values for w seem equally likely; this is the case $F(x) = constant$ and gives us the special case:

$P(H(m)|H(1) \& \ldots \& H(m-1))$

$$= \frac{P(H(m) \& \mathrm{H}(m-1) \& \ldots \& H(1))}{P(H(m-1) \& \ldots \& H(1))}$$

$$= \int_0^1 w_m\mathrm{d}F \Big/ \int_0^1 w^{m-1}\mathrm{d}F$$

$$= \frac{w^{m+1}}{m+1}\Big|_0^1 \Big/ \frac{w^m}{m}\Big|_0^1$$

$$= \frac{1}{m+1}\Big/\frac{1}{m} = \frac{m}{m+1}$$

which is just BE2.

We must now ask two questions: Exactly what interpretation of boson behaviour does this suggest? And, would this interpretation remain tenable if we looked at other aspects of boson behaviour?

In framing the interpretation, let us consider M disjoint cells and N particles, for the reconstruction in general. First I shall state three postulates that give us the classical Maxwell–Boltzmann statistics.[11]

Postulate I: The particles are individually distinguishable, and for each particle the objective chances of being found in cell C_i is the same number w_i.

Postulate II: (*Independence*) The objective chance of particles $1, \ldots, N(1)$ being in cell $C1$, and $\ldots$, and particles $k+1, \ldots, k+N(M)$ in cell

C_M—for any ordering of particles—equals the product $w_1^{N(1)}, \ldots w_M^{N(M)}$.

Postulate III: The number w_i is the same for each cell C_i, namely $1/M$.

These three postulates pick out a single probability function. If we accept them, we profess no ignorance of what the objective chances are. To get Bose–Einstein statistics we must replace Postulate III by a much weaker one:

Postulate III(BE): The numbers w_i, are constrained solely by the total probability condition $\sum w_i = 1$.

This postulate insists that all objective chance functions remain possible, as far as our theoretical model goes. If we treat them all as equally likely, therefore, we form the mixture which treats as equally likely (*a*) all orderings of the sort described in Postulate II, and (*b*) all possible assignments of chances to cells.

The precise representation is then as follows. Let X stand for the state of affairs of there being $N(1)$ particles in cell C_1, and $\ldots$, and $N(M)$ particles in cell C_M. Let N be the sum of the numbers $N(1), \ldots, N(M)$. Next, let p be an index parameter which ranges over the set of all possible chance functions (corresponding to the sequences $w_1^p, \ldots, w_M^p$ whose elements are non-negative and sum to 1). Our ignorance about objective chance is then represented by a density function $F(p)$ defined on this index set. Taking into account the possible orderings, we have then the personal probability function P such that:

$$P(X) = \int \frac{N!}{N(1)! \ldots N(M)!} (w_1^p)^{N(1)} \ldots (w_M^p)^{N(m)} F(p) \mathrm{d}(p)$$

When our ignorance is such that we treat all the objective chance functions as equally likely, the density function $F(p)$ is constant. In that case, the same sort of calculation as before yields

$$P(X) = \frac{N!(M-1)!}{(N+M-1)!}$$

which is the correct general BE assignment. It can be expressed briefly as: $P(X) = 1$ divided by the number of partitions of $N(M)$ into M ordered summands.[12]

The technical result is impressive and interesting. We have

seen the Bose statistics appear as the natural probability function, for a perfectly simple and natural combination of ignorance and chance. We have now seen that it really is just the
same statistics as we would use for a magician who chooses a
coin or die to be tossed from an unlimited supply of differently
biased ones. But should we conclude that boson behaviour can
be interpreted entirely in purely classical terms? Should we try
to think of bosons as classically conceived, independent, indeterministic devices?

The answer to this seductive question, we can quickly show, is
no. The two pictures indeed agree under conditions of maximal
ignorance. Under those conditions, both use a uniform statistical
distribution—and that is the sole point of contact between the
quantum models and the suggested classical model. Success for
the latter would require that the agreement· persist under
conditions of accessible information, of diminished ignorance.
Suppose for example that we prepare an ensemble of identical
particles in a pure state—a single state of form
$(x \otimes y) + (y \otimes x)$. Then the quantum-mechanical treatment
ceases to use the uniform distribution over pure states—the
statistical operator $(1/K)I$—so that sole point of contact between the two pictures is lost. But the quantum model still
incorporates correlations of individual behaviour, without any
assumption of interaction.

First, let the particles be individually distinguishable, with the
state of the two-particle system $x \otimes y$, and let cell C_1 be
represented by observable (operator) B. Then we have

$$Pr(\text{1st particle in } C_1) = (x \cdot Bx)$$

$$Pr(\text{2nd particle in } C_1) = (y \cdot By)$$

$$Pr(\text{1st and 2nd particle both in } C1) = (x \otimes y) \cdot B \otimes B(x \otimes y))$$

$$= (x \otimes y \cdot (Bx \otimes By))$$

$$= (x \cdot Bx)(y \cdot By)$$

So the joint probability is indeed merely the product of the
individual probabilities.

But in a symmetric state $(x \otimes y) + (y \otimes x)$ we find 'probability interference terms':

$$(x \otimes y + y \otimes x) \cdot B \otimes B (x \otimes y + y \otimes x))$$
$$= (x \otimes y + y \otimes x) \cdot (Bx \otimes By + By \otimes Bx)$$
$$= 2[(x \cdot Bx)(y \cdot By) + (x \cdot By)(y \cdot Bx)]$$

I did not normalize, so the number 2 is an irrelevant proportionality factor. The term $(x \cdot Bx)(y \cdot By)$ is just the product of the individual probabilities, but the other term $(x \cdot By)(y \cdot Bx)$ is additional, an 'interference term'.

If at this point someone were to suggest that we could continue to look for a classical model—with the two particles having independent but unknown chances of being found in C_1—he would run into the usual obstacle for all hidden-variable interpretations. For we would ask him to let the interpretation cover various divisions into cells, to correspond to non-commuting observables. The 'no hidden variable' theorems would show that this could not be done, with any single probability function.

So, despite the interest of the results, they cannot be cited as reason to look for a classical interpretation after all. What is established is only that the use of the uniform distribution in the maximum ignorance case can lead to Bose statistics in both classical and quantum models. It is of no help in understanding boson behaviour in the quantum case of minimum ignorance (the pure state), which needs to be understood as well.

Proofs and illustrations

Suppose S is a state-description which assigns no more than one named individual to any one cell. Then its isomorphism class equals just the number of permutations of the individuals. Hence isomorphism classes are equinumerous, whenever m^F does not assign them *zero*. Therefore m^F results equally from the m^+ policy or the m^* policy for assigning probabilities.

If we have K cells, and $N \leq K$ particles, the exclusive state-descriptions are produced by choosing one of K cells for the first, ..., one of $K - (N - 1)$ cells for the last. The total number of exclusive state-descriptions is therefore $K!/(K - N)!$. Dividing that by the number $N!$ of permutations of particles, we get the number of isomorphism classes; the result is usually written $\binom{K}{N}$ for short.

If we drop the 'no duplication' requirement, we have K choices for each particle, so a total of K^N state-descriptions. Permutation affects a state-description now only if it interchanges particles in different cells, so the isomorphism classes are not equal. In this case the number of structure-descriptions turns out to be $\binom{N + K - 1}{N}$. For 2 cells and 3 particles these numbers K^N and $\binom{N + K - 1}{N}$ are 8 and 4, with no state-descriptions which are exclusive; for 3 cells and 2 particles, there are 9 state-descriptions, 6 structure-descriptions, and 6 exclusive state-descriptions which are isomorphic in pairs (with each pair uniquely determined by which cell is empty).

8. THE MODAL INTERPRETATION APPLIED TO AGGREGATE BEHAVIOUR

Let us now turn from the challenges posed by the quantum statistics for *any* attempt at interpretation, to those it presents for my favourite, the modal interpretation. I shall show how permutation invariance for the dynamic states places constraints also on the values observables can have in actuality. But these constraints are not entirely rigid. It is possible to utilize the Exclusion Corollary for fermion states of Section 3 above to produce models in which fermions *are* individuated and the Identity of Indiscernibles is not violated. In the case of bosons, however, that is not possible — at least, not if we take the quantum-mechanical description of reality, as understood in the modal interpretation, to be complete.

9. POSSIBLE WORLD-MODELS FOR QUANTUM MECHANICS

For what follows it will be best to consider the modal interpretation in the format in which it can be cast to secure maximal precision.

A possible world-model consists of a set ('the worlds') with certain relations and functions defined thereon. One such function gives to each world another set: its domain, the (ordered) set of things that exist in that world (or, of which that world consists). We shall look here only at the simple sort of model in which the domain is the same set of entities for each world. Thus, each world represents one way that these entities could be.

Another function, call it s, assigns to each world a history. Thus, if x is a world and t a time, $s(x, t)$ is the condition or configuration of x at time t. But x consists of a set of entities, its domain—for example, a set of particles, or some particles and some measuring apparatus. This configuration of x must therefore be the configuration of a total physical system, possibly quite complex.

According to the modal interpretation, this configuration—which encapsulates all information being given—has really two aspects: the *dynamic state* (which is the main subject of physical theory) and the *value state* (which specifies the values of the observables). It is also part of the interpretation to say that the value states are describable by *formally speaking* the same mathematical representation as the pure dynamic states—and in addition that the value states at any given time are constrained to some extent by the dynamic states.

The description of a possible world-system so far therefore looks as follows. There is a set W of worlds, a (n ordered) set D of entities, a set K representing the states, an interval T of time—these are the building blocks. The set K is, for example, the set of all statistical operators on a Hilbert space, each of these operators representing a possible instantaneous state of the total aggregate D. Next, there is a function s which assigns to each world x, at each time t, a value $s(x, t)$—something which represents what x is like at time t. This something splits into parts—say the dynamic state $s_d(x, t)$ and the value state $s_v(x, t)$, both of which are members of the set K of possible states. (And indeed, since these pertain only to the total aggregate D, we need still further functions, assigning dynamic and value states to the parts of D.) This is only the beginning of our description: we have not yet introduced, e.g., the Hamiltonian, which governs how the dynamic state varies with t, or any probabilities.

Some functions must assign probabilities. If the model is one of an isolated quantum-mechanical system, which starts off in a pure state, then the dynamic states are the same in each possible world at all times—for their evolution is deterministic.

But the dynamic states of the subsystems are sometimes mixtures, and these leave open various options for the value states (of these subsystems at those times). Each option is realized in one of the possible worlds.

If this were the model of a classical physical system, we would ask physical theory for a single probability function, developing deterministically in time, for these options. That is, we would ask for a single function P which gives us, for each time t, a measure ('proportion') for the set of worlds x such that subsystem X_i has value state λ_j in x at time t. In Part I, the section on ergodic theory shows in principle what such classical models look like. But according to the modal interpretation, quantum mechanics gives us probability assignments of this sort only for special times and special subsystems—namely, at the end of measurement interactions, for the systems involved in those interactions. The theory does nothing else, in the way of assigning probabilities. For other times and/or other subsystems, we have only a description of the possibilities, with no probabilities attached by the theory.

We need to keep in mind here that the function(s) which assign the dynamic states do not give all the information about the system—even from the quantum-mechanical point of view. To classify one of the subsystems as an A-measurement apparatus is to say that that is the sort of thing for which a certain kind of interaction is always governed by a certain sort of Hamiltonian. This goes beyond saying what happens in just those interactions which happen in *this* dynamic process presented by *this* possible world-system. And indeed, even the Hamiltonian for *this* dynamic process cannot be uniquely inferred from the succession of dynamic states in it. The description of the Hamiltonian—equivalently, of the dynamic group—is again an independent bit of information. By keeping this in mind, we also keep at bay the idea that the assignment of Born's probabilities for measurement outcomes refers to consciousness or some other subjective element. It does not: it refers to the classification of certain subsystems as measurement apparatuses, and that is a classification of the sort of interactions in which they can take part.

A logician sums all this up in the form: *a possible world-model of type* ... is an ordered sequence $\langle W, D, K, \ldots; s, p, \ldots \rangle$ which satisfies the following *conditions* ... I will just take for granted that readers can put everything into that form if they so wish. It is important however to inquire further into those conditions which characterize the admitted models.

A proposition is identified with the set of worlds in which it is true. In the possible world-models described above, we see special sets of worlds describable in rather simple ways. Corresponding propositions are denoted as follows:

Where X is a set of states:
 (i) the proposition $[X, t]$ is (true in) the set of those worlds x such that dynamic state $s_d(x, t)$ of x at t is in set X;
 (ii) the proposition $\langle X, t \rangle$ is (true in) the set of those worlds x such that the value state $s_v(x, t)$ of x at t is in set X.

One main condition we place on the possible world-systems is that $[X, t]$ implies $\langle X, t \rangle$, whenever the proposition $\langle X, t \rangle$ can be read as saying that a certain observable has a certain value at t. That is, one of the conditions such an ordered sequence $\langle W, D, K, \ldots; s, p, \ldots \rangle$ must meet, in order to belong to the type of possible world-models we are describing, is that the set $[X, t]$ is part of the set $\langle X, t \rangle$ for such an X.

I could only sketch this condition so far; it is to be given precise content. When a certain possible world-model meets that Permutation Invariance condition for dynamic states, will it automatically meet an analogous condition for the value states? This is the same question as: can we think of these particles, which are certainly not individuated by the attributed dynamic states, as individuated by the actual, unpredictable, values of observables? From a purely formal point of view, that is the main question investigated in the remainder of this chapter.

I have already announced the guess that we should regard fermions, but not bosons, as individuated by their quantum-mechanical description. This is actually not as easily tenable as it looks at first sight. I propose to do three things. First, I shall discuss the question of individuation for fermions. Here we have to cope with an argument due to Margenau, which says that electrons violate the Principle of Identity of Indiscernibles (PII), if quantum-mechanical description is complete. Then I shall discuss the question for bosons. There we have to consider Reichenbach's discussion of identity over time and material identity. The general conclusion I shall maintain is that it is tenable to hold PII, though only with the implication that there are characteristics which are not covered in quantum-mechanical description, but which have no empirical effect. The exact philosophical status of PII, and of possible violations, is to be discussed in the next chapter.

10. THE EXCLUSION PRINCIPLE IN THE MODAL INTERPRETATION

The idea of a connection with Leibniz's Principle of the Identity of Indiscernibles was perhaps first suggested by Herman Weyl, who referred to the Exclusion Principle as 'the Pauli–Leibniz principle of exclusion'. The nomenclature suggests the connection. Consider two distinct orbital electrons in an atom. Each is characterized by certain constants, definitive of electrons, plus a state of motion. The quantum-mechanical description admits nothing further, so if we assume that description to be complete, then the identity of indiscernibles requires their states to be different. This is the content of PII: any entities which are numerically distinct must differ in some significant respect.

The assumption of completeness here may have a 'metaphysical' air, but it is involved in the very application to atomic structure for which the Exclusion Principle was introduced. This can be seen in its use in the theory of atomic structure, which I described in Section 1 above. PII plus a list of quantum numbers determines the maximum number of atoms for each orbit. It appears therefore that the Exclusion Principle can appear simply as a corollary, if we assume that we have, in this list, a *complete* set of characteristics that could individuate the particles.

In the light of this, it is with some surprise that we find Margenau (in 1944 and again in 1950) citing the electron as violating Leibniz's principle, in writings specifically devoted to the Exclusion Principle. But looking more closely at the Permutation Invariance Principle, we can reconstruct an argument that leads to Margenau's conclusion.[13] Consider a two-particle system in the simple anti-symmetric state $\varphi = (1/\sqrt{2})$ $[(x \otimes y) - (y \otimes x)]$. What states, if any, can we attribute to each particle? The answer is given by the Principle of Special

Reduction, and is the same for each, namely:

$$W = \tfrac{1}{2}I_x + \tfrac{1}{2}I_y$$

The uniform superposition in the state of the whole gives us a
uniform distribution (mixture) for each part. That each component must receive the same reduced state is clear for any
anti-symmetric or symmetric state of the whole.[14]

Partly at issue here is the so-called 'ignorance interpretation'
of mixed states. It is true of course that, if I am sure that a
given particle was prepared either in state x or in state y, but I
have no idea which, I can adequately represent the situation by
a mixed state, a half-and-half mixture of these two pure states.
In that case my ascription of a mixed state reflects my ignorance. But mixed states are encountered also in a different
context. Sometimes a complex system has a pure state, and it is
different from any pure state we would ascribe it, in view of,
say, the spatial separation of its components, on the basis of any
supposition of pure states for those components. This happens
typically after past interaction; Schroedinger called it 'the'
distinguishing feature of quantum mechanics. In that case,
however, predictions about observables relating to *one* component can be based on a 'reduction of the density matrix', which
ascribes a mixed state to the component part. One view is that
the component has no state at all (of its own). Another view,
which I attribute here to Margenau, is that the mixture is the
state of the component; and, correlatively, that mixtures *are* the
possible states of motion, with pure states representing only an
unprivileged special case.

But on this view, the two particles discussed in the second-last
paragraph (which are, as an aggregate, in a superposition of
exclusive product states) are not themselves in different states at
all. Hence they are literally indiscernible, at least if this description is complete—though they are not identical.

The violation of identity of indiscernibles, at which we have
now arrived, is implied in part by a certain interpretation of the
difference between superpositions and mixtures of states.

Let us begin with the notion of state itself. This is connected
with two concepts of operational procedures: preparation and
measurement. Our notion of state must be such that, for each

possible state, it is in principle possible to prepare a system in that state, and that distinct states result from distinct state-preparation procedures, though not necessarily conversely (see Beltrametti and Cassinelli 1981*b*; 1981*a*, sects, 2.4, 43; also De Muynck and van Liempd 1986). But secondly, this notion must be such that, given the state, we have a determinate probability for each possible outcome of any measurement of an observable pertaining to the system. These requirements are met by both pure and mixed states.

Given an aggregate of identical particles, can we ascribe a state to an arbitrary single member of the assembly? First, we have a range of possible states available that a single particle can have, so, logically speaking, the answer is at once affirmative. Secondly, given our knowledge of the state of the aggregate, we can also predict with determinate probabilities the outcomes of any measurement made on any arbitrary member of that aggregate. It makes no physical sense to choose one member rather than another. Whatever we can attribute to any member, we must attribute to all. Thus, Margenau is quite correct in saying that, if a state is to be attributed to any member of the aggregate, it must be the same state that is attributed to each. This state must be, in effect, the summary of the predictions that can be made about any one member, on the basis of the state of the aggregate. And so the only candidate is the mixture described above.

At first sight we have only two options. The first is to insist on the ignorance interpretation. With a little strain, it will allow us to say, in the above case, that one particle is in state x and the other in state y, but we don't know which, and that this truth is in any case irrelevant when we predict their joint behaviour. I say 'a little strain', but actually a gratuitous addition is made, to rule out the case that both are in state x or both in state y, which the unaided ignorance interpretation would certainly allow. The second option is Margenau's, which insists equally that the states of the components are irrelevant to predictions of joint behaviour, but entails in addition that PII is violated, if the description is complete.

However, the modal interpretation allows us a reconciliation. In its possible world-models for such a complete system, the dynamic states are assigned exactly as by Margenau. But in addition, each subsystem S' (including the whole system) has a value state $\lambda[S']$ which assigns actual values to the observables. It is true that in some of the possible worlds PII will be violated if we let $\lambda[S']$ be whatever it can be, subject only to the general constraints. But if we eliminate, or classify as unreal, those worlds in which this happens, the system of possible worlds is not rendered defective. All the same dynamic states can still occur, so the basis of prediction is left untouched.

As an example, let us take again the simplest two-particle fermion aggregate $X + Y$ in dynamic state $\varphi = (1/\sqrt{2})$ $(x \otimes y - y \otimes x)$. The dynamic state of X and Y is the same, namely $W = (\frac{1}{2})(I_x + I_y)$. If we ignore the fact that this is a fermion aggregate, we associate (at least) four distinct possible worlds with this dynamic state:

Value states	$X + Y$	X	Y
In world 1	φ	x	y
In world 2	φ	y	x
In world 3	φ	x	x
In world 4	φ	y	y

But to represent a fermion aggregate, this is a 'bad' system of possible worlds, according to the suggestion that fermions are individuated by their value states. So we must rule out the last two worlds. Happily, the image space of W has dimension 2, so there is no need to assign the same states to X and Y; indeed, we can assign them orthogonal states. In Section 3 we saw that, in any fermion aggregate state, the reduction to the single-particle space has a dimension at least as great as the cardinality of the aggregate. Therefore, as far as that is concerned, there are indeed enough possible states for all the individual particles.

Note well that, in accordance with our discussions of PII, the particles need only be distinguished in some way, two by two. Thus it suffices if, for every two particles, there is an observable which has distinct values. The value states assigned to the particles need not be an orthogonal set, therefore; it suffices that a distinct value state is assigned to each particle. The Hermitean operators which represent those observables that pertain to a single particle will separate these value states.

11. A FERMION MODEL WITH INDIVIDUATION

What I have said now does not yet establish that the modal interpretation can be carried through in this fashion. For, as we saw in an earlier chapter, that interpretation needs to meet certain criteria especially applicable to many-particle systems. The proof that these can be met here too follows below. This proof builds on the more general consistency proofs given in Chapter 9, Sections 8 and 9.

To substantiate the modal interpretation of the aggregate states of fermions, we need to fulfil four desiderata for a *possible* choice of value states:

(a) All three general criteria (a)–(c) for the attribution of component states, listed in Chapter 9, Section 8, must be satisfied.

(b) The value state assigned to each sub-aggregate must itself also be anti-symmetric.

(c) The value states assigned to distinct individual particles must be different.[15]

The tenability of the interpretation requires that some such choice be possible for each anti-symmetric total state of a fermion aggregate.

To fulfil (a) most easily, it is best to begin with the construction used in Chapter 9, Section 9. If either (b) or (c) is not automatically satisfied, we should then modify the construction. We start again with the three-body example $\varphi = \Sigma a_{ijk} (x_i \otimes x_j \otimes x_k)$, with for example the value state attributed to subsystem $(X_a + Y_b)$:

$$\lambda_{ab} = \sum a_{ij2}(x_i \otimes x_j)$$

To say that φ is anti-symmetric means that, if $(x_{i'} \otimes x_{j'} \otimes x_{k'})$ is a permutation P of $(x_i \otimes x_j \otimes x_k)$, then $a_{i'j'k'} = s(P)a_{ijk}$ where $s(P)$ is the signature of P. If that is so, will λ_{ab} also be anti-symmetric? We need only look here at the (odd) permutation which exchanges the first two vector terms. By hypothesis, if $(x_j \otimes x_i \otimes x_2) = x_{i'} \otimes x_{j'} \otimes x_2$, then $a_{i'j'2} = -a_{ij2}$, as is required. Thus λ_{ab} is indeed anti-symmetric. Thus desideratum (b) is fulfilled in this special case.

To prove that in general, we go to the general notation, with $\varphi = \Sigma f$, where $f = a_{i(a)...i(n)}(x_{i(a)} \otimes ... \otimes x_{i(n)})$. By hypothesis, if the term $x_{i'(a)} \otimes ... \otimes x_{i'(n)}$ denotes permutation P of the vector denoted by $x_i(a) \otimes ... \otimes x_{i(n)}$, then

$$a_{i'(a)...i'(n)} = s(P)a_{i(a)...i(n)}.$$

If we now look at the value state

$$\lambda(S') = \sum f^{(S-S')}|(S - S')$$

we see that the only relevant permutations are those exchanging vectors keyed to index subset S'. But these are clearly covered correctly by the hypothesis, in just the same way as happened in our special example above. So we conclude that desideratum (b) is fulfilled in general.

This really concerned only the subsystems which are aggregates of at least two particles. A one-particle state is vacuously and trivially both symmetric and anti-symmetric. Hence, if we alter the general construction in a way that affects only the single-particle value state $\lambda_a, \lambda_b, ...$, the above conclusion will remain intact.

To consider desideratum (c), we begin again with the three-particle example. This includes value states:[16]

$$\lambda_a = \sum a_{i32}x_i$$

$$\lambda_b = \sum a_{5j2}x_j$$

$$\lambda_c = \sum a_{53k}x_k$$

Could these be the same? We note that the anti-symmetry of φ requires that there be no repetitions. Thus, $a_{i32} = 0$ if $x_i = x_3$ or $x_i = x_2$. Since all three bases are the same, this means that $a_{i32} = 0$ equals zero unless x_i is orthogonal to $[x_3, x_2]$. Therefore, λ_a must be orthogonal to that little subspace.

The integer constants were chosen carefully: a_{532} is not zero, so none of these vectors is the zero vector. But then λ_b includes the summand $a_{532}x_3$, which is clearly not orthogonal to $[x_3, x_2]$. Therefore we conclude that $\lambda_a \neq \lambda_b$. By exactly similar reasoning, we conclude that all three vectors are distinct.

Again, the reasoning transposes easily to the general context. There we find the individual particle value states

$$\lambda_a = \sum f^{(S-a)} | (S - a)$$

and so forth, where $(S - a)$ is the index set minus index a. Thus, more concretely we could have, e.g.,

$$\lambda_a = \sum_{i(a)} b_{i(a)235\ldots K(n)} x_{i(a)}$$

By the anti-symmetric decree against repetitions, we deduce as above that

$$\lambda_a \text{ is orthogonal to } [x_2, x_3, \ldots, x_{K(n)}]$$

and that λ_k is not orthogonal to it, if $k \neq a$. Thus $\lambda_a \neq \lambda_k$. Similarly for every other pair—as we were required to prove.

12. BOSONS AND GENIDENTITY

Boson aggregates are capable of symmetric states, and not of anti-symmetric ones. Margenau's argument can be given for such a symmetric state:

$$\varphi = \sum a_r (P_1 \psi_r + \ldots + P_{N!} \psi_r)$$

where each Ψ_r has form $(x_1^r \otimes \ldots \otimes x_N^r)$, in which repetitions are now allowed (no restriction to exclusive product states). It has the same conclusion: we can assign reduced states to the N particles, but they will all be the same. Hence the assignment of states does not individuate the boson. If this description is complete, then PII is violated.

Now even the modal interpretation cannot help in all cases. For let $\varphi = x \otimes x$; then both particles are assigned x as dynamic state, and hence also as value state.

However, we have restricted our discussion so far to the state of the aggregate at a given moment. Should we not, now that we are stymied, bring in the time-variable, which certainly has a role in quantum mechanics as well? Then we can follow Aquinas's course with respect to disembodied souls and say that each boson may be individuated by its history. We would have to add, though, that the historical individuation is to be regarded as empirically superfluous for the statistics (in the sense I explained above), if it goes in any way beyond the description of (dynamic and value) states. And surely it will go beyond, if we are allowed to image a boson aggregate created in state $x \otimes x$ and annihilated before any change of state. Nothing in the theory forbids that.

It has been emphasized and clearly explained by Peter Mittelstaedt that re-identification over time has no empirical significance in quantum mechanics. Mittelstaedt (1983) illustrates this specifically with a boson aggregate: a class of free He^4-atoms with a conserved momentum. When Mittelstaedt constructs possible world-models, he interprets only definite descriptions, and does not allow singular terms (such as names) without descriptive content. The same issue is examined by Dalla Chiara and Toraldo di Francia (1983), and is graphically described for both electrons and photons. They propose the even more radical departure of a theory of 'quasets', which are abstract objects which have a cardinality but no order type. For the reasons given above, I am more sanguine about the applicability of classical logic and set theory without presupposing PII. But one main conclusion we must accept from these discussions: identity through time—history or, in Reichenbach's terminology, 'genidentity'—loses at least its *empirical* significance in quantum mechanics.

It was Reichenbach who tied the discussion of individuation of fermions and bosons on the question of identity through time. Reichenbach reformulated the traditional questions concerning *genidentity* (his term for identity across time), first in connection with relativity (1957, sect. 43) then for quantum mechanics (1956, sect. 26). The classical particle/wave distinction is used as illustration. A floating cork bobs up and down when a wave reaches it; thus we see that no water moves laterally, although the wave moves across the surface. If the individual water droplets, or better its molecules, are entities persisting in time, the wave is merely a changing configuration of these entities. In Reichenbach, a particle has *material identity* (its temporal stages, or the events involving it, are *genidentical* with each other) and the wave does not. Hence questions of individuation or identification of waves (which may form super-positions) are either misplaced, or can be settled by convention.

The suggestion may fit photons (which are in various ways atypical among elementary particles, and even among bosons) especially well. Let us here just address the general suggestion that bosons are 'not genidentical'. This means that, where intuitively we have, say, an assembly of N photons, each persisting in time, we really have only at each moment N photon-stages (temporal slices), and there is no objectivity of any sort to the classification of one of these photon-stages at time t belonging to the same photon as one or other of the stages at time $t + d$. A photon-stage at a certain time is really no more than an event—the *being-occupied* of a certain photon-state. But now we recall that the boson aggregate states are symmetric, which entails that several or even all of these events may correspond to the same pure state at once. Hence all N events may have exactly the same character—there are n *being-occupieds* of the same photon-state. If these events are not individuated by their historical connections to previous such events, or in any other way, then we have a clear violation of PII.

Reichenbach does not draw this conclusion, for he does not discuss PII at all. But he gives an argument for why we should regard bosons as not genidentical. If we think of bosons as entities persisting through time, then the correlations evident in Bose–Einstein statistics mean that these particles tend to go into the same states. They exhibit this tendency even though there is no non-negligible interaction between them in the past. Hence this is correlation without common cause. Reichenbach had, most of the time, the same attitude toward interpretation as I espouse. Hence he writes in conclusion:

We see now how the thesis concerning indistinguishable particles is to be qualified. In precise language we cannot simply say: The particles are indistinguishable. We must say: Either the particles are indistinguishable, or their behavior displays causal anomalies. We are left the choice of selecting the one or the other interpretation. Neither interpretation is 'more true' than the other; the two are equivalent descriptions.

However, only one of the two descriptions supplies a normal system, that is, a system free from causal anomalies; this is the description according to which the particles are indistinguishable. When we follow the usual rule of employing a normal system whenever it is possible, we may therefore say, without hesitation, that the particles are indistinguishable. (Reichenbach 1956, 234–44)

But Reichenbach was writing a decade or so before Bell's Inequalities were published, and well before the Common Cause Principle had been clearly seen to contradict the empirical predictions of quantum mechanics. No normal interpretation, in his sense, is possible anyway. Secondly, to regard m^+ as more natural than m^* for empirical statistics, when the particles are historically individuated (as he clearly does, when he concludes that Bose statistics exhibits an anomaly prima facie), can only rest on an arbitrary application of the flawed Principle of Indifference.

Much more puzzling perhaps is the question: why aren't all particles fermions? The answer in the modal interpretation can be: because fermions represent the special case of particles individuated not historically, but by their description, in quantum-mechanical language, at any given time. We can consistently add to this: and the bosons are individuated, possibly through genidentity or through some other difference, in a way not represented in the quantum-mechanical description of states and observables.

This inversion of Reichenbach's classification (which had instead allowed fermions to be genidentical) was suggested in a paper I gave in 1969. It was challenged in a paper by Cortes (1976), who argued that it is better to reject PII than to admit empirically superfluous factors into one's interpretation. There were a series of further papers. Barnette (1978) accused Cortes of confusing metaphysics and epistemology. Ginsburg (1981) showed that, in quantum field theory, Barnette's reasoning looked much less plausible (see also Teller (1983*b*). Aerts and Piron (1981), as I have noted before, took exactly the position that bosons are distinguished by some feature ignored in the physical description.

In conclusion, I would maintain the tenability of PII only in the form of this non-specific assertion by Aerts and Piron. The example I gave of a state $x \otimes x$ for a system which exists only briefly allows of no 'internal' historical differentiation. Hence individuation, if it is insisted on, is by characteristics not describable in quantum-mechanical terms, as well as being empirically superfluous. That seems fine to me, for an interpretation. The intelligibility it conveys derives from the way it leads us in detail from one aspect of the story to another, always guided by the same thread. That is all; but it is still worth making explicit.

12

Identical Particles: Individuation and Modality[1]

> If our question is directed simply to a yes or no, we are
> well advised to ... consider what we would gain as the
> anwer is in the affirmative or in the negative. Should we
> then find that in both cases the outcome is sheer nonsense,
> there will be good reason ... to determine whether the
> question does not itself rest on a groundless presupposition
> ...
>
> I. Kant, *Critique of Pure Reason*, A485/B514

THE most satisfying way to end a philosophical dispute is to find
a false presupposition that underlies all the puzzles it involves.
Contrary to my own, as well as others', previous writings on
indistinguishability and quantum statistics, I shall argue that the
'loss of identity' dispute can be so dissolved. The questions rest
on a mistake—or, more precisely, on a metaphysical position
which has already been moribund for centuries.

1. ARE THERE INDIVIDUAL PARTICLES?

How shall we make intelligible the passing of the Maxwell–
Boltzmann statistics' primacy, and the hegemony of the new
quantum statistics? One response was: we must derive consequences for statistics from assertions of indistinguishability.[2] The
other main response, which tended to point meaningfully to the
quantum field formalism, was this: we must eliminate the idea
of individual particles altogether. If we do so, quantum statistics
will be derivable or made intelligible.

This thesis and antithesis have much in common. Both entail
that we must turn to ontology; a new conception of substance
must replace the old, and will provide a foundation for the new
statistics. They share a common diagnosis of the problem: that

the very concept of an individual, with its own identity, so to say, forces us into Maxwell–Boltzmann statistics. Despite their opposition, they agree that we must discard a cherished concept of individual, deeply embedded in the structure of our thinking so far. They differ only in the final step: either we must arrive at a concept of individual particles which can be many and yet have no identity (in some sense), or we must say that there are no individual particles. In the second case the replacement will be fields, and while this move is reminiscent of aether theories, those fields fall under no classical concept either. Like the particles which have no individuality or identity of their own, these fields are to have a mode of being entirely different from anything dreamt of in Descartes's, Newton's, or even Einstein's philosophy.

One philosophical illusion concerning science is absolutely perennial: that a theory wears its content on its sleeve, written unambiguously into the shape of its formalism. Elementary quantum theory is written in a way that uses labels apparently associated with individual particles. The quantum field formalism does not. Does that mean that the one is a theory of particles, and the other not? Or is the second a streamlined and improved formulation of the first—allowing finally for the description of systems, in which the number of particles varies with time? Or, thirdly, does it bring a new 'ontological insight'; do we have here an improved formalism that shows that the original theory was not really a theory about particles after all? The illusion is that those questions can be answered simply by looking at the formalism.

Even if that illusion is rejected, the questions can be taken seriously on the basis of an assumption. We may assume that we have two different world-pictures on offer as basis for interpreting science: in the first the world is a world of particles with their own individuality, so to say, and in the second there is no such thing. The opposition is subliminally supported by the connotation of 'particle', that particles are spatially localized. But no idea of space or locality is essentially involved in that of individuation. The assumption is that, even if the empirical phenomena might well turn out to fit both ways of thinking, the two pictures are incompatible; the two worlds depicted are genuinely different.

What I have just said is necessarily obscure; it refers to a puzzle in metaphysics. Let me give a more blatant example. Consider the two statements *There is more than one cow* and *The species COW is multiply instantiated*. Can we understand the second in such a way that it does not imply the first, and could be true in a world in which there are no individual cows? But if so, can we not understand the first in the same way? There must be some metaphysical doctrine about individuals or individuality that we presuppose when we say the two are genuinely different. But are they?

In the next section I will explain the field formalism. What I shall argue for is that, despite the improvement it incorporates, it is not a different theory. It is equivalent to a somewhat enriched and more elegantly stated theory of particles. That we can take it as a description of a world which is particle-less only masquerades as an incompatible alternative. We will have to look at the general metaphysical puzzle, and see if we can escape its possessively loving clutches. In subsequent sections I shall argue that the loss of individuality is illusory, since there is no individuality to be lost.

2. BRIEF EXPOSITION OF SECOND QUANTIZATION[3]

In Section 3 below I shall discuss coordinate-free formalism in general. Whenever we have a symbolism designed for a general case, and then impose a symmetry to define a special case, the symbolism can be simplified. The invariance imposed by the symmetry makes some aspect of the symbolism redundant. In the quantum-mechanical treatment of particle assemblies, permutation invariance makes individual particle labels redundant. This is the reflection that leads to 'second quantization'.

2.1. *Describing bosons*

To be concrete, let us consider the case of an assembly of identical bosons. As a beginning we specify a definite cardinality N for this assembly. Let H be the Hilbert space for a single such particle, and H^N its Nth tensor power. This space is too large; we look at the symmetric subspace H_+^N:

Let PP be the group of permutation operators on H^N, and define PP^+ to be the linear operator on H^N which satisfies

$$PP^+(x_1 \otimes \ldots \otimes x_N)$$

$$= (1/N!) \sum \{P(x_1 \otimes \ldots \otimes x_N) : P \in PP\}$$

for $x, \ldots, x_N \in H$. Then PP^+ is a projection operator; the subspace H_+^N on which it projects is the subspace of symmetric vectors in H^N.

For example, if $N = 2$ and $\varphi = (x \otimes x)$, $\varphi' = (x \otimes y)$, then $PP^+ \varphi = \varphi$, and $PP^+ \varphi' = [(x \otimes y) + (y \otimes x)]/2$. This subspace is the state-space for such a boson assembly, as we have discussed before.

The division by $N!$ is there not to normalize the result, but to ensure the idempotency of PP^+. If we omitted it we would have $PP^+ \varphi = 2\varphi$, $PP^+(PP^+ \varphi) = 4\varphi$, etc. Given an orthonormal basis $\{x_i\}$ for H, we obtain a basis for H^N which is also orthonormal. If we then apply PP^+ we obtain a basis for H_+^N of which the members are mutually orthogonal, but not of norm 1. This is the point at which we can begin to simplify our symbolism.

Take the basis vector $\psi = x_{i(1)} \otimes \ldots \otimes x_{i(N)}$, and let us define the *occupation number* function n—which really depends on ψ, so n is short for n_ψ—as follows:

$n(j)$ = the number of indices $i(k)$, $k = 1, \ldots, N$, which equal j

For example, if our original basis was $\{x_1, x_2, x_3\}$, then for $\psi = x_1 \otimes x_2 \otimes x_1$, the occupation numbers are: $n(1) = 2$, $n(2) = 1$, $n(3) = 0$. Obviously, those numbers have to sum to 3; in general, for H^N they will sum to N. Each of these occupation numbers $n(j)$ defined for ψ is linked to a specific pure state x_j in the original space H. Hence for a fully explicit terminology we should say that $n(j)$ is the *occupation number of ψ for x_j*, so of x_j in ψ. At this point, we may still think of ψ as the state of an assembly consisting of $n(j)$ particles in pure state x_j.

Lemma 1: If $n_\psi(j) = n_\psi(j)$ for $j = 1, \ldots, N$, then $PP^+ \psi = PP^+ \varphi$.

This is quite easy to see, because if P is any permutation, then $PP^+(P\varphi) = PP^+(\varphi)$. Hence in the case imagined we permute ψ and φ first so that all the indices 1 occur first, then all the indices 2, and so on, to produce something of form

$$x_1 \otimes \ldots \otimes x_1 \otimes x_2 \otimes \ldots \otimes x_2 \otimes \ldots$$

The subsequence of $x_j \otimes \ldots \otimes x_j$ must have length $n(j)$, so we get the same result in both cases. But then the two projections by PP^+ are also the same.

Lemma 2: With ψ as above, $|PP^+(\psi)|^2$ equals $\Sigma n(j)!/N!$

This I won't bother to prove; it merely shows what the normalization factors must be to turn $\{PP^+(x_{i(1)} \otimes \ldots \otimes x_{i(N)})\}$ into an orthonormal basis for H_+^N, which is what we wanted. More important are the implications of Lemma 1. Given that lemma, we can write the basis vectors of H_+^N as

$$\psi^+ \langle n(1), \ldots \rangle$$

(which has last member $n(\dim H)$ if H is finite-dimensional), in terms of the occupation numbers rather than the individual particle labels. Now we have a coordinate-free—i.e. in this case an individual-label-free—notation for the vectors in this space. As the quick proof of Lemma 1 shows, this is exactly because of the permutation invariance in the symmetric subspace.

It is not difficult to see how this construction can be modified for an assembly of fermions. As projection operator, we define

$$PP_-(x_1 \otimes \ldots \otimes x_N)$$
$$= (1/N!) \sum \{s(P)P(x_1 \otimes \ldots \otimes x_N) : P \in PP\}$$

where $s(P)$ is the signature of permutation P. This projects on to the anti-symmetric subspace. For concreteness I will continue mainly with the boson case, with side-remarks like these to cover fermions.

2.2. *Number as observable*

Recall that the above construction was carried out beginning with a specific one-particle space basis $\{x_i\}$. The basis elements can be the eigenvectors of a non-degenerate, i.e. maximal Hermitean operator A, with $x_i = |a_i\rangle$. Now the phraseology can be made still more explicit:

$n(j)$ is the occupation number of the jth eigenvalue of A in the state ψ,

where ψ is a basis vector $x_{i(1)} \otimes \ldots \otimes x_{i(N)}$ of H^N as above.

Let us now ask whether there is an operator N^j on the symmetric subspace H_+^N such that $N^j PP^+ \psi = n(j) PP^+ \psi$. That would be what before I called a *counting observable*, in our discussion of quantum statistics, and the discussion there should show that the answer here is indeed *yes*. That is easy to see because of the following lemma:

Lemma 3: If ψ and φ are basis vectors of H^N constructed from $\{x_i\}$ as above and $n_\psi(k) \neq n_\varphi(k)$, then $\psi \perp \varphi$ and $PP^+(\psi) \perp PP^+(\varphi)$.

If the antecedent holds, and $\psi = z_1 \otimes \ldots \otimes z_N$ while $\varphi = y_1 \otimes \ldots \otimes y_N$ with z_j, y_k all in the basis $\{x_i\}$, then we conclude at once that, for at least one index j, $z_j \neq y_j$. But since they are chosen from the same basis, they must then be orthogonal, so $\varphi \perp \psi$. The same holds for any permutation of either, since that operation preserves the occupation numbers. Hence also $P\varphi \perp P'\psi$ for P, P' in PP and $PP^+\varphi \perp PP^+\psi$.

Therefore, the equations $n(j) = 0$, $n(j) = 1$, $n(j) = 2$, ..., $n(j) = N$ partition the basis vectors of the space H_+^N, and we can define the linear operator N^j as required. Note then that each N^j is a function of maximal operator A, and should really be written N_A^j.

Now we can go a real step forward, beyond the mathematical representations so far explicitly utilized in this book. For we can think of an open system, an assembly with a variable number of elements. Particles can come in from outside or leave; they can also be emitted (created) or absorbed (annihilated). The reason we have not come to this before is simply, I think, that most philosophical discussions—e.g. of perfect correlations, Schroedinger's Cat, measurement—focus especially on the case of an isolated closed system.

Definition: The *direct sum* of the finite or countable indexed collection $\{H_r\}$ of Hilbert spaces is the space $\oplus H_r$ whose vectors are the sequences

$$v = \langle v_0, \ldots, v_r, \ldots \rangle, \qquad v_r \in H_r$$

such that $\Sigma |v_r|^2$ is finite and the vector operations are defined by

$$(v \cdot w) = \sum (v_r \cdot w_r)$$
$$v + w = \langle v_r + w_r \rangle$$
$$av = \langle av_r \rangle$$

This direct sum $\oplus H_r$ is again a Hilbert space; the norm is as is to be expected:

$$|v|^2 = \sum |v_r|^2$$

Now we return to the collection of which the arbitrary Nth member H_+^N has been holding our attention. We need a special *vacuum* state, all of whose occupation numbers are *zero*, for technical reasons below; we give it convenient slot H_+^0.

Definition: The *Fock space* for boson assemblies is $\oplus H_+^N : N = 0, 1, 2, \ldots$

What do the elements of this space represent?

Above we focused on space H with basis $\{x_i\}$, and the basis vectors $\psi = x_{i(1)} \otimes \ldots \otimes x_{i(N)}$ of H^N, and then basis vectors which are the vectors $PP^+\psi$ (normalized) of H_+^N. The latter are specified uniquely by their occupation numbers, so we could also write them as

$$P^+ \psi = \psi \langle n(1), \ldots \rangle$$

This vector reappears in the Fock space, as the sequence whose $(N + 1)$th member is $\psi \langle n(1), \ldots \rangle$ and whose other elements are zero vectors. Two such vectors, characterizing N-particle and N'-particle systems, reappear thus in Fock space as mutually orthogonal when $N \neq N'$. For example, x and $x \otimes x$ reappear as components in the scalar product

$$\langle x, \varnothing, \varnothing, \ldots \rangle \cdot \langle \varnothing, x \otimes x, \varnothing, \varnothing, \ldots \rangle = 0$$

Therefore these vectors so produced form an orthonormal basis for the Fock space. Each is characterized jointly by the total particle number N and the sequence $n = \langle n(j) \rangle$ of occupation numbers. But since $N = \Sigma n(j)$, the latter sequence alone suffices even here. Let us therefore introduce the perspicuous notation: $\psi(\langle n(1), n(2), \ldots \rangle)$ in H_+^N reappears in Fock space as unit vector

$$|n\rangle = |n(1), n(2), \ldots\rangle$$

with $N = \Sigma n(j)$. By the same reasoning as for Lemma 3, they form an orthogonal and complete basis.

Two cautionary remarks. Of course, this means also that each number operator N^j can also be defined, as before; and as the above notation shows at once, these operators form a complete set for the space. Remember that they are not the only such set: N^j is short for N_A^j, and A was a specific operator whose basis we selected originally to characterize H. Because that basis can be transformed, yielding a new description of the space in terms of an operator which does not commute with A, we must definitely *not* think of the above vector as the state of a simple collection of particles with $n(1)$ in pure state x_1 and so forth. This is a point about the holism of the quantum-mechanical total state, and we shall return to it below.

The second remark to be made here is that we must not confuse the sequence $v = \langle v_0, v_1, \ldots \rangle$, which appears in the definition of Fock space, with either the sequence $n = \langle n(1), n(2), \ldots \rangle$ or the unit vector denoted via that sequence as $|n\rangle$. So for example, $|n\rangle + |n'\rangle \neq |n + n'\rangle$ and $k|n\rangle \neq |kn\rangle$. All the vectors $|n\rangle$ for distinct sequences n are orthogonal to each other.

Creation and annihilation. Since Fock space is a Hilbert space, we can define Hermitean operators in the usual ways, by specifying how they transform individual vectors. The interesting new possibility is to change an occupation number $n(j)$ to $n(j) + 1$—*creation* of a particle in such a way that state x_j occurs one more time—or to $n(j) - 1$—*annihilation* of one particle so that this state occurs one less time. One is tempted to say here 'create an extra particle in state x_j'. That temptation had better be resisted, as we noted above.

The unit vector $|n\rangle$ has a single non-zero element which is a unit vector that belongs to the Hilbert space H_+^N where $N = \Sigma n(i)$—the space for N-particle assemblies. If we change the sequence n by adding 1 to its rth element:

$$n = \langle n(1), n(2), \ldots, n(r), \ldots \rangle$$
$$(r +)n = \langle n(1), n(2), \ldots, n(r) + 1, \ldots \rangle$$

and treat this new sequence as specifying the occupation numbers, then that uniquely identifies a new unit vector in the space $H_+^{(N+1)}$ and a corresponding one in Fock space. Of course, that is the vector we call $|(r +)n\rangle$. The linear operator a_r^*, defined by its effect,

$$a_r^*|n\rangle = \sqrt{(n(r) + 1)}|(r +)n\rangle$$

on the basis vectors is a *creation operator*. The operator a_r corresponds to a similar numerical operation on sequence

$$(r -)n = \langle n(1), n(2), \ldots, n(r) - 1, \ldots \rangle$$
$$a_r|n\rangle = \sqrt{n(r)}|(r -)n\rangle$$

and is an *annihilation operator*. (Note that the result is the null vector if $n(r) = 0$.) The operator $a_r^* a_r$ is Hermitean.

2.3. *The label-free theory*

At this point we can leave the construction behind, and think of Fock space as an abstract object in itself, ignoring the way in which we came to it. Looking at the creation and annihilation operators, we see that they are each other's adjoint:

$$a_r^*|n\rangle \cdot |n'\rangle = 0 \text{ unless } n' = (r +)n$$
$$|n\rangle \cdot a_r|n'\rangle = 0 \text{ unless } n = (r -)n'$$

Either scalar products both are *zero*, or both are $\sqrt{(n(r) + 1)}$, because if any occupation number is different, the vectors are orthogonal.

Next, we note for the same reason that we have the commutation rules

$$[a_r, a_s] = [a_r^*, a_s^*] = 0$$
$$[a_r, a_s^*] = \delta_{rs}$$

where the commutator $[b, c] = bc - cb$. The operator $N_r = a_r^* a_r$ satisfies

$$N_r|n\rangle = a_r^* \surd n(r) |(r-)n\rangle$$
$$= \surd n(r) \surd (n(r) - 1 + 1)|(r+)(r-)n\rangle$$
$$= n(r)|n\rangle$$

so we can use that as the definition for the (*r*th *occupation*) *number* observable. The *total number* observable is just $N = \Sigma N_r$.

Several times above I made a point of emphasizing that we should not read an equation like $n(1) = 3$ as saying that we have here an assembly of particles, three of which are in pure state x_1. The reason is not that there is anything wrong with the concept of a particle in this context; the reason is the holism of the total state. If $N = 2$, for example, and the assembly is in pure state $(x_1 \otimes x_2) + (x_2 \otimes x_1)$, then neither particle is in a pure state—each is in a 50/50 mixture of states x_1 and x_2. The two occupation numbers $n(1)$ and $n(2)$ each equal 1, but that does not count the number of particles which are in specific pure states, for there are indeed $2 = n(1) + n(2)$ particles, but both are in the same mixed state.

As usual, this holism has to do with the principle of superposition. We made the transition from 'labelled' to 'label-free' formulation by choosing initially a one-particle maximal observable A, and its basis $\{x_i\}$ of eigenstates, and eventually ended up with the creation and annihilation operators a_r^* and a_r. But we can choose another maximal observable B, incompatible with A, and its basis $\{y_j\}$ of eigenstates. Then we must arrive similarly at creation and annihilation operators b_r^* and b_r. How are the two modes of creation and annihilation related to each other?

The bases $\{x_i\}$ and $\{y_j\}$, and the operators A and B, are related by a unitary operation:

$$x_i = Uy_i = \sum c_{mi} y_m$$

where the unitary matrix $[c_{mn}]$ is a matrix representation of unitary operator U. From the basis $\{y_j\}$ of H we can get to a basis of unit vectors of H_+^N as we did above, and we can define the associated occupation numbers $\hat{n}(1)$, $\hat{n}(2)$, ..., which count occurrences of y_1, y_2, in the state so expanded. Then, as before, the occupation numbers specify the symmetric state uniquely, and we can write

$$\psi = \psi(\hat{n}) = \psi(\langle \hat{n}(1), \hat{n}(2), \ldots \rangle)$$

Next, we can build up the Fock space in which Ψ reappears as

$$|\hat{n}\rangle = |\hat{n}(1), \hat{n}(2), \ldots \rangle$$

All this goes as before and, because we started by merely redescribing the same situation in terms of a different basis, *we must here have arrived at the same Fock space.* But the new basis $\{|\hat{n}\rangle\}$ is not the same as the earlier basis $\{|n\rangle\}$. Indeed, $|\hat{n}\rangle$ must be expressible as a superposition of the vectors $\{|n\rangle\}$. What exactly is their relationship? The picture of a set of $\Sigma\hat{n}(j)$ particles neatly arranged in cells 1, 2, 3, ... occupied by $\hat{n}(1)$, $\hat{n}(2)$, ... particles respectively certainly makes no sense, if the depicted entity is also to be a superposition of other ways in which these particles could have been arranged. Beware, as usual, the lures of the classical imagination!

Think for a moment about the one-particle state $|n(1) = 1, 0, 0, 0, \ldots \rangle$. That is the sequence $\langle \emptyset, x_1, \emptyset, \emptyset, \ldots \rangle$ in the construction of Fock space. By the above equation, that is $\langle \emptyset, \Sigma c_{m1} y_m, \emptyset, \emptyset, \ldots \emptyset \rangle$; but by the definition of the operations on the constructed Fock space, that is $\Sigma c_{m1} \langle \emptyset, y_m, \emptyset, \emptyset, \ldots \rangle$. And here the components are the one-particle states related to the B basis. Hence

$$|n(1) = 1, 0, 0, \ldots \rangle = \sum c_{m1} | \ldots, \hat{n}(m) = 1, \ldots \rangle$$

If we now think of the one-particle states as produced by creation operators from the same vacuum state, we write this as

$$a_1^*|0\rangle = \sum c_{m1} b_m^*|0\rangle$$

Therefore on the vacuum state, a_1^* acts as $\Sigma c_{m1} b_m$, and more generally a_i^* acts as $\Sigma c_{mi} b_i^*$. This argument can be continued for the adjoint annihilation operators and for other states; let us accept this as a proof by illustration of the equations

$$a_i^* = \sum_m c_{mi} b_m^* \qquad a_i = \sum_m c_{mi}^* b_m$$

Hence the creation and annihilation operator families transform exactly, as do the bases in the 'label' formalism, namely via unitary matrices.

Proofs and illustrations. Let us take an even briefer look at two subjects which will not enter the main line of argument. The first is what the relativistic quantum field formalism looks like, and the second is the relation between second quantification and the thesis of Dichotomy discussed in the preceding chapter.

The last item above was the kind of transformation that corresponds to changes of basis in the original Hilbert space.

The trick we used there with the vacuum state is of course always possible. Indeed, each unit vector ψ in Fock space belongs to some orthonormal basis, hence there is some suitable one-particle observable A such that we can write ψ in terms of the associated occupation numbers, as $\psi = |n\rangle$. Now that means that ψ can be produced from the vacuum state by repeated applications of creation operators:

$$a_1^*|0\rangle = |1, 0, \ldots\rangle$$
$$a_1^*|1, 0, \ldots\rangle = \sqrt{2}|2, 0, \ldots\rangle$$
$$a_2^*|2, 0, \ldots\rangle = |2, 1, \ldots\rangle$$
$$a_2^*|2, 1, \ldots\rangle = \sqrt{2}|2, 2, \ldots\rangle$$
$$a_2^*|2, 2, \ldots\rangle = \sqrt{3}|2, 3, \ldots\rangle$$

and so forth. But this means in turn that we can think of any linear operator, specified through its effect on a given basis, as defined in terms of repeated creation and annihilation. For example, suppose that $R|n\rangle = 3|\hat{n}\rangle$ where $n(1) = 1 = n(2)$ and $n(j) = 0$ for $j > 2$, and where $\hat{n}(2) = 1$ and $\hat{n}(i) = 0$ for $i \neq 2$. Then we notice that the effect of R on $|n\rangle$ can also be produced as follows:

$$a_1 a_2|1, 1, 0, \ldots\rangle = |0\rangle$$
$$b_2^*|0\rangle = |0, \hat{n}(2) = 1, 0, \ldots\rangle$$

So

$$R|n\rangle = 3b_2^* a_1 a_2 |n\rangle$$

This is not a good format for defining R, but it shows that any one operation can be reconstructed as a linear combination of sequential creation and annihilation operations.

But these reflections will be a good preparation for the formalism of relativistic quantum field theory. In classical relativity, a model will consist of a space–time, which may have a field defined on it—that is, a function which takes a certain value at each space–time point. The function is a tensor, which represents a physical quantity that varies over space–time, the *field strength*. In the quantum case, of course, physical quantities are represented no longer by scalars, vectors, or tensors, but rather in terms of operators on the states which allow us to calculate measurement outcome probabilities. Accordingly, a field assigns to each space–time point a linear combination of creation and annihilation operators. These operators act on the quantum-mechanical state, in the way we have described, and the role of the old field strengths is taken over by the calculable expectation values. That it can be sufficient for a quantum field to have this simple structure should be plausible given the outline above of how the effect of any observable on any given state is duplicated by the effect of a certain linear combination of creation and annihilation operators.

Returning now to the non-relativistic context, let me first fill in the story as it applies not just to bosons but to particles in general (see e.g. Merzbacher 1970, ch. 20, sect. 3). The transformations that relate the different creation and annihilation families are in all cases by unitary matrix in the format that was exhibited above:

$$(1) \qquad a_i^* = \sum_m c_{mi} b_m^* \qquad a_i = \sum_m c_{mi}^* b_m$$

However, the commutation relations I gave above,

$$(2) \qquad a_r^* a_s^* - a_s^* a_r^* = 0$$

hold only for bosons. We can derive the possibilities in the following way. If we apply a_r^* and a_s^* sequentially to a state, we increase the numbers related to pure single-particle states x_r and

x_s by one each. The result should not depend on the order of application, therefore, except perhaps by a constant factor which will be removed by the normalization required to yield a unit vector. Hence we conclude:

(3) $$a_r^* a_s^* = k a_s^* a_r^*$$

where k might depend on r and s. By (1) above we deduce from (3):

(4) $$0 = (a_r^* a_s^* - k a_s^* a_r^*)$$

(5) $$0 = \sum_m c_{mr} b_m^* \sum_n c_{ns} b_n^* - k \sum_n c_{ns} b_n^* \sum_m c_{mr} b_m^*$$

(6) $$0 = \sum_{m,n} c_{mr} c_{ns} (b_m^* b_n^* - k b_n^* b_m^*)$$

However, in (1) the unitary matrix $[c_{mn}]$ is subject only to the constraint that $\sum_m c_{mr}^* c_{mq} = \delta_{rq}$ for all r and q. Accordingly, (6) will not hold unless, for all m and n, individually,

(7) $$b_m^* b_n^* - k b_n^* b_m^* = 0$$

(8) $$b_m^* b_n^* = k b_n^* b_m^*$$

Since this holds for all m and n, we can add the consequence:

(9) $$b_n^* b_m^* = k b_m^* b_n^*$$

(10) $$b_n^* b_m^* = k^2 b_n^* b_m^* \qquad \text{from (8) and (9)}$$

(11) $$k = \pm 1$$

which means that the commutation relations must take either the form (2), which says that the commutator $[a_r^*, a_s^*] = a_r^* a_s^* - a_s^* a_r^* = 0$, or the form

(12) $$a_r^* a_s^* + a_s^* a_r^* = 0$$

which says that the anticommutator is 0. The first characterizes bosons, and the second fermions.

So now it appears that we have a new deduction of Dichotomy, in the non-relativistic quantum field formalism. In the relativistic theory Pauli's Spin and Statistics theorem does indeed have bearing on Dichotomy, though it does not establish it.[4] Do we really have something much stronger here? Of course the answer is *no*; the deduction we have just given has the same status as Blokhintsev's. Equation (3) above follows from the preceding remarks only if it is assumed that two vectors cannot represent the same state unless they are parallel. That is to say, the pure states are represented by rays or one-dimensional projection operators, and therefore there are no superselection rules present. But that is an assumption holding only for the simplest case.

2.4. *Field representation, and the proper conclusion*[5]

Starting with elementary quantum theory, we arrived above at (elementary, non-relativistic) quantum field theory by *construction*. I use the term 'construction' here in its mathematical sense. Beginning with any set, we can construct a group, for example the group of one-to-one mappings of this set on to itself. There is a converse procedure. It can be proved that this method of construction suffices for groups: every group is isomorphic to a group of transformations. That result is a *representation theorem*: every group can be represented as a transformation group. Construction is never enough to draw much of a conclusion. Representation as well, however, means that we really have just one mathematical concept, realized in different ways, more or less abstractly.

The question we must now ask therefore is: does the above way of constructing those quantum field models suffice? Can we make the journey back, so to say, to the 'label' formalism if we are given an abstract presentation, the 'label-free' formalism in its unencumbered self-sufficiency?

The answer is *yes*. This question has been investigated by several writers, notably by De Muynck, who carried through a reformulation of quantum field theory with the 'individual particle labels' reinserted. All models of (elementary, non-relativistic) quantum field theory can be represented by (i.e. are isomorphic to) the sort of Fock space model constructions I have described above. Since the latter are clearly carried out within a 'labelled particle' theory, we have a certain kind of demonstrated equivalence of the particle—and the particle-less—picture.

What does this entail? That depends obviously on exactly

what kind of equivalence we have found. It is a strong kind; but we must be careful not to jump to philosophical conclusions. I quote the remark on this point by one strong defender of a particle-less interpretation, Allen Ginsberg (see also Ginsberg 1981):

Cognoscenti may object to the position taken in this paper on the following grounds. It is a well-known fact that the non-relativistic QFT [quantum field theory] formalism is mathematically equivalent to the many-particle Schroedinger equation formalism of EQM [elementary quantum mechanics]. Since the two theories are equivalent how can they 'really' say anything different from one another? (Ginsberg 1984, 348)

Ginsberg has two answers. The first is that the fundamental theories of contemporary physics are relativistic, and there is no similar demonstrated equivalence there. This is not in my opinion a relevant point. There has not been, as far as I know, an equally thorough investigation of what is possible at that level, and Ginsberg's own interpretation of the subject contains some at least controversial philosophical elements. Obviously, the puzzle about continuants versus events characteristic of space–time theories will also be involved there (see Section 2.1 below). But in any case, we should not take the attitude that, if a more recent or rival theory avoids a certain problem of interpretation, the problem disappears. Otherwise we shall always be in the position of saying that (*a*) we can't interpret the newest theory yet because it is still incomplete and in flux *qua* developing science, and meanwhile (*b*) we do not understand the older theory, however recent, but it has been supplanted anyway! The questions of interpretation at issue here concern the non-relativistic quantum theory.

Ginsberg's second reply, however, relates directly to the significance of that equivalence.

Secondly, I do not agree that mathematically equivalent theories always 'say the same thing'. Newtonian gravitational theory [for example] can be formulated either in terms of bodies acting on each other at a distance, or in terms of bodies interacting locally with fields. I see no reason to say that these mathematically equivalent formulations are also equivalent with respect to the model of reality they embody. (Ginsberg 1984, 348)

Ginsberg is making a point here, I think, which I also consider valid and important. But the point is obscured in various ways, because we need to be precise first of all on the difference between rival theories and rival interpretations of one theory, and secondly on the possible meanings of 'mathematically equivalent'. The example he gives hinders rather than helps if we are not told the precise relations between Newton's theory and the two purported formulations.

The equivalence between the field formalism and the many-particle formalism which we found above is neither very weak nor the strongest one could have. This sort of equivalence holds between the abstract theory of groups and the theory of transformation groups, as demonstrated by the representation theorem. It is the equivalence, I think (though there have been demurrals) between Schroedinger's wave mechanics and Heisenberg's matrix mechanics and von Neumann's abstract Hilbert space formalism. Each of these pairs is also inequivalent in a certain way; similarly, in terms of our example, there are groups which are not transformation groups. But is this inequivalence significant?

As a result of the demonstrated equivalence—the representability of one sort of mathematical object as another sort—a weaker but philosophically more interesting equivalence holds: the theories are *necessarily* empirically equivalent. Any possible phenomena that can be accommodated in the one sort of model can also be accommodated in a model of the other sort. This in turn entails several still weaker but philosophically interesting equivalences, such that all actual, or all known, phenomena are such that, if the one theory can model them, then so can the other theory. But the demonstrated equivalence is stronger than any of these.

Schroedinger's wave mechanics, Heisenberg's matrix mechanics, and von Neumann's abstract quantum theory may well have been differentially instrumental in suggesting the different interpretations of quantum mechanics being advocated today. Because of their strong equivalence, however, any interpretation tenable for any one of the three is *mutatis mutandis* tenable for the others. This is exactly what must also be said about the field formalism and the many-particle formalism which we have discussed here. Indeed, each may suggest a different sort of interpretation; perhaps even they are communicated in ways that involve drawing incompatible pictures of the world. But we must add about the models which they offer us for representing the world: their structure *cannot rule out* the rival interpretation of their content.

Let me give one other example. The normal way, post-set-theory, of conceiving of geometry is as a theory of points and relations among points. Spheres, for example, are definable sets of points. But it is also possible to write an equivalent theory (in the above sense of equivalence) in which the elements are spheres and relations among spheres. Points are then identifiable with a definable set of spheres (intuitively, the point *is* a set of spheres which contain it; points are introduced by limit construction). Obviously we are here offered two rival world-pictures for our consideration. But if someone likes to talk in terms of spheres, I can reconstrue his every assertion *salva veritate* (and saving also all valid inferential relations) as an assertion about points. And vice versa! This is not to deny that it is possible for a person to believe that points are the only real concrete individuals—what we cannot do is to say that geometry forces this view on us.

3. THREE PARALLEL DEBATES IN METAPHYSICS

Philosophical puzzles do not have just one native habitat; they tend to crop up, sometimes overtly and sometimes disguised, in many places. A preliminary look at some other metaphysical debates will equip us with some of the concepts needed for the discussion of the identity of identical particles: *genidentity*, coordinate-free descriptions, indiscernibility as symmetry, reference with or without individuation.

3.1. *Substance and event in the relativistic universe*

'That is a very *Earthling* question to ask, Mr. Pilgrim. Why *you*? Why *us* for that matter? Why *anything*? Because this moment simply *is* . . . . All time is all time. It does not change. It does not lend itself to warnings or explanations. It simply *is* . . . . 'You sound to me as though you don't believe in free will,' said Billy Pilgrim. 'If I hadn't spent so much time studying Earthlings,' said the Tralfamadorian, 'I wouldn't have any idea what was meant by "free will". I've visited thirty-one inhabited planets in the universe, and I have studied reports on one hundred more. Only on Earth is there any talk of free will.'

Kurt Vonnegut, *Slaughterhouse-Five*

Billy Pilgrim is kidnapped by a flying saucer, by beings who, so to say, live in space–time; they can witness what occurs at any space–time point by focusing their attention there. So there is for them no question about what to decide to do; what they will do they can see by looking. The concept of decision does not have any applicability for them. Billy realizes that, space–time being as they (and we in this century) conceive of it, he must accept their conclusion that decision and freedom of will are an illusion. Do you like this story? If you do, you will love twentieth-century medieval metaphysics.

Relativity theory was the great impetus for the development of logical empiricism—it was for the Vienna and Berlin Circles the paradigm of science's progress when metaphysics is cleansed away. As they saw it, the 'operational' analysis Einstein gave of simultaneity revealed the element of convention and the base-lessness of absolute synchronicity. That is what made relativistic electrodynamics possible; and it was also the paradigm for logical empiricist philosophy of science. In England, on the other hand, relativity was instrumental in a return to realist metaphysics, as part of the revolt against idealism. Samuel Alexander, Alfred North Whitehead, and Bertrand Russell developed new systems of the world, in which the relativistic view of nature was the 'natural' one. The simplest of these was Russell's, which had perhaps as much debt to Hume as to Einstein. Individuals persisting in time, he held, were mere logical constructs from events. What we think of as the history of a single entity which exists, say, both now and a year ago is a class of events linked by such purely external relations as contiguity and similarity.

How exactly is this metaphysical view related to relativity in physics? In just the way that can keep us endlessly fascinated without profit. It is perfectly possible to accept that metaphysics while remaining loyal to classical physics. It is also possible to take the contrary view while embracing Einstein. But there is a suggestive link. That is what feeds the desire of metaphysics to find support in science, and the desires of those puzzled by science to find peace. After Einstein, the relations among events which have the objectivity of invariance—the ones that are not perspectival, or relative to frames of reference—are neither spatial nor temporal. The distance between my outstretched hands is nothing absolute, nor is the time that will have elapsed between my first love and last rites. Doesn't that make my identity an illusion as well? Is it not the events, whose only proper location is a spatio-temporal one, that are the relata of 'real' relations, such as the space–time interval:

Events located at coordinates (x, y, z, t) and (x', y', z', t') in any frame are separated by a space–time interval with magnitude $[(x - x')^2 + (y - y')^2 + (z - z')^2 - (t - t')^2]$, and this is the same in every inertial frame of reference,

so doesn't modern physics do away with entities which are not so related? In the non-perspectival, *coordinate-free* description of nature, the remarks I like to make about my armspan do not even appear in the formalism.

Note especially here the emphasis on the new formalism: it omits notation for what is not invariant, which has all dropped out of this description of nature, and hence(?) physics no longer acknowledges its reality.

But this is spurious. Certainly my birth and death are events between which there is an invariant relation, the space–time interval, and indeed, it is measurable as my *Eigenzeit*. That they are related by being the birth and death of the same person is also an invariant relationship, the same in every moving observer's frame of reference, *unless* we assume already that persons are unreal. That my armspan is of puny extent in your frame of reference is amusing to remark, but it certainly does not threaten my identity. That the coordinate-free formalism leaves out such remarks altogether we may attribute to a selection of what is of interest *sub specie* physics, without implying that what interests me is an illusion. Physics as such has no interest in where *I* am *now*, but I do, and if I did not, I would not be in a position to apply science at all.

Does such a Humean philosophy of nature, as Russell once espoused for example, make Einstein more intelligible? If it gives a clue to a possible, and possibly adequate, way to interpret physics, *yes*. But in the end, while it does help to realize that a certain form of language is feasible, preference for it does not help. Reichenbach discussed the issue in several books, using the term *genidentity* for the relationship between events which belong to the history of a single persistent individual. Since he developed a relational theory of space–time — roughly siding with Leibniz against Newton on the issue of Absolute Time and Space, and with Einstein against certain neo-Kantians — he could not very well use spatio-temporal relations to define genidentity. Instead, he attempted to use causal and genidentity relations to define spatio-temporal order. Nevertheless, as he pointed out, a description of nature in terms of events, causation, and genidentity does not imply (nor does it contradict) either that persistent entities are real, or that they are merely logical constructs. In his terms, the *object-language* and the *event-language* can both be systematically developed, and a preference for either must rest on convenience or convention, if not on superfluous metaphysics. Science certainly does not rule either way.

There is something very heady about the idea that a new development in physics can settle or inspire truly philosophical questions (or, less excitingly, answers). Inspiration and new enthusiasm — without which no insight is perhaps ever achieved — are indeed welcome; but they are not honoured by logical fallacy. Whether persistent individuals are real, or only events, or some third sort of miasma, is not the question. Which forms of language are and are not adequate is an objective matter, and then, only relative to the criteria of adequacy we impose — that is all.

3.2. *Incongruous counterparts and absolute space*

When two figures are congruent — i.e. are related by a Euclidean transformation — but cannot be 'moved into coincidence' — i.e. are not related by a proper motion — they are called *enantio-morphs* or *incongruous counterparts*.[6] Intuitively, they are each other's mirror image. So, being congruent, they are exactly the same, geometrically speaking; yet they stand in a geometric relationship which precludes identity! We are theoretically torn: if they are geometrically the same but numerically distinct, does it not follow that the geometric description of the world is incomplete? But on the other hand, the non-identity is entailed by a geometric relationship; so why suspect incompleteness in geometry; why not suspect rather their geometrical sameness?

Kant attacked the question in this way: what is the difference between a pair of two left hands and a pair consisting of one left hand and one right hand? Let us idealize the question by assuming that the spatial regions occupied are perfectly congruent. That means that any spatial relations internal to any of these hands—length of thumb, angle between thumb and forefinger, and so on—are exactly the same. But while you can then move the left hands so that one comes to occupy the position of the other exactly, you could never do this for the right and left hand.

Imagine next a universe containing a single hand. Kant insists that it must be *either* a left hand *or* a right hand—but the internal spatial relations do not suffice to answer the question, and they are the only physically instantiated relations in that universe. So spatial relations among physical entities are not all there is to space!

One way out of the impasse is certainly this: space is real, an independent entity in its own right, and left-handedness is a relation to space itself. Another way out is to say that left-handedness is an irreducible property of the object, not consisting in relations either to other bodies or to space. This second move would be analogous to the medieval flirtation with the *ubi*, the 'where-ness', of an individual, as basic spatial property. And the third reaction is to say that, if the universe contains only a hand, then it is neither a left hand nor a right hand. In this last option, there is then a definite 'loss of handedness'!

The parallel to discussions of identical particles becomes even more striking when we look at a universe U containing a left hand and a right hand. It is easy to distinguish a universe U' with two left hands from U—in the second case, Euclidean reflection in a certain plane is a symmetry of the figure, and in the first case it is not. But now let us label one hand A and the other B. We use these labels in our description; we cannot actually go into that world and attach the labels. Isn't it objectively true either that *A* is the left hand or *A* is the right hand? In other words, if we imagine the universe *rU* produced from *U* by that Euclidean reflection, don't we have this problem: $U \neq rU$ but no internal spatial relations are different? There are no people in either, but if *we* had an ontological telescope to look at them one at a time, we would see no difference. If through that ontological telescope we look at both together, we see that they are distinct.[7] But what is it to look at *U* and *rU* side by side if not to look at a third universe U^+ containing two pairs of hands? The very idea of looking at two possible universes through one ontological telescope is absurd.

Could one assert then that $U = rU$? It is egregious to insist that these two universes are really different, and then let that insistence drive us to accept either Absolute Space or primitive left-handedness. But in this universe $U = rU$ itself, how are hands *A* and *B* different? Each has the property that it exists in a universe where there is another hand that 'would not fit the same glove', so to say. Indeed, anything we can say truly about *A* we can also say about *B*. Now it appears that we are in trouble with the Principle of Identity of Indiscernibles (PII). That is a curious tangle: the predicate

> . . . *exists in a universe containing only one other object*

can be true of an object only if it is not the only one in the universe. But in addition, it is true of an object only if it is equally true of a different object. So why should we think that the two objects must be different from each other in any respect?

3.3. *Quine: to be is to be the value of a variable*

Returning to universe *U* which contains only the congruent left and right hand, we notice now a difficulty with our description. In our universe, I can designate as left whichever hand is closest to the heart. But the symmetries of *U* prevent us from attaching the word 'left'—or any other label—determinately to one hand on the basis of a description. This is not just the point that the labelling 'left' and 'right' is conventional. Let them be called 'left' and 'right' by convention; *now* describe to me which of them is called 'left' by your convention! Obviously you cannot do that. Similarly for: let them be called A and B.

We should however not jump to an unforced conclusion. Suppose I begin a geometric proof with 'Given a pair of enantiomorphs of type H, call them A and B, . . .': will I be reasoning invalidly? Not at all. For the mathematical assertion *that there are two*, when explicated set-theoretically, becomes equivalent to *there exists a function which maps this collection of enantiomorphs one-to-one onto the class* $\{'A', 'B'\}$. The point of the preceding paragraph is therefore *not* that labelling is impossible, but only that the different possible labellings cannot be distinguished by description.

W. V .O. Quine made famous in philosophy the slogan 'to be is to be the value of a variable'. By this he meant the following. Suppose I say 'Some Greeks believe that Zeus exists.' Here we see the name 'Zeus', but the sentence does not carry the implication that there is such an entity as Zeus. But if we write this in logical language, it becomes 'There is some entity x such that x is Greek and x believes that Zeus exists', and this does carry the implication that there exists at least one Greek. The appearances of the bound variable x, and not the name, carries the 'ontic commitment'.

Quine added to this the thesis that names are entirely eliminable. We can first rewrite anything with a name using the identity relation as follows:[8]

'Z is thus and so' is equivalent to 'Every entity x such that $x = Z$, is also such that x is thus and so',

then introduce a single predicate Z^* which will henceforth take the place of the complex phrase '$= Z$'. When Z is the name 'Zorba' then Z^* is 'Zorba-izes' and is true of an entity if and only if that entity is Zorba.

Now we have a logically designed language which has, it seems, lost no powers of expression, but by the systematic use of variables carries its existential—or rather, ontic—commitments on its face.

An *interpretation* of this language uses the above insight that a labelling consists in the existence of a function—which need not be descriptively specifiable in order to exist. We call f an interpretation if it maps each variable x, y, z, ... to an entity in the world. Then

> 'x is happy' is true under f exactly if $f(x)$ is happy. 'There is an entity x such that x is sad' is true under f exactly if there is some interpretation f' which is like f for all variables other than 'x', and such that 'x is sad' is true under f'.

We can see now that the universe U can be described in this language which is indeed a regimented version of the description we gave. The curious features of U, which contains just two objects which are exactly alike and yet distinct, do not prevent such description.

But we must also note that what we called here an interpretation—i.e. an arbitrary labelling—corresponds to what in the discussion of relativity we called a coordinatization. For that too was an arbitrary labelling which consisted in choosing an X-axis, a positive direction, and so forth. And now Quine's idea of ontological commitment, written on the face of a perfect language, begins to suffer. For descriptions in terms of a given chosen coordinatization may be convenient—but they can also be turned into coordinate-free descriptions.

To put it another way: Quine's elimination of names can be carried further; we can also eliminate the variables! Quine himself eventually noticed and discussed this. An elegant and precise reformulation of Quine's perfect language in variable-free (coordinate-free) form was given by Svenonius (1960). I will sketch this here, though only just enough to show the relevant conclusions to be drawn. Take a sentence of form

1. Everybody loves somebody who admires him or her.
2. Every entity x is such that there is some entity y such that x loves y and y admires x.

The regimentation leaves out a few details of humanity and gender, of course. Sentences like 'x loves y' and 'x admires y' we assign the form Rxy. Now we introduce the following devices (not a complete list but sufficient for our example):

Converse	$Rxyz$ ... if and only if $Conv(R)yxz$ ...
Conjunction	$[Rxyz$ ... *and* $Sxyz$ ...$]$ if and only if $K(R, S)xyz$ ...

Universalization [Every entity x is such that $Rxyz \ldots$] if and only if $U(R)yz \ldots$

Existentialization [Some entity x is such that $Rxyz \ldots$] if and only if $E(R)yz \ldots$

Using these devices, we now rewrite sentence 2 as

3. *UEK (Loves, Conv (Admires))*

during which rewriting both variables dropped out.

Now how shall we read this logically coordinate-free language? A sentence like 3 says of certain complex predicates that they are or are not *instantiated*. Thus, if F is a one-place predicate like 'is red', then EF says that F is instantiated and UF says that the negation of F is *not* instantiated. Moreover, we can introduce counting into this language (see Svenonius for details) and so we will also get sentences that say that certain complex predicates are multiply (2-fold, 3-fold, ...) instantiated.

We are now only a little step away from terminology familiar from physics. Let us say 'cell' rather than 'complex predicate' and 'occupied' rather than 'instantiated'. Then we conclude: the logically coordinate-free statements do exactly one thing: *they specify occupation numbers of cells*.

Does the use of one language rather than another, when we know that each can be translated perfectly into the other, carry any ontological implications? Of course not. Quine's programme, to deduce ontology from syntax, was just a mistake.

4. THE IDENTITY OF INDISCERNIBLES

So far I have argued that quantum mechanics, including its formulation in the field formalism, is properly interpretable as a theory about particles. Now I shall argue that this sort of interpretation need not bring along a conception of particle which makes quantum statistics one whit less intelligible. Indeed, the 'loss of identity' is not a new feature of the quantum world, and does not yield a presumption in favour of any particular statistics. The only way we will be able to see that is by looking at identity in a purely classical context, and asking whether the situation was really different then, in this respect, from what it is now.

The notions of individual, individuality, individuation, and qualitative and numerical identity form a traditional subject cluster in philosophy. If I am right, we are still bedevilled by certain metaphysical concepts of substance which were actually already surpassed in philosophy centuries ago. I shall argue specifically that:

1. The principle of the identity of indiscernibles is not a matter of logic;
2. nor is the identity of indiscernibles inherent in the conception of an individual;
3. nor does individuation force us into a dilemma between identity of indiscernibles and metaphysical realism;[9]
4. nor does the completeness of physics require that its language be less than purely general.

It is equally important to realize what these conclusions do not imply. They do *not* say that physics is or even can be complete, or that real individuals do not always differ in some characteristics, or that all singular propositions can be translated without loss into purely general ones. They just open up a manifold of possible interpretations, in principle all equally tenable and capable of doing justice to physics.

4.1. *Leibniz and the Aristotelian knot*

Central to the philosophical discussions of identity is Leibniz's Principle of the Identity of Indiscernibles. We must distinguish numerical identity from qualitative identity, or indiscernibility. The latter means, roughly or intuitively, having all properties in common. Two bosons of the same type, forming a compound system in a state of motion of form $x \otimes x$, are certainly indiscernible as far as quantum-mechanical description goes. Leibniz's Principle appears to entail then that either this case cannot arise, or quantum-mechanical description is incomplete. In the latter case, what is left out is some distinguishing characteristic of the bosons that 'individuates' them. This position is indeed found in the foundational literature.[10] But so is the position that Leibniz's Principle has been violated and indeed proved false.[11] In this section I shall explain the principle's rationale, its vulnerability, and the prospects for a semantics without it.

Would it be a logical mistake to violate Leibniz's Principle? Although of venerable history in philosophy, it has certainly been disputed. Leibniz used it polemically in his disputes with Newton (via Clarke) over absolute space and time. Later writers have sometimes tried to save at least the conceptual admissibility of absolute space either by trivializing Leibniz's Principle or by denying it (see van Fraassen 1985*b*). What is inconsistent, even by the strictest classical standards of logic, in the idea of a world consisting simply of two spheres alike in every respect; or, more strikingly, with the universe U, which we examined above, consisting of two congruent enantiomorphs?

There is not much profit in putting a question in this way. In logical analysis you can shift to ever higher levels of abstraction, and for any assertion which appears a priori on one level there will be another level on which it does not. Consider for example the assertion that nothing can be warmer than itself. Surely this is a priori; you have to assent to it if you understand 'warmer'. But taken a little more abstractly, the assertion has the form 'Nothing can bear relation R to itself'—and no statement is true by virtue of having *that* form. Similarly, if we treat identity in the abstract, as an arbitrary relational predicate, none of the principles peculiar to identity will look a priori. Logic, in all its purity, will also never single out a particular level of abstraction as privileged or appropriate. It simply does not deign to settle such disputes. Reflection on logic does not therefore answer the above question; it does not yield the answer that there is something inconsistent in that supposition, nor that it is consistent. Instead, it leads us to correct the question: 'consistent' is an elliptical term, and logic does not remove the ellipsis.

To understand PII we must return to the philosophical context in which it emerged. That was the *problématique* of seventeenth-century metaphysics, in the historically conscious form it took in Leibniz. Unlike scornful contemporaries, Leibniz reflected on the Aristotelian–Scholastic tradition which had, in historical fact, and in many ways, shaped the philosophies that emerged from its ruins.

The reasons Leibniz had for PII go back to an essay he wrote at the age of sixteen, concerning the medieval problem of individuation. This problem was originally encountered in Aristotle: there can be many men, but there cannot be many Socrates. What accounts for this? What gives Socrates his uniqueness? There are two questions here, and we should not be too hasty to conflate them. It is a logical principle—before PII is accepted—that each individual is identical with only one thing, namely itself. Is that not enough to answer the first question? But the Aristotelian tradition searched for an individuating principle, something which by its presence would identify Socrates and individuate him, separate him out uniquely.

The same question may be more perspicuous for us with respect to the universe U, which contains only two hands (enantiomorphs), which are congruent but not related geometrically by a proper motion. We may imagine them also structurally and qualitatively entirely alike, in the sense of the correspondence by geometric reflection. If this universe is indeed a possible one, we must ask (if we follow Aristotle): *what distinguishes the one hand from the other?*

Aristotle said that a material substance is individuated by its matter; it could have its form (its properties) in common with another material substance, but its matter is peculiar to it (*Metaphysics*, vii, 1034d. 5–8). Socrates has *this* flesh and *these* bones; Callias has not. Each bone, when taken out of the skeleton, is in turn individuated by its matter, and not by its properties, each of which another bone could share. But how is the matter of the bone individuated? Have we landed in a regress? Aristotle blocks any such regress because for him there will come a level at which the entity's matter is not a separate or separable substance. Then the question of individuation does not arise for that matter. But this way of closing the regress has to raise eyebrows. How could Socrates and Callias be individuated by each possessing something for which the question of individuation cannot even arise? Both have matter—that fact does not distinguish them. What do we do when we say, but Socrates has *this* matter? How can we designate something specifically, or even speak about *it*, refer to *it* and not to another thing, when it is not a separable individual at all?

That Aristotle's relatively short discussion had not laid the questions to rest is clear from medieval writings. Some introduce special properties which only one individual can have — notably the *ubi*, something like absolute spatial position — or discuss the *haecceity* (this-ness) as if it were such a property. Aquinas, in *De Ente et Essentia*, carries on this discussion and brings in the problem examples natural to his Christian Aristotelianism.[12] There are two especially: Socrates and Callias after death, and immaterial intelligences, such as angels. (Reader, stay with me: this is crucial to the genesis of Leibniz's PII, and we will see important similarities to e.g. Reichenbach's discussion of bosons!)

Aquinas took the soul to be what Aristotle called the form, and this is the same in all humans; it comprises the essential properties which delimit the *infima species* of humanity. Socrates' soul is, you might say, his humanity or his personhood. But now, after death, Socrates' soul still exists and his matter does not. The same is true for Callias. If matter is what individuates, how are Socrates and Callias distinct after death? The answer according to Aquinas shows how he has changed Aristotle's notion of form — for he says that Socrates' soul is distinguished by its history, in which it was erstwhile received in *this* flesh and *these* bones. Two souls after death are qualitatively entirely alike, and are not then distinguished by being received in distinct designated matter — but are still distinguished, namely by having different histories behind them. Having a past is counted as a real property and distinguishing mark, even if not present today in any sort of material trace or record.

Angels are immaterial intelligences, and never were received in matter. So they cannot be distinguished by past material individuation. Therefore, Aquinas concludes, there are as many species of angels as there are angels: any two of them have a different form, different essential properties. Their individuation is not by matter but by form.

Why could this not have been asserted for men and trees as well? Surely any two men or trees are distinguished by some property? However, the two men share all essential properties and differ only in their accidental ones. An essential property of Socrates, such as his ability to laugh, is inalienable; any accidental property, such as his snubnose, he might have shared or lacked without imperilling his identity. Callias might have had a snubnose.

Leibniz very clearly perceived the chink in this tightly woven chainmail. It is true for any accidental property of Socrates' that Socrates could have shared it with Callias. But it does not follow that Socrates and Callias could have shared all of them. It can be asserted that the individual is individuated by the totality of all his properties. On this view, Leibniz later says with a little exaggeration, it is for all things the way Aquinas says it is for angels.[13] (Yes, it is—if you ignore the essential/accidental distinction.)

But this assertion is exactly the Principle of Identity of Indiscernibles. It allows us to say: Socrates has a snubnose and Callias does not, and it is not necessary that Socrates have a snubnose, nor that Callias lack one. However, it is not possible that Socrates and Callias have all properties in common. Thus, in any possible situation in which both exist, one has a property that the other lacks. As corollary, if entities A and B have all properties in common, then $A = B$.

I have dwelt on this history in order to make two points. Leibniz's PII was not simply an arbitrary addition to the logic of identity. It solved a problem in a certain philosophical tradition. On the other hand, the principle definitely was an addition to the logic of identity. It was surely the simplest and cleanest way to capture the sense that, if individuals are distinct, there must be something about them that makes them distinct. But the historical perspective shows that it is less than incontrovertible even if it was not capricious.

Let us now see how these reflections should guide us in modal semantics. I will argue here that they should not make us wary of referring to entities in possible worlds, even when we have no uniquely identifying descriptions handy. It is necessary only to insist that the family of worlds is closed under permutations of individuals—and that any *significant* proposition is truly general, i.e. is also true exactly in such a closed set.[14] In the next subsection I shall discuss this in detail. Later I shall also show how different conditions of individuation can be used to define the various statistics in an interesting way.

In recent literature there has been considerable worry about whether the indistinguishability of identical particles does not undermine any usual form of reference.[15] There are very interesting suggestions to revise semantics, quantum logic, and even set theory so as to eliminate assumptions about the very applicability of the identity predicate.[16] I will try to show that a much more conservative approach can also be satisfactory.

4.2. *Permutation invariance in semantics*

Almost every puzzle in metaphysics corresponds to a problem in the philosophy of language. The question 'What distinguishes one individual from another?' (What is it that individuates?) has as correspondent the question, 'In virtue of what does a term refer to one specific individual and not to another?' Given Leibniz's PII, we could answer the latter with: for every individual, there is (in a sufficiently rich language) a description which is satisfied by that individual alone, so all reference can proceed via description. Alternatively, not being metaphysicians in the Aristotelian tradition, we could dismiss the 'problem' of individuation—but if we reject the PII, we would then still have to produce a new answer to the question about reference.

If our language has no simple (primitive) singular terms, i.e. none which purport to refer to a specific individual alone, that question about language does not arise for us any more either. In that case, however, we must defend the assertion that *everything*, the world, can be completely described by entirely general propositions. All the puzzles we have looked at so far then come back to haunt us—for how can we say that the sort of description we offer is complete, when it allows for two individuals which are numerically distinct and yet are alike in all respects accessible to our sort of description?

'Complete' is not univocal. Perhaps (*a*) all factual description can be completely given in entirely general propositions, and yet (*b*) our language would not be adequate if its only function were to express those propositions. Let us give the position that (*a*) is correct the name *semantic universalism*. Let us add to it the understanding that completeness in the sense intended requires at least that the whole body of physics can consist of such propositions. What exactly the word 'general' means here is to be explained below. That something may be left out in some way even if (*a*) is correct appears when we reflect on our own concern to be able to *apply* science.

We shall not be able to apply science unless we can refer to specific things in the world. So we need, even *qua* scientists, something more than the body of physical theory. This is the simple point that you need to know who you are, where, and what time it is if an ordinary factual statement (such as 'The bomb in 1879 Hall will explode at 3 pm local time on 1 January 1999') is to be a guide for action. You might know a complete description of the situation in the language of physics, and not know *that*. The objective descriptions furnished by physics do not coordinatize the world automatically with you at the centre. (For note: you and I can both know of all the same objective descriptions, whether they are true or false; see further Lewis 1979.)

Semantic universalism can be maintained if we add that our language needs more than completeness with respect to factual description. The 'view from nowhere' (to adapt Thomas Nagel's phrase) leaves out no fact, but we—persons, agents—need an enriched language, in which (to adapt another philosopher's insight) the 'I think' can accompany every thought. This means indexical language, whose characterization is not possible within semantics, but only within pragmatics: The utterance of 'I am --- now' by person X at time t expresses the proposition that X is --- at t. Note that this principle does not use, but only mentions, the indexical words 'I' and 'now', so that knowledge of this principle does not suffice to give a person the ability to use and understand indexical language, either. The incompleteness of semantics as a study of language consists in part in the fact that we language users do know how to use such indexical language nevertheless. This *knowing how* is irreducible to (factually descriptive) *knowledge that*. Having made this point, we leave pragmatics, and return to semantics again.

How does the position of semantic universalism enter into semantics? To put it briefly, the requirement that all propositions be general imposes a symmetry principle—Permutation Invariance—on semantic representation. These representations, the models of the phenomena of inference which are constructed in semantics, can take many forms. In the typical form, such a model comprises a *domain* of individuals and a *universe* of possible worlds. That is because semantic study is typically focused on the sort of syntax in which variables (individual coordinates) do appear. Knowing the translation into variable-free (coordinate-free) syntax discussed above, it is easy to see that we could shift to models in which the domain of individuals is replaced by a set of properties and relations. That point has been settled. For the sake of familiarity, let us look at the usual formulation. To quicken our interest, consider the following assertion:

(P) There are two entities, call them x and y, such that x and y are entirely alike in every qualitative respect in actuality, but for x it is possible to be G, and for y it is not possible.

Will the models proper to semantic universalism allow for the possible truth of this assertion? Note that at least one version of the PII is violated. A weaker version is not violated if we allow factual description to contain modal elements irreducibly. But does not the removal of *haecceity* from semantics entail that what is possible for any individual must be possible for all? How could (P) be true unless there is a description-independent identification of individuals across the boundaries between alternative possible worlds? And if (P) cannot be true, does that mean that all modal features and differences supervene on actual, qualitative features and differences?

To construct our semantic models, we need first of all a set H of *cells* to represent qualitative distinctions. A philosopher passionately interested in language can think of these cells as bits of language — *predicates*. You could however also think of these cells as the members of a partition of a 'logical space' or the subspaces of a state-space, whatever you like. Next, we need a *domain* of individuals; call this set D. Now a particular *world* is specified by a mapping w of all of D into those cells. (The ones which intuitively do not exist in that world may be mapped into a special cell, the trash heap of the universe.) Finally, a *universe* W (on H and D) is a set of these worlds. At this point we may add global structure to the universe, say an *access* relation R between worlds.[17]

A *proposition* is characterized if we specify in which worlds it is true. So it may be identified with the set of those worlds. We add more global structure if we like by designating a special class of significant propositions. These propositions are closed under Boolean operations, finitary and infinitary:

$$-A = W - A$$

$$\cap X = \{w \in W : w \in A \text{ for each } A \text{ in set } X\}$$

correspond to negation (complementation) and conjunction (meet). Moreover, there are special *modal* operations which relate to the global structure:

Absolute necessity $\quad \Box A = W$ if $A = W$
$\qquad\qquad\qquad\qquad\quad = \Lambda$ otherwise

Absolute possibility $\Diamond A = \Lambda$ if $A = \Lambda$
$\qquad\qquad\qquad\qquad\quad = W$ otherwise

Relative necessity $\quad \boxed{R}A = \{u \in W : w \in A \text{ for all } w \text{ such that } uRw\}$

Relative possibility $\langle\!\!\boxed{R}\!\!\rangle A = \{u \in W : w \in A \text{ for some } w \text{ such that } uRw\}$

Such an operation is significant if it maps the family of significant propositions into itself. Since a world in W has been identified with a map of D into H, we can introduce a defined relationship as well:

A *permutation* of D is a one-to-one mapping of D onto D.
If g is a permutation of D and w a world, then $g(w)$ is the world defined as $g(w)(x) = w(g(x))$ for all x in D,

and such a world $g(w)$ may then be called a permutation of w, of course. Consider now as intuitive examples:

$A \quad x$ is brave and y is cowardly.

$B \quad x$ is brave or not brave and y is cowardly or not cowardly.

Syntactically, A and B are similar, in that they mention the same two individuals. But there is no problem of reference for B, because it is true regardless of which individual x and y really are. This is not so for A, and indeed the permutation which interchanges x with y would turn A into a falsehood, if it were true to begin with. So A is indeed about specific individuals.

Call *A peculiarly about x* in *D* (*about* subset *E* of *D*) exactly if there is some world *w* and a permutation *g* which permutes *x* with $y \neq x$ (which is not the identity on *E*) such that *A* is true in *w* but not in *g(w)*—while no permutation has that effect if it leaves *x* (the members of *E*) fixed.

Now this is still a notion which concerns individual reference. But notice how it affects propositions which *look* general. Let *F* and *G* be cells, and $-F$, $-G$ their complements in *H*, and suppose that universe *W* is a set of three worlds with the following character:

	F	−F	G	−G
w_1	x	y	x, y	
w_2	y	x	x, y	
w_3	x, y		x	y

Consider now the following proposition:

C There is an entity, call it *z*, such that *z* is *F* and □ (*z* is *G*)

Worlds w_1 and w_2 are permutations of each other, but w_3 is different from them in another way. It is easy to see that in w_1 the proposition *C* is true (the entity in question is *x*), but in the permutation w_2 *C* is *not* true—and that is due to the presence of world w_3 in the same universe.

So proposition *C* looks general, for it has only a bound variable in it. Indeed, it could be written:

(C′) Cell *F* has at least one occupant which is necessarily in cell *G*.

But by our previous definition, it is peculiarly about the individual *x* (and also about *y*) and so is not purely general. Recall that a proposition is a set of worlds:

A proposition is *purely general* in universe *W* if its truth-value is invariant under permutation.

Such a proposition need not be necessary. For example, *cell F has exactly 17 occupants* is purely general by this definition, but it need certainly not be necessary. The family of purely general propositions is closed under operations $-$, $\cap$, $\square$, $\diamond$. What is required for this family to be closed under $\boxed{R}$, $\langle\!\!\!\!\;\text{R}\;\!\!\!\!\rangle$? Suppose that A is a purely general proposition, and let us identify it as usual with the set of worlds in which it is true. So A contains all the permutations of all its members. Now we also have:

w is in $\boxed{R}A$ iff, for all w', if wRw' then w' is in A.

w is in $\langle\!\!\!\!\;\text{R}\;\!\!\!\!\rangle A$ iff, for some w', both wRw' and w' is in A.

So in order for $\boxed{R}A$ to be purely general, what we need is that, if w is in $\boxed{R}(A)$, then any permutation $g(w)$ of w must have the same status with respect to (A):

If w is in $\boxed{R}A$, then, for all w', if $g(w)Rw'$ then also w' is in A.

We can guarantee that by requiring that the set of worlds accessible by R from $g(w)$ is determined exactly by the three elements R, g, and w, as follows:

(PERM)   If wRw', then $g(w)Rg(w')$.

Once stated, it is obvious that semantic universalism requires that, since nothing can be significant unless it is permutation-invariant. This condition (which implies its own converse as well) does indeed suffice to ensure that $\boxed{R}A$ and $\langle\!\!\!\!\;\text{R}\;\!\!\!\!\rangle A$ are purely general if A is. (See *Proofs and illustrations*.)

It is very important to recognize what (PERM) does *not* imply. It does not say that $R(w)$ is closed under permutations. If w is our world, and u is possible relative to us, it does not follow that $g(u)$ is also possible relative to us. Let us look very carefully at what permutation invariance does and does not require.

Let $M = \langle H, D, W, R \rangle$ be called a *full* model if H, D, W, R are as described above and both W and R are closed under permutation. This means that W contains all permutations of all its members, and R satisfies condition (PERM) above. Semantic universalism clearly demands that all models be full; if a sub-family of propositions is singled out as significant, they must all be purely general. I will let F, G, ... stand for disjoint cells of the space H, and examine three propositions. The first will be a version of (P) and the other two, variants thereof:

(P1)   There are two individuals a, b such that both are in F and $\Diamond_R$ (a is in G) and not $\Diamond_R$ (b is in G).

(P2)   There are two individuals a, b such that both are in F and $\Diamond_R$ (a is in G and b is not in G).

(P3)   There are two individuals a, b such that both are in F and yet not $\Diamond_R$ (both a and b are in G) although $\Diamond_R$ (a is in G) and $\Diamond_R$ (b is in G).

Readers familiar with this sort of discussion will notice at once that all three 'quantify into' the modal contexts governed by R. Since being in the same cell—in logical terms, satisfying all the same primitive predicates—is the nearest we come, in precise terms, to complete actual qualitative sameness, each of these appears to imply that qualitatively similar individuals can be modally different. Are any or all of these ruled out by semantic universalism?

The first answer is: (P1)—and hence our original (P) is definitely ruled out. We can prove this by Fig. 12.1. There the initial supposition is that wRw'. To begin to satisfy (P1) we must depict w so that individuals a, b are both in cell F in world w, but only b is there in w'. That way we have represented the information that 'a and b are both in cell F, but $\Diamond_R$ (a is in cell G)' is true in w. Now we look at a permutation g which exchanges a and b. The effect of g is represented by the broken arrows downward. Principle (PERM) requires us to *close the diagram*, that is to draw the double arrow, indicating that $g(w)$ bears R to $g(w')$.

But now, let g do nothing except exchange a and b, leaving other individuals fixed; then w and $g(w)$ are one and the same world. So then $wRg(w')$, and hence $\Diamond_R$ (b is in G) is also true in w. This shows that (P1) is false in w. Since our argument must apply to any model, that proposition is not satisfiable in a full model.

Below I shall argue that (P2) and (P3), on the other hand, are satisfiable. Despite their 'de re' look, they are not peculiarly about any two individuals. They really describe 'global' properties of a world, which are permutation-invariant. Before showing this, however, we must wonder a little whether (PERM) is not too weak. To rule out (P1), we only needed (PERM), but now let us consider

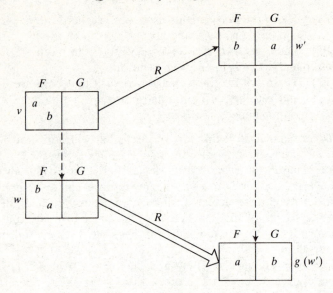

Fɪɢ. 12.1 Concerning (P1)

(PERM*?) For any world w, the set of worlds accessible to
w via R is permutation invariant.

In other words, if w bears R to w', then w must also bear R to
$g(w')$, regardless of what g does to w. Does semantic universalism require this? That question reduces to: does (PERM*?) rule
out anything which we must retain, if we are to do justice to
science?

The answer is that (PERM*?) is indeed too strong. It rules
out as unsatisfiable the further proposition:

(P4) There are individuals a, b such that a and b are in
distinct cells F and G and ⟨Ⓡ⟩ (a is in cell H) and not
⟨Ⓡ⟩ (b is in cell H).

It is immediately clear that (PERM*?) rules this out. A relatively possible world, accessible via R, in which a is in H will
spawn another world, by permutation of a and b, in which b is
in cell H. By (PERM*?) that permuted world would also be so
accessible.

Now I add that (P4) should not be ruled out. To show my

reasons, let me now exhibit the quantum-theoretical analogues of these propositions. (They are analogues, of course, on the modal interpretation of quantum mechanics.)[18] There the actual properties of an individual in a given world are values of observables, while its quantum-mechanical state is reflected not only in those actual values but also in what is and is not possible. In addition, impossibility implies, and can generally be equated with, zero probability (for discrete observables), while possibility divides into all the positive degrees of probability. We get four sorts of analogues, but the first three start in the same way. Note that we can construct such analogues in two ways, by replacing 'possible' either with 'has positive probability', or with 'has probability x', for some positive number $x < 1$.

(QP1–3)   There are systems X and Y which are in the same quantum-mechanical state and are such that:

(P1*)   the probability that a measurement of B will yield value r is positive (or, for example, $= 0.5$) if performed on system X and *zero* (or, for example, $\neq 0.5$) if performed on system Y;

(P2*)   the probability that measurements of B performed on X and on Y will yield value r and value $s \neq r$ is positive (or, for example, $= 0.5$);

(P3*)   the probability that measurements of B performed on X and on Y will both yield the same value r is *zero* (or, for example, $\neq 0.5$), although the probability that such a measurement will yield value r if performed on X is positive (or, for example, $= 0.5$), and is also positive if the measurement is performed on Y.

(QP4)   There are systems X and Y which are in distinct quantum-mechanical states, and the probability that a measurement of B will yield value r is positive (or, for example, $= 0.5$) if performed on X while it is *zero* (or, for example $\neq 0.5$) if performed on Y.

There is no doubt that (QP1) runs counter to quantum theory, while (QP2)–(QP4) are correct in certain situations. In the case of (QP3), one such possibility is the typical one of perfect anti-correlation in EPR–Bell situations.

Reflecting on the possibility of (QP4), and the consequent need to reject (PERM*?), however, we see that any plausibility it has concerns the special case of indiscernibility with respect to modal as well as actual qualitative properties. That is an equivalence relationship which we can define as follows:

Definition: Individuals a, b, c, ... are *modally indiscernible* in world w if and only if $R(w)$ is closed under permutations which affect at most these individuals.

In quantum-theoretical analogue, this concerns sameness of state.

Finally, then, I shall show the satisfiability of (P2) and (P3) in a full model. In addition, I shall use a model construction which shows that (P2) and (P3) can still hold even if the individuals are modally indiscernible.

Consider the following full model. We need not specify everything about it, but say that D contains distinct entities a and b; H contains the three cells F, G, G'; and W contains worlds w_1, w_2, w_3:

	F	G	G'
w_1	a, b		
w_2		a	b
w_3		b	a

If g is the permutation which interchanges a and b (and leaves any other individual fixed), we can take it without loss of generality that $w_3 = g(w_2)$ and $w_2 = g(w_3)$, while $w_1 = g(w_1)$. Let the relation R be such that each world bears R to itself, and in addition

$$w_1 R w_2 \qquad w_1 R w_3 \qquad \text{not-}w_2 R w_3$$

which implies via (PERM) that not-$w_3 R w_2$. Given this information, we can consistently add that R satisfies (PERM), and that w_1 does not bear R to any other worlds. Note also that all this is compatible even with (PERM*?), so at the very least we can specify a and b to be modally indiscernible.

This construction was designed to satisfy (P2) and (P3), for a case in which the relevant individuals are modally indiscernible. If R is reflexive—as it should always be, if R is worthy of the reading 'Necessarily'—then modal indiscernibility entails qualitative sameness as well.

We can still have a full model, with the above situation in it, without entailing that these propositions are peculiarly about any given individual. That is so even if there are in the world some entities which are modally discernible from them. For $\{w_1, w_2, w_3\}$ can be one cluster of worlds, not related by R to any worlds outside that cluster, while somewhere else in the model we will see another cluster $\{h(w_1), h(w_2), h(w_3)\}$ where h exchanges a and/or b with certain other entities c and d. There, in that new cluster, c and b or d play the role which we depicted for a and b above, while a plays an entirely different role. All this may sound rather strange, for it sounds as if I'm saying, for example, that it would have been *possible* for all the things which are electrons (actually, in our world) to have been photons instead. In one sense of 'possible', *yes*, but not in the sense of 'physically possible', which is represented by ⟨Ⓡ⟩—only in the more *outré* ultimate sense of verbal possibility.[19]

This has been a lengthy discussion, but it was needed to show how far semantic universalism does and does not go. The cells of our language can carry information both about modal properties (the dynamic states) and about qualitative differences (the value states). This requires that, in one sense, all modal features *supervene* on qualitative characteristics; that is, (P) or (P1) is ruled out. But that does not imply that these modal distinctions are lost, not even all those that have often been classified as *de re* because they are stated with quantifiers reaching into modal contexts.

And now, we return finally to the question of worlds with and without individuals. The models required by semantic universalism are exactly those which can be described *equally* on either view. So far we have described them in terms of individuals. But each world—a mapping of individuals into the cells of a logical space—can be characterized simply as a set of occupation numbers for the cells. Closure under permutation of the access relation R entails that the R-modalities operate on fully general propositions without losing the generalities. Therefore every significant proposition can be restated entirely in terms of occupation numbers. This means that we can 'abstract' an individual-free model:

A world is a mapping of cells into natural numbers,

and this abstracted model corresponds to many models of the more ordinary sort, but (up to isomorphism) to a *unique* full model among them.[20]

Now not even the ontology can be 'read off' the semantics, let alone the syntax. Ontology can find no purchase.

Proofs and illustrations. We prove first that (PERM) implies its own stronger version:

(PERM+) wRw' iff $g(w)Rg(w')$,

and is therefore equivalent to (PERM+). For assume (PERM) and suppose that $g(w)$ bears R to $g(w')$. Each permutation has an inverse, so by (PERM) itself we argue

$g(w)Rg(w')$, hence $g^{-1}g(w)Rg^{-1}g(w')$, hence wRw'.

We wish to prove now that the set of purely general propositions is closed under operations $\boxed{\text{R}}$ and $\langle\!\!\langle\text{R}\rangle\!\!\rangle$:

> *Definition*: $R(w) = \{w' : wRw'\}$
> (*) if $R(w) \subseteq A$ then $R(g(w)) \subseteq A$
> (**) if $R(w) \cap A \neq \Lambda$ then $R(g(w)) \cap A \neq \Lambda$

So suppose $g(w)$ is not in $\boxed{\text{R}}A$. Then we have some world u such that $g(w)Ru$, and u is not in A. But $u = g(v)$ for a certain v, so $g(w)Rg(v)$, and hence wRv. But if v were in A then u would be too, so v is not in A. Therefore w is not in $\boxed{\text{R}}A$. This proves condition (*). Similarly, if w is in $\langle\!\!\langle\text{R}\rangle\!\!\rangle A$, let wRz and z in A. Then $g(w)Rg(z)$; since z is in A and A is purely general, $g(z)$ is in A too. Therefore $g(w)$ is in $\langle\!\!\langle\text{R}\rangle\!\!\rangle A$. This proves condition (**).

The non-diagrammatic argument that (P1) is ruled out by (PERM) is as follows. Suppose x, y are in F in world w and that x is in G in world w' while wRw'. Now let $g(x) = y$ and $g(y) = x$, and g leaves all other individuals fixed. Then $g(w)Rg(w')$, which means $wRg(w')$. But $g(w')$ is in the model since it is full, and so y is in G in $g(w')$.

4.3. *Classical statistics in possible worlds*

Now let us revisit statistics. Here is a proposal which may be suggested (perversely?) by the relative success and relative failure of the classical reconstruction via De Finetti's theorem which we examined in the preceding chapter. Let us follow Carnap in taking the measure m^* described in the preceding chapter as the basic probability function. For a metaphysician this could only be the proposal that Carnap's m^* *is* objective chance, and he or she must immediately face the question: what makes it so?[21] But as a proposal for interpretation, it raises only the question: if we adopt this as basic probability function, can the systems of possible worlds be so fashioned as to represent one way things could be, which is still in accordance with science?[22] Our main problem then becomes: under what special conditions will m^* reduce to (make the same assignments as) m^+ and m^F? Intuitively, the answer should be: respectively, when the entities are distinguishable, and when an Exclusion Principle is in effect. Upon proper precise construal, that is indeed correct.

Recall that for Carnap the cells are characterized by families of predicates; thus, in the study of an urn-problem, the predicate might be *cubical* or *red*, the cells being cubical–red, cubical–nonred, noncubical–red, and noncubical–nonred. The four complex predicates representing the cells are called *Q-predicates* (logically strongest consistent predicates in the language). Let us say that a family of predicates *individuates* a set of individuals if no two of them can be alike with respect to all these predicates (i.e. if no Q-predicate formed from this family can characterize more than one such individual). We now have three and only three possible situations:

1. The family does not individuate the individuals. (Every logically possible state-description can be true.)
2. The family as a whole individuates. (Only those state-descriptions in which each individual satisfies a different Q-predicate can be true.)
3. A proper subfamily individuates. (Each individual satisfies a different Q-predicate *of that subfamily*, in each state-description that can be true.)

Cases 2 and 3 can hold only relative to some postulates, on which the ur-probability is to be conditionalized; for *logically* speaking, of course, every state-description could be true. In fact, it is easy to formulate the relevant postulates. Let $a_1, \ldots, a_n$ be individual constants and let $\{F_1, \ldots, F_k, G_1, \ldots, G_m\}$ be the total family of predicates. Let $\{Q_1, \ldots, Q_q\}$ with $q = 2^{k+m}$ be the set of Q-predicates for the whole family, and let $\{Q'_1, \ldots, Q'_r\}$ with $r = 2^m$ be the set of Q-predicates for $\{G_1, \ldots, G_m\}$ alone. Then those postulates are as follows:

Situation 2: $Q_h(a_i) \supset {\sim} Q_h(a_j)$ for $h = 1, \ldots, q$

Situation 3: $Q'_h(a_i) \supset {\sim} Q'_h(a_j)$ for $h = 1, \ldots, r$

in both cases for each $i \neq j$ from 1 to n. Here $\supset$ is the logical symbol of material implication: $A \supset B$ is false if A is true while B is false; and otherwise $A \supset B$ is true.

In Section 5 it was already illustrated that m^F is m^* conditionalized on the situation 2 postulate. That is, with H, T as the two Q-predicates, we have

$$m^F(--) = m^*(-- \,\|[Ha_1 \supset Ta_2] \,\&\, [Ha_2 \supset Ta_1])$$

This is easily checked by noting that the odds between the *remaining* state-descriptions are the same in both cases $(1/6 \div 1/6, 1/2 \div 1/2)$. This is a trivial case but the argument is general: the situation 2 postulates rule out all *structure*-descriptions exhibiting multiple occupancy of cells. The remaining structure-descriptions each contain the same number of state-descriptions, so all remaining state-descriptions (as well as, separately considered, all remaining structure-descriptions) are treated as equiprobable by m^*—just as by m^F.

To illustrate the effect of the situation which 3 postulates, we need a bigger table (Table 12.1). Let $k = m = 1$, so we have only four cells (and $G_1, {\sim}G_1$ are the Q-predicates of the relevant subfamily), and let $n = 2$. Here cases 1–4 are mutually non-isomorphic; case i is isomorphic to case $4 + i$ (for $i = 1, 2, 3, 4$). All other cases are non-isomorphic to these and are ruled out by the situation 3 postulates. So we have here four of the original structure-descriptions, in fact. By m^* all structure-descriptions are equiprobable, and conditionalization leaves the 'internal odds' the same, so the remaining four are now still equiprobable (1/4 each). Now we notice that *state-descriptions* in the *other* subfamily $\{F_1\}$ correspond to those remaining struc-

TABLE 12.1.

	F_1	$\sim F_1$	G_1	$\sim G_1$	
Case 1	a_1, a_2	—	a_1	a_2	1/8
Case 2	a_1	a_2	a_1	a_2	1/8
Case 3	a_2	a_1	a_1	a_2	1/8
Case 4	—	a_1, a_2	a_1	a_2	1/8
Case 5	a_1, a_2	—	a_2	a_1	1/8
Case 6	a_2	a_1	a_2	a_1	1/8
Case 7	a_1	a_2	a_2	a_1	1/8
Case 8	—	a_1, a_2	a_2	a_1	1/8

ture-descriptions in the whole family $\{F_1, G_1\}$, and hence are equiprobable. But that means that m^*, so conditionalized, *coincides with m^+ on the other subfamily* (on the remainder of the overall family, after we remove the subfamily which indi-viduates—in this case just $\{G_1\}$). An example is: we have two coins (a_1, a_2), one of which is scratched (G_1) and the other not; each can be heads (F_1) or tails $(\sim F_1)$ independently of the other. If we now look at heads v tails *alone*, m^* conditionalized on the relevant postulate $(G_1 a_1 \supset \sim G_1 a_2)$ gives us the effect of m^+.

To sum this up, then, we can see the three statistics as special cases of the same principle (prior equiprobability for structure descriptions) for situations of different extents of individuation by the predicates considered. Again we see here the relevance of completeness and PII. A claim of individuation is a complete-ness claim for a family of predicates; PII entails that there must always be some family of predicates which individuate. Thus, in this perspective, situation 1, the boson case, provides the challenge to PII.

However, one could also take the following point of view. Certain characteristics are empirically significant; their presence or absence affects the observable phenomena. Others are empir-ically superfluous. The proposal that m^* is basic is meant to apply to the family of all empirically significant properties. It ,may be therefore that PII is satisfied by distinguishing character-istics outside the family which defines the domain of m^* in our system of possible worlds. Being outside the domain, m^* cannot be conditionalized on them, and so the argument relating to case 3 does not apply. The question that can be answered definitively is therefore not whether science entails a violation of PII, but only whether its description *can be regarded* not only as complete of all that is empirically significant in its domain, but also sufficient to individuate. Our present guess, before looking more closely at quantum mechanics in this light, is that the answer is *yes* for fermions and *no* for bosons. The case of the classical particles may be regarded as one in which individuation is by empirically significant characteristics that could only be *temporarily* omitted from the scientific description for some limited purpose.

5. CONCLUSION: GOOD-BYE TO METAPHYSICS

What I used to find seductive about metaphysical realism is the idea that *the way to solve philosophical problems is to construct a better scientific picture of the world.*

Hilary Putnam (1988, 107)

Why should Putnam have found this seductive? Because a philosopher, feeling that he or she has come to understand the structure of science, and admiring its intellectual reach, will then be naturally inclined to engage in activity similar to what he or she admires. But then, what results? Descartes's or Leibniz's attempt to add 'more basic' principles to science, on which it can 'rest' or 'be supported', is one result. This yields a larger body of science, a larger theory of the world. But if there are philosophical questions about science *as such*, they will equally arise no matter how it is enlarged, or in what direction, if the enlargement is still science. Carnap's and Ayer's conclusion that those metaphysical forays were mistaken in intent, and that there really are no meaningful questions outside science and mathematics, is another result. That implies that there are no questions about science *as such*, only e.g. scientific sociological questions about parts of science not identical with present sociology.

There is something right in both reactions, and also something very wrong. What is right about the first is that philosophical puzzles about science tend to involve hidden scientific questions—we must carefully disentangle conceptual difficulties from questions about what the theory itself implies. What is right about the second reaction is that none of this can lead to an extra-scientific foundation for science. The philosopher *qua* philosopher is not contributing to science. Separating out scientific questions entangled in the philosophical puzzles is for the philosopher only brush-clearing, removal of impediments. And—here is perhaps the main point—when all those factual, theoretical, and conceptual questions have been disentangled and the impediments removed, we are in a position to understand; we are free.

What then is the seductive temptation of metaphysical realism? It consists in the idea that removing those impediments leaves a residue of factual questions of a different order, which the philosopher can answer speculatively by postulating abstract, unobservable, or modal realities. Strangely enough, when that idea exercises its lure, it tempts us also to stop disentangling scientific and metaphysical issues and instead to start 'identifying' those supposed residual factual questions at once, and to postulate the answers. To be fair, we must admit that many philosophers explicitly engaged in metaphysics have given us valuable insights as well. But it is possible to re-evaluate these as insights into the manifold of tenable interpretations, and not at all to take them as clues to 'the' uniquely right interpretation.

The phenomena underdetermine the theory. There are in principle alternative developments of science, branching off from ours at every point in history with equal adequacy as models of the phenomena. Only angels could know these alternative sciences, though sometimes we dimly perceive their possibility. The theory in turn underdetermines the interpretation. Each scientific theory, caught in the amber at one definite historical stage of development and formalization, admits many different tenable interpretations. What is the world depicted by science? That is exactly the question we answer with an interpretation, and the answer is not unique. Perhaps no interpretation ever finishes the task of answering all questions about the depicted world it displays as the theory's content. To have even a sketch of one interpretation is valuable and brings understanding. To appreciate, however dimly, its horizon of alternative possible interpretations brings more insight. There cannot be in principle, but only as historical accident, convergence to a single story about our world.

NOTES

Chapter 1

1. There are also philosophical views, not discussed here, which disagree with some of that, e.g. logical positivism and the 'non-statement' or 'structuralist' view of theories due to Sneed, Steg-müller, and Moulines.
2. See Suppes (1967). The impact of Suppes's innovation is lost if models are defined, as in many standard logic texts, to be partially linguistic entities, each yoked to a particular syntax. Here the models are mathematical structures, called models of a given theory only by virtue of belonging to the class defined to be the models of that theory.
3. For the history see Suppe (1974, 1989); see also van Fraassen (1970), Suppe (1972), and Suppes (1967, 1974).
4. This problem was raised by Jill Sigman.
5. As an (aspirant) empiricist, I should say that I take only the second question seriously (since the first appears to assume uniqueness of correct interpretation). By the same token, I envisage rational life as not necessarily committed to a world-picture yet vitally concerned with what world-pictures might be tenable for us, and what forms of life each might open to us.
6. I draw here on Sigman (1989).
7. See further van Fraassen (1989, 257–8).
8. For a summary and references for these different treatments of mass, see van Fraassen (1980*b*, ch. 3, sect. 8).
9. See Eco (1979), especially the chapters 'The Poetics of the Open Work' and '*Lector in Fabula*'.
10. Giere (1988, ch. 5) is to be recommended here, though written in the service of another debate. Recent history of science has some beautiful studies, such as Peter Gallison's (1987); sociology of science, though always in philosophical hot water, is very helpful to enrich our view of actual scientific practice (Knorr-Cetina 1981; Latour 1986). For further reflection see also Hughes (1989, 79–82).
11. Cartwright maintains that real science is not concerned to produce theories which have *that* feature, or to maintain any claim as far-reaching as that. I do not wish to dispute this here; I think it

affects not the form we attribute to the theory, but the aim we ascribe to science (in the context of an account of theories).

Chapter 2

1. I am indebted here to A. F. Chalmers's instructive article (1970), which also analyses possible applications of the principle in physics.
2. See van Fraassen (1989), in which this tactic, its uses, successes, and limits, are treated at greater length.
3. See the characterization of what propositions are about in van Fraassen (1978*a*), in terms of the permutation group; and see further Ch. 12 below. The metaphysics I have in mind here is the anti-nominalism which posits an objective distinction between 'natural' classes ('natural kinds') and arbitrary classes.
4. See also the analysis of this sort of reasoning by Mach (1974, 516–20, 549, 456–9). See further Redhead (1975).
5. This is a point of view defended in van Fraassen (1989), but denied by philosophers who espouse realism with respect to laws and/or modalities.
6. This subsection draws especially on Lloyd (1988, ch. 2; ch. 3, sect. 3.2; ch. 8).
7. See e.g. Blumenthal (1961, ch. VII) for this view of geometry.
8. As general sources, see Martin (1982), Weyl (1952), Yale (1968). See also van Fraassen (1985*b*, ch. IV).
9. Compare this to Butterfield (1987, 1989) on definitions of determinism for space–time theories.
10. Would it make sense to say that a system *became* deterministic at a certain time? Only in the derivative sense that its possible trajectories from then on are just the final segments, starting at an arbitrary time, of the possible trajectories of a deterministic system.
11. This has an interesting and welcome corollary, which we should in any case require in a state-space model. There is no criterion of simultaneity for merely possible events; accordingly, we should allow clocks and calendars to be reset however we like. That means: If s is a possible trajectory, and s' is definable by the equation $s'(t) = s(t + b)$, where b is any real number, then s' is a possible trajectory also. Of course we can call s and s' equivalent descriptions of what happens, in terms of a different clock.
12. Wilson (1989), a review of Earman (1986), suggests that we should perhaps understand classical mechanics this way, in view of the possibility of unstable equilibria, where the theory does not predict what will happen if a certain kind of force is applied, and the possibility of spatial paths which are not everywhere differentiable.

13. See Reichenbach's collection of early articles (1959).
14. For extensive discussion of classical mechanical systems along these lines, see Lefschetz (1977).
15. More generally, a Borel set. I restrict the discussion here to intervals because the notion of Borel set is to be discussed in the next Chapter.

Chapter 3

1. It does have the troubling feature of separating past from future in a way that after Einstein we take to be relative. There are indeed conceptual obstacles to the combination of indeterminism and relativity, though they are not disabling. See van Fraassen (1978).
2. In van Fraassen (1979*a*; 1980*b*, ch. 6) I have defended the modal frequency interpretation of probability in physics; in van Fraassen (1989, chs. 8, 12, 13), I have argued for an empiricist view of how our own probabilities (expressing our opinion) are to be related to the probabilities found in the scientific theories we accept. Cassinelli and Lahti (1989*a*) have shown how the modal frequency interpretation can be used to illuminate certain issues in the foundations of quantum mechanics.
3. For further details see Moore (1982).
4. There is a large literature on the shift from the classical account of probability, summarized in, and thereafter dominated by, Laplace's *Théorie analytique des probabilités* (1814), to the new conception for which probability cannot be determined a priori. See Krueger *et al.* (1987) and especially Kamlah (1987); see also van Fraassen (1989, ch. 12).
5. See Jaynes (1973, 477–92), which has references to preceding discussions.
6. Jaynes (1973) uses this form of reasoning to disarm Bertrand's chord paradox.
7. Rosenkrantz (1981, 4.2-2). See also sect. 3.5 of Perey (1982), which develops this approach to data reduction in general form. Note that the numerical tables discussed here are e.g. entries for subjects arranged alphabetically. In the case of a table of logarithms, or entries generated by any continuous function, Poincaré proved that the even digits in the nth place are as frequent as the odd digits. See Poincaré (1905, 193–4).
8. See the concise, perspicuous exposition in Dawid (1983). The main figures in the search for 'invariant priors' besides Jaynes were Jeffreys (1939) and Fraser (e.g. 1961).
9. Cf. Dawid (1983, 235): the programme 'produces a whole range of

choices in some problems, and no prior free from all objections in others'.

10. See Jaynes (1968); Villegas (1971, 1977, 1981). I want to thank Dr F. G. Perey, of the Oak Ridge National Laboratory, for letting me have a copy of his excellent and insightful presentation (1982) of this approach.

11. Required are certain topological properties for the group. If it is locally compact and transitive (for any y and z in the domain there is a group element g such that $g(y) = z$), then the left Haar measure is unique up to a constant. However, the left and right Haar measures need also to be the same; this is so if the group is compact.

12. See De Finetti (1964) and the exposition in Jeffrey (1983, ch. 12). For an insightful discussion, see Zabell (1988). For both a survey of how this subject relates to classical and quantum-statistical mechanics and important new results, see Bach (1987, 1988, 1989).

13. A criterion for extensibility was proved already by De Finetti; see Diaconis (1977) and Bach (1989, sect. 3.2), who connects this with correlation in the sequence of random variables and with Fermi statistics.

14. For a philosophical introduction to ergodic theory see von Plato (1988).

15. A transformation which satisfies the Basic Condition is called an *endomorphism*.

16. Birkhoff's theorem, as well as Poincaré's, is proved in almost every book on ergodic theory; see e.g. Halmos (1956, 10–11 and 18–21), or Parry (1981, 19 and 25). Birkhoff's theorem itself, rather than its corollary for ergodic systems, is usually called his ergodic theorem.

Chapter 4

1. See note 4 below about the word 'causal'. If this term is rejected, the main point of this chapter remains intact: no model with certain features (which have often been associated with demands for causal explanation) can fit these phenomena. This shows why quantum theory must lack those features, and must take its peculiar form, if it is to accommodate such phenomena.

2. Approximately: the most stable isotope has a half-life of 1620 years. The immediate disintegration product is radon; its most stable isotope has a half-life of 3.825 days.

3. The most complete exposition of this point of view appeared in his posthumous work (Reichenbach 1956). As a logical empiricist,

however, he presented this mainly as a methodological rule, which we could settle on by decision to guide theory construction, rather than as an insight into the structure of the world.

4. Reichenbach's statistical relationships are not sufficient to establish a causal relationship; to that extent 'causal' is a misnomer. That will not obstruct our discussion, provided only that a truly causal account must at least have the features here used to define 'causal model'. For the claim at issue will be the negative one that certain phenomena admit no causal model (and hence no truly causal account a fortiori). But various writers have also proposed amendments to Reichenbach's criterion, which do not allow his original one to capture even necessary conditions for causality. The result, and perhaps the point, of these amendments is so to weaken the empirical import of the criterion that even quantum mechanics can no longer be said to violate causality. See the discussion of Salmon's later theory of explanation in van Fraassen (1980*b*, 1985*c*); see further Cartwright (1988). The latter begins with a generalized notion of causal model based on a set of linear equations, and then introduces special variables which reflect whether a true singular cause is present or not in a given individual event. Unfortunately, there exists no generally accepted pre-theoretical analysis of the concept of cause to legislate such proposals, and it is not clear what is to be preserved beyond the pleasing sound of the word 'cause'.

5. The remainder of this section follows van Fraassen (1982).

6. Principle III is also called *Factorizability* (A. Fine), *Completeness* (J. Jarrett), and *Outcome Independence* (A. Shimony).

7. Principle IV is also called *Locality* (J. Jarrett) and *Parameter Independence* (A. Shimony).

8. Thanks to my student Ned Hall for showing this. Since independence is symmetric, V entails $P(Li \& Rj|Aq) = P(Li \& Rj)$, and this can be used to remove the factor Aq from IV to yield II.

9. That I and III entail determinism in some respects became clear in Wigner's (1970) reformulation of Bell's argument. For a rigorous and general treatment, see Suppes and Zanotti (1980).

10. The early experiments, such as by Clauser and Holt, are described and discussed at length in Belinfante (1973). For the later experiments, see Aspect, Dalibard, and Roger (1982); Aspect, Grangier, and Roger (1982); Aspect and Grangier (1985).

11. Both parts of the assumption are crucial. In the discussions of the quantum-logical 'revolution' below we shall note the importance, and dubitability, of the assumption that the quantities must always have some value (belonging to the set of possible measurement outcomes). Jon Dorling (1981) has constructed a striking illustration

of how, if that assumption holds, systematic but indeterministic discrepancies between actual and apparently revealed values can operate.

12. What I have called here simple models are what Accardi calls Kolmogorov models (for references see next chapter), which suggests—misleadingly, in my opinion—that these are the only sort allowed by classical probability in its standard formulation.

13. This is one way of summing up Bell's Inequalities in a general form. There are other elegant ways to do that; see e.g. Itamar Pitowsky's description of what he calls the Bell–Wigner and Clauser–Horne polytopes of probability functions, in ch. 2 of Pitowsky (1989).

14. See for example De Muynck (1986*a*, esp. 998). See further Dieks's (1986) discussion of this argument; Dieks and Hoekzema (1984); Rastall (1983); and the discussion between Stapp and Rastall in Stapp (1985) and Rastall (1985).

15. The many operative senses of locality, contextuality, and separability, as well as the logical relationships between them, are explored in Redhead (1987). See also D'Espagnat (1976, 1984).

Chapter 5

1. In Chapter 10 we shall briefly consider such a ploy, the 'de-occamization' of observables via a pertinent theorem of Stanley Gudder.

2. To see this consider the simple case $P(A|B \text{ or not } B) = P(A) = P(A \text{ and } B) + P(A \text{ and not } B) = P(B)P(A|B) + P(\text{not } B) P(A|\text{not } B) = $ some average of $P(A|B)$ and $P(A|\text{not } B)$.

3. The above discussion was arrived at in response to the 1986 lectures at Princeton by Professor Luigi Accardi of the University of Rome. Despite our disagreement, I thankfully acknowledge my debt to Professor Accardi's writings and lectures; this debt will be apparent in the remainder of this chapter.

4. I think here especially of the programme of Jauch and Piron, and the more recent programme of Ludwig. See Jauch (1968), Piron (1972, 1976), Varadarajan (1968), Ludwig (1967, 1983/5, 1985/6), Mittelstaedt (1978, 1981). See also work of Marlow (1978*b*, 1978*c*).

5. $P_s^m(1) = [\cos(\psi - \theta)]^2 = [\cos\psi\cos\theta + \sin\psi\sin\theta]^2 = \cos^2\psi \cos^2\theta + 2\cos\psi\sin\psi\cos\theta\sin\theta + \sin^2\psi\sin^2\theta.$

6. See also Beltrametti and Cassinelli (1981*a*, 204–7).

7. On the modal interpretation (Copenhagen variant) that I shall advocate below, only the statistical distribution of outcomes in a class of measurements gives information about the state of the system.

8. A similar, entertaining discussion in terms of pupils passing exams is found in D'Espagnat (1983, sect. 4.7).

9. See also Harper *et al.* (1981) and Hajek (1989) for the recent history of this subject.

10. The critique in this section is mainly that of van Fraassen (1981*b*). The subject has been thoroughly treated since in Hellman (1982) and Skyrms (1982). See also D'Espagnat (1983, ch. 12).

11. See the discussion of non-historical propositions and chance in van Fraassen (1980*c*, sects. 1.2, 2.4, and 3.2).

12. The easiest way this pair of equations could be consistent is perhaps this: there is a certain observable n_Y such that n_Y has a certain value $v(y, n_Y)$ when Y is in state y—and $Y(E)$ is just the set $\{y \in S_Y : v(n_Y, y) \in E\}$. In that case n_Y is the 'pointer observable' whose values are the outcomes of the measurement, and equivalently the sets $Y(E)$ are the characters of states of Y that could *equivalently* be called outcomes of the measurement. As I said, that is the *easiest* way in which this story could be consistently elaborated, but perhaps not the only one.

13. For the later development I refer to here, see Dalla Chiara (1983).

14. The set must be a Borel set, to ensure that the function is defined.

15. See Jauch (1968, sect. 5.8); for further discussion see Ch. 9, Sect. 7 below.

16. This is a variant of Desargues's theorem, and was first formulated by Alan Day; see Greechie (1981).

17. For a complete exposition of these results see Varadarajan (1968, ch. VII, sect. 5).

18. For the quantum-logical interpretation of quantum mechanics (implicitly contested here), see the bibliography entries for Bub, Bub and Demopoulos, Demopoulos, Friedman and Putnam, Putnam, and Stairs.

Chapter 6

1. J. van Aken (1985, 1986) has correctly emphasized the limits of this assumption.

2. In quantum logic and its associated approaches to foundations of quantum mechanics (QM), such notions are conceptually distinguished, and assumptions introduced one by one in order to explore exactly how much each contributes to the structure of QM models; see van Fraassen (1981*a*, sect. 1–6); Mittelstaedt (1981).

3. In the next Section (see especially its *Proofs and illustrations* and the cited work by Lahti and Cassinelli) it will become clear that eigenvalues and possible values are not to be equated in general.

The restriction to discrete spectra allows us to ignore this complication.

4. For a clear account of the whole subject, see Jordan (1969). A philosophical discussion of observables with continuous spectra is found in Teller (1979). Recent work has brought out further qualifications concerning non-discrete observables, which I do not discuss here: see Ozawa (1984) and Lahti (1988).

5. See Cassinelli and Lahti (1989a), where this issue is coupled with the connection between observables and their discretizations, and the modal frequency interpretation of probability.

6. As a geometric illustration of the case in which the condition is not satisfied, think of the z-axis in space and two lines m and n through the origin, in the $x-y$ plane. The projection on either line commutes with the projection on the z-axis, because doing one after the other turns a vector into the null vector. But if those two lines are not orthogonal to each other, the projections on them do not commute. Take a vector in the $x-z$ plane which is orthogonal to m but not to n; if we first project it on n we obtain a vector of positive length, and we turn this into another vector of positive length by projection on m.

7. The representation of mixed states by means of statistical operators is clearly presented in von Neumann (1955); the seminal paper for more recent practice was Fano (1957). For a general discussion which covers also continuous mixtures and some questions of interpretation, see Ochs (1981). Mixed states are also referred to as incoherent superpositions.

8. It follows at once that Tr is linear. That it is order-independent is not equally obvious; see e.g. Jordan (1969, ch. V, sect. 22).

9. Note that by my definition a statistical operator has necessarily a pure point spectrum. We could have proceeded conversely with the more usual definition of a statistical operator as a positive self-adjoint operator with trace 1, and then the proof that these have a discrete spectrum (Jordan 1969, ch. 5, sect. 22). Hence there is here no loss of generality in our discussion.

10. This is a subject with a history, starting with an inspired guess by Schroedinger (1935a; 1936) and ending with the fully general results of Hadjisavvas (1981). The result for the finite-dimensional case was already proved in Jaynes (1957).

11. Remember that our discussion is restricted to separable Hilbert spaces; the dimension of a subspace is always less than or equal to that of the whole space.

12. There is an exposition in Appendix 2 of Hughes (1989).

13. Bub (1977) draws the connection between conditionalization and Lueders's rule explicitly; see further Beltrametti and Cassinelli

(1981*a*, sects. 25–6), and Hughes (1989, sects. 8.2, 8.4, 8.8).

14. For 'learns' we may read 'finds out by measurement'.

15. Lueders (1951). This rule is discussed in Furry (1966), and its connection with conditionalization is very clearly exhibited in Bub (1977). For a recent discussion see also Martinez (1989).

16. See also Beltrametti and Cassinelli (1981*a*, ch. 26).

17. For this theorem, see Herbut (1969, 291).

18. There are no other cases; see Wigner's Theorem, discussed in Jordan (1969, sect. 29).

19. For a discussion of its exact meaning, see Jordan (1969, ch. III, sect. 15). See also Beltrametti and Cassinelli (1981*a*, sects. 6.1, 6.2 and further 6.4, 7.5, 23.1, 23.2).

20. This is the usual assumption, but see n. 21 and Ch. 8, Sect. 8 below for complications.

21. This reasoning rests on an assumption, namely that H does represent an observable. At the moment, I am assuming that all Hermitean operators do. When this assumption is discarded, as in the presence of superselection rules (see below), we must say more carefully: by definition, H represents the energy, if it represents an observable at all.

22. The simplest Hilbert space of this sort uses one spatial dimension and its vectors are the square integrable, complex-valued functions of a real variable. That means: functions $f(x) = c$ such that x is real and c complex, with $\int |f(x)|^2 \, dx$ finite, and with the inner product defined by $(f \cdot g) = \int f(x)^* g(x) \, dx$.

23. The term was introduced in Wick *et al.* (1952, esp. 103).

24. A simple example is given by Beltrametti and Cassinelli (1981*a*, sect. 6.4, exercise 7).

Chapter 7

1. For a more abstract treatment, see Beltrametti and Cassinelli (1981*a*) and the references therein especially to the work of Aerts. The *locus classicus* is von Neumann (1955, ch. VI, sect. 2).

2. I shall ignore superselection rules in this section; see Beltrametti and Cassinelli (1981*a*, sect. 7.6).

3. See Finkelstein (1962/3, esp. 630) and Bub (1976, esp. 7–9). Cf. also Lahti (1988).

4. Communicated by J. Bub.

5. This fact, which follows from the more general Polar Decomposition Theorem, has been playing a significant role in some recent interpretations of quantum mechanics.

6. For the abstract see Beltrametti and Cassinelli (1981*a*, ch. 24 and references, esp. to Aerts); for the recipe see Beltrametti and Cassinelli (1981*a*, ch. 7, sect. 2) and van Fraassen (1972, app.).

7. For general discussions of the interpretation of mixed states, see Ochs (1981); Hellwig (1981); D'Espagnat (1976); and Peres (1978).

8. For discussion see especially Hooker (1972); see also Grossman (1974); Ochs (1981); Helwig (1981); Hadjisvvas (1981); Mittelstaedt and Stachow (1983).

9. See Cartwright (1974*a*); see also Beltrametti and Cassinelli (1981*a*, sects. 2.3, 2.4, and 4.3).

10. It may be that the process cannot be initiated except under favourable circumstances, which constrain S. For example, its temperature should not be such that the apparatus melts. In that case we should speak of an apparatus that is an A-measurement apparatus only under those conditions, which is a *derivative* notion of measurement. I shall leave this qualification aside here.

11. These will be essentially what Lahti *et al*. call pre-measurements. But I cannot use the same term, for they characterize the distinction between pre-measurement and measurements quite differently, in terms of 'objectification', so a different contrast is connoted.

12. The word 'calculable' can be inserted so as not to include arbitrary Borel functions, directly accessible only to inhabitants of Cantor's Paradise. The metatheorem will remain provable; see Ch. 6, sect. 3.3.

13. As is indicated by the title 'Pointer basis of quantum apparatus: Into what mixture does the wave packet collapse?' I shall recount his discussion without the assumption that there is any real collapse of the wave packet (which I shall discuss in the next Chapter).

14. To see that (1) allows U to be unitary, consider the simple example of the tensor product of two Euclidean spaces with orthonormal bases $\{x_0, x_1, x_2\}$ and $\{y_0, y_1, y_2\}$ respectively. Then (1) says that $U(x_0 \otimes y_i)$ simply acts on the first component, rotating x_0 around axis x_k into x_i, where k is a value which is neither 0 nor i (unique if $i \neq 0$; indifferent when $i = 0$). To define U entirely, let it act on the other basis vectors by means of a similar rotation on the second component instead. This means that for $j \neq 0$, $U(x_j \otimes y_i) = (x_j \otimes R_j y_i)$ where R_j is the rotation around axis y_j which sends y_i into y_k, when $i \neq j$, with k being the third remaining value. It is clear that U preserves length and is one-to-one on the basis vectors; therefore, it is indeed an isometry. See Sects. 4.5 and 4.6 below for general results.

15. This difficulty was brought to my attention by Jon Dorling; see van Fraassen (1974*a*).

16. I want to thank David Albert and Barry Loewer for making me

think about this. See Albert and Loewer (1990).

17. See van Fraassen (1989, ch. 8, sects. 4–5) for a fuller empiricist account of acceptance of irreducibly statistical scientific theories.

Chapter 8

1. There is an insightful survey of such answers in Earman (1986, ch. XI), with pessimistic conclusions. He omits however, *inter alia*, what I call the modal alternative. Most of the historically important papers on the subject have been reprinted in Wheeler and Zurek (1983).

2. More precisely, the observable A such that all and only the observables $f(A) : f$ a Borel function are effectively measurable by this process.

3. A good deal more could be said here; the subject is clearly linked to the voluminous philosophical literature on whether any or all attributions of qualities can amount to statements about what (normal) observers would experience under (standard) conditions.

4. The crucial role of the Projection Postulate was pointed out by Henry Margenau (1936), who I believe coined the name.

5. For recent philosophical analyses see Sneed (1966), van Fraassen (1974*a*), Teller (1981, 1983*a*, 1984).

6. The weakest form of acausal transition appears in the superselection account of measurement (see Sect. 8 below), in which the initial state evolves in a determinate manner, but into a mixture only one (unspecified) component of which represents the actual new state. As I shall try to show, this is the most promising way to try and make the Projection Postulate intelligible, but we shall see that it too is subject to serious difficulties.

7. My critique of the Projection Postulate (van Fraassen 1974*a*) essentially followed Margenau; here I am indebted also to De Muynck's discussion of the postulate (1986*b*), and the discussion note by De Muynck and J. van Stekelenborg (1986).

8. See the discussion at the end of this section of the point made by David Albert; see also Redhead (1987, 53–5).

9. Albert made this point at a workshop on the philosophy of quantum mechanics at Princeton University, 1988.

10. As noted in the preceding chapter, caution is required in the discussion of repeatable measurement: the measured quantities cannot be continuous. However, they have arbitrarily fine discretized counterparts which do admit of repeatable measurement.

11. In addition to the references below, see the text by Gottfried (1966); Cartwright (1974*a*); and Belinfante (1976, ch. 1).

12. Beltrametti and Cassinelli (1981*a*, ch. 8); see also the exposition above, Ch. 6, Sect. 8.
13. Furry (1936*a*, 1936*b*); cf. the analysis in Hooker (1972, esp. 114–20).
14. See also the discussion in Hughes (1989, sect. 9.8).
15. See Hepp (1972) and Bub (1989*b*); this point is emphasized in Bub (1989*d*, 143).

Chapter 9

1. The first version of the modal interpretation which I presented (van Fraassen 1972*a*) was rightly criticized by Nancy Cartwright (1974*c*). The version developed in this chapter is an elaboration of what I called the Copenhagen variant of the modal interpretation in van Fraassen (1973*b*), which I developed further in van Fraassen (1974*a*, 1981*c*). For discussion see Hardegree (1976, 1979); McKinnon (1979, 1981); Healey (1979), and my reply (1979*b*); Burghardt (1984*a*, 1984*b*); Redhead (1987, 135–7). Some of these discussions relate to the anti-Copenhagen variant, to which I shall turn only very briefly, in Ch. 10.
2. See Wheeler's (1957) comments on Everett's initial paper.
3. There are several versions, as I shall note below; only one (the 'Copenhagen variant') will be developed here. In comparison with the first two options listed above, it allows for isolated systems in the scope of quantum theory, but classifies the indeterminism as reality and not mere appearance.
4. It is important to emphasize this, for the imagery of the Projection Postulate tends to suggest a much more direct revelation of the initial state in the measurement outcome. Within the general quantum theory of measurement, we can pose the question: how can the initial state of the object system be inferred from measurement outcomes? The answer is then that it can be inferred uniquely only from information about incompatible observables. Of course, measurements to obtain such information cannot be performed on a single system without mutual interference, but only on different samples from a relevant ensemble. There is a fascinating discussion of this topic already in Reichenbach (1944). For a contemporary general account see Lahti (1988, sect. 6); Busch and Lahti (1989).
5. There is more to this than I can spell out at once. We shall look at its implications again in Sect. 4, and at more basic reasons for it in the quantum-logical development of the modal interpretation below.

6. Kochen's use of the Polar Decomposition Theorem could be adapted here; and it leads to a general and comparatively simple rule of this sort. I described this essentially, though without appeal to that theorem, in the last section of van Fraassen (1981*c*), but did not make it part of the interpretation. See further Sect. 7.5 and n. 19 below.

7. Although it is not necessary for our little discussion here, there will be more general constraints on value states for N-body systems with $N > 2$. See Sect. 8 and 9 below.

8. It may be an interesting project to explore what additions can be made consistently to the modal interpretation to guarantee that certain values of observables remain the same during a given time interval. The first condition is clearly that a measurement designed to check such agreement over time gives a positive outcome with certainty. A further condition that could be used is Zurek's (see Ch. 7 above), which concerns the interaction Hamiltonian for apparatus and environment. A third item of use may be the Polar Decomposition Theorem, along the lines of Kochen's interpretation. Whatever additions are made will certainly be such as to apply to some set of compatible observables only. Personally, I feel that such additions would only be aesthetic, for the explication of what the Born Rule says is complete without them.

9. We might put it this way: the many-worlds interpretation 'modalizes' the collapse of the wave packet, while the modal interpretation 'modalizes' the ignorance interpretation of mixtures.

10. See the discussion of the 'canonical' description in the *Proofs and illustrations* of Ch. 7, Sect. 2. In the earliest variant I made up of the modal interpretation (van Fraassen 1972*a*), I followed Everett in this; I eliminated that because of those asymmetries pointed out by Cartwright (1974*c*).

11. This section and the next follow van Fraassen (1981*a*, 1981*c*).

12. For expositions by some main contributors themselves, see Piron (1972, 1976), Jauch (1968), and Varadarajan (1968). For a retrospective that covers a number of alternative approaches, see Beltrametti and Cassinelli (1981*a*). For the severe theoretical limits on the project see Goldblatt (1984); for philosophical discussion see e.g. Hardegree (1981), McKinnon (1981). See also the recent thorough discussion of major developments and their significance in Hughes (1989, ch. 7).

13. See also Hardegree (1981), Hardegree and Fraser (1981), Kalmbach (1984).

14. See Jauch (1968), Varadarajan (1968), Piron (1972, 1976).

15. See Ch. 6, Sect. 4. (See also the discussion there of the amendment needed for the infinite-dimensional case.)

16. Several of the papers in Beltrametti and van Fraassen (1981) deal with that topic, specifically Aerts (1981) and Zecca (1981).

17. I give this postulate here in its strongest form; appropriate weakenings introduce the set of superselection rules by adding a clause of form 'and if P commutes with every operator in set SSR'.

18. See Ch. 6, Sect. 5; reference is to Gleason (1957).

19. The discussion of the possible addition of PQ4 was prompted by Kochen's use of the Polar Decomposition Theorem in his interpretation; see Kochen (1979, 1985a, 1985b). Unlike in the modal interpretation, probabilities are assigned by Kochen also outside measurement situations. Similar attempts, along somewhat similar lines, have been made more recently in interpretations proposed by Healey and Dieks.

20. As James Cushing has pointed out to me, the ignorance interpretation would have to meet still one more criterion. For if the system to which we assign a mixed state is really in one of the component pure states, then that pure state must be developing in time in accordance with quantum mechanics. Nothing in what follows establishes that this further criterion can be met (or even exactly how it should be explicated). The modal interpretation, however, is not subject to this further criterion: only the dynamic state, and not the value state, is directly subject to dynamic principles about evolution in time. On the contrary, the changes over time in the value state (as allowed by the dynamic state, via criteria (a) and (b) above), constitute the ineliminable element of indeterminism in the quantum world, according to this interpretation.

21. I want to thank Wm. De Muynck for substantial help with the proof. Leslie Ballentine independently sent me essentially the same proof shortly after.

22. There will be some judicious confusion of use and mention; the proof consists in the construction of a consistent notation for the sought states.

23. Suppose we started with a particular basis $\{x_i \otimes y_j \otimes z_k\}$ and vector φ is not orthogonal to e.g. $x_2 \otimes y_3 \otimes z_5$. Then choose three new bases for the component Hilbert spaces $\{x_i'\}$, $\{y_j'\}$, and $\{z_k'\}$ such that x_2, y_3, and z_5 are respectively non-orthogonal to all members of those bases. It follows that ϕ is not orthogonal to any member of the basis $\{x_i' \otimes y_j' \otimes z_k'\}$.

24. I apologize to my logician friends who may be driven wild by the way I ride roughshod over the use/mention distinction in this paragraph (with a bow to those perspicacious authors who once began a paper with the footnote: 'with respect to use and mention we follow the conventions of *Principia Mathematica*').

Chapter 10

1. This chapter is an expanded version of my paper by this name (van Fraassen 1985d), presented at Joensuu, June 1985. For subsequent discussion see Kronz (1988), Halpin (1987). Compare also the paper by A. Fine in Cushing and McMullin (1989), and, for views contrary to mine and to Fine's, the papers by e.g. Shimony, Teller, and Wessels in that volume.

2. I am indebted here to various analyses of the argument, notably those of McGrath (1977) and Wessels (1981).

3. Considerable ingenuity has been spent, for example by James McGrath and Linda Wessels, to uncover exactly and precisely what those premisses might be; see McGrath (1977) and Wessels (1981, 1989).

4. I want to thank Professor Erhard Scheibe for bringing this question forcefully to my attention, and for our subsequent correspondence in which he showed in detail that the divergence from the classical case cannot be shown using only two pairs of variables. His results are reported in Scheibe (1991). I also want to thank my student Ned Hall for his help with the proofs in this section.

5. Although I shall not pursue this point here, I conjecture that it has been the lure of this 'perfect' solution that initiated and maintains the so-called quantum-logical interpretation of quantum mechanics; see e.g. Bub (1973, 1974, 1976, 1989a), Glymour and Friedman (1972), Friedman and Putnam (1978), Stairs (1983a, 1983b, 1984).

6. For the quantum-logical interpretation, see Putnam (1969), Bub (1973, 1974), Friedman and Putnam (1978), Stairs (1983a, 1983b, 1984). My two conjectures, about this motive as lying behind the quantum-logical interpretation and also behind Bohr's reply to EPR, support each other in Bub's (1989d).

7. As pointed out in Ch. 9, the Copenhagen variant of the modal interpretation presented there implies (K–S), in the form of the Identity of Observables principle. I have also described (but definitely not advocated) a variant of the modal interpretation ('anti-Copenhagen' variant) which rejects (K–S) and consistently assigns sharp values to all observables (see van Fraassen 1973b, 1979b). There is an illuminating discussion of this alternative in Redhead (1987, 135–7; also 139–52).

8. There is a good deal of literature on this topic, and I can only list a selection. See Jordan (1983); De Muynck and van den Eijnde (1984); De Muynck (1986b); Redhead (1987, ch. 4, sect. 6). I am furthermore indebted to the discussions of K. Kraus in Lahti and Mittelstaedt (1985), Rastal (1985), and Dieks (1982). For a recent

discussion and review of some of the literature see K. Kraus (1989). Concerning apparent conflicts with relativity theory, see the next subsection.

9. See e.g. Mittelstaedt and Stachow (1983), Dieks (1986), Smith and Weingard (1987); for a recent survey and diagnosis, see Eberhard and Ross (1989). This literature differs from most of what I cited earlier in being directed specifically to relativistic formulations of quantum mechanics.

10. See e.g. Eberhard (1978); Ghirardi *et al.* (1980); Jordan (1983); Redhead (1987, ch. 4).

11. Despite all this, the topic remains alive. See Krauss (1985) and Stapp (1985); also Kronz (1988, 1990), and the reply by Halpin (1987).

12. For an illuminating history of the attempted forms of explanation of distant action in connection with Newtonian physics, see McMullin (1989).

Chapter 11

1. For this chapter I want to express grateful debt especially to Simon Kochen, and also to the lectures and articles of Peter Mittelstaedt, Maria Luisa Dalla Chiara, Giuliano Toraldo di Francia, and Willem De Muynck. Correspondence and discussion with Dennis Dieks has been of significant help. Specific debts to each of these are indicated below.

2. At least in the general form here given; Dennis Dieks (1990*c*) argues that there is a natural equivalence relation which is needed to integrate dynamics and statistics.

3. This point has been clearly made by Redhead (1983), and also by Kochen (1985*a*).

4. I shall discuss the similar arguments by Kaplan (1976) and Sarry (1979) below. For critique see Messiah and Greenberg (1964), De Muynck and van Liempd (1986).

5. Cf. Jauch (1968), sect. 15-4 and problems, especially problem 6, which deduces the dichotomy for a system with abelian superselection rules.

6. There is a recent, startlingly ingenious, proof of Dichotomy by Simon Kochen, on the basis of his interpretation of quantum mechanics. In his formalism, subaggregates are assigned states relative to the remainder, and the Permutation Invariance is required for all the relative ('witnessed') states. I do not know whether his assumptions could be rephrased as concerning the family of observables.

7. I want to thank Robert Weingard for the details.

8. Recalling Ch. 2, we note of course that the choice is not *logically* informationless, for the choice of a Hilbert space model is presupposed.

9. This is disputed by Dieks (1990).

10. See also the sequel (Bach 1985), and see further the quantum generalization of De Finetti's Theorem by Hudson (1981), and the recent work by Costantini and Garibaldi (1988, 1989).

11. I have tried to write Postulate II as perspicuously as I could; it must be kept in mind that the cells are disjoint, so the description must be understood to assign a particle to only one cell.

12. Compare this to the Maxwell–Boltzmann function which results from Postulates I–III. For a special ordering it gives $(1/M)^N$; but because the particles are not required, by proposition X, to be ordered in any specific way, $P(X)$ is then $(N!/N(1)! \ldots N(M)!)(1/M)^N$. I want to thank John Paulos and John Collins for carrying out the detailed calculations for these cases, for my seminar at Princeton.

13. Which I did not appreciate when I wrote my 1969 article.

14. It does not follow from Permutation Invariance alone. Recall the discussion of Kaplan's argument, which derived Dichotomy from Permutation Invariance, via an assumption about the reduced states.

15. This implies that, for any two particles, some observable has different values. It does not imply that there is a single observable with a different value in each particle.

16. Each index letter $a, b, \ldots, n$ is assigned a constant numerical term $K(a), K(b), \ldots, K(n)$. When I use specific numerals, they are chosen by way of example.

Chapter 12

1. In my thinking on this topic I have been especially indebted to the articles by Redhead, Teller, De Muynck, and Ginsberg in quantum field theory, by Dieks on identity and individuality of identical particles, by Costantini, Galavotti, and Garibaldi on classical indistinguishability, and finally by David Kaplan and Robert Stalnaker on modal semantics. Work presently in progress by Teller and Redhead will undoubtedly give new impetus to the questions addressed here.

2. For critical discussions see Costantini (1987); Costantini and Garibaldi (1988); Costantini *et al.* (1989); Bach (1987, 1988, 1989); Dieks (1990).

3. In this section I am much indebted to De Muynck (1975). See also Redhead (1983, 1988).

4. Ginsberg (1984) misstates the result as 'particles of half-integer spin must be Fermions, particles of integer spin must be Bosons' (Ginsberg 1984, 335). See the discussion in Ch. 11, Sect. 2, *Proofs and illustrations*.

5. See Robertson (1973) and De Muynck (1975, sect. 4–5).

6. See the discussion in van Fraassen (1985*b*), Nerlich (1976), and Weyl (1952).

7. The idea of an ontological telescope is of course an impossible fiction, for many reasons. Here the story of its use implies that looking through it at the universe does *not* introduce an orientation, does *not* introduce one perspective (the observer's) into the universe. A less metaphysical discussion could replace this ontological 'view from nowhere' with a complete description of the universe by means of entirely general propositions. See further below.

8. I'm riding roughshod over the use/mention distinction, but it would be too pedantic to do otherwise.

9. Specifically, notions of individuating essence or of haecceity, to be explained below.

10. See Aerts (1981, esp. 402: 'They can be distinguished, but the observer does not see the interest in distinguishing them'); also Aerts and Piron (1981).

11. There is an extensive literature on this subject. In chronological order, see Margenau (1944), Reichenbach (1956), van Fraassen (1969), Salmon (1969), Cortes (1976), Barnette (1978), Teller (1983*b*), van Fraassen (1984*b*, 1985*e*).

12. See Aquinas, *Being and Essence*, esp. ii. 32–3; iv. 44–5; v. 52; also *Summa Theologicae*, I, q. 11, a.3c.

13. *Discourse on Metaphysics*, IX; see also his *New Essays*, II-xxvii.

14. For a more technical treatment see van Fraassen (1978*a*). The view about significance requiring permutation closure is one form of anti-haecceitism in modal semantics, and is not uncontroversial. See also Stalnaker (1979).

15. Very helpful discussions of the language of physics in this connection are provided by De Muynck (1975); also De Muynck and van Liempd (1986).

16. Cf. esp. Mittelstaedt (1985), Dalla Chiara (1985), Dalla Chiara and Toraldo di Francia (1983).

17. The cells correspond to monadic predicates; so what about relations? I omit them to keep the discussion perspicuous. However, the formal treatment is not curtailed. The simplest formal treatment

of relations is to treat them as properties of infinite sequences of individuals—see van Fraassen (1973*a*). See also, for this section, the discussion of modal semantics in van Fraassen (1978*a*) and Stalnaker (1979).

18. But it is only an analogue still, since we are here discussing classical modal semantics in which the non-classical complexities of quantum theory may not be reflected. I wish to infuse our enterprise with ideas from modal semantics, not to reduce it.

19. To this I add the assertion elsewhere that all modalities are indeed reducible to verbal modalities, but only on the level of pragmatics; see van Fraassen (1977).

20. See van Fraassen (1978*a*), which also compares this treatment with other approaches to *de dicto* and *de re* modalities.

21. He or she might find comfort in the reflection that if there is no primitive this-ness, and numerical identity reduces via PII to qualitative identity, then there is indeed no real difference in the world if two entities are permuted. But does that mean then that there can be no permutation? Hence no more than one entity?

22. Of course, there is no question of reducing quantum-statistical mechanics as a whole to a classical probabilistic theory—this we saw clearly in the preceding chapter—but only of a way of looking at Bose–Einstein and Fermi–Dirac statistics.

BIBLIOGRAPHY

ACCARDI, L., and FEDULLO, A. (1982), 'On the Statistical Meaning of Complex Numbers in Quantum Mechanics', *Lettere al Nuovo Cimento* 34: 161–72.

ADDISON, J., HENKIN, L., and TARSKI, A. (1965), *The Theory of Models* (Amsterdam: North-Holland).

AERTS, D. (1981), 'Description of Compound Physical Systems and Logical Interaction of Physical Systems', in Beltrametti and van Fraassen (1981), 381–404.

—— (1986), 'A Possible Explanation for the Probabilities of Quantum Mechanics', *Journal of Mathematical Physics* 27: 202–10.

—— and PIRON, C. (1981), 'Physical Justification for using the Anti-symmetrical Tensor Product', preprint TENA, Free University of Brussels.

AGAZZI, E. (1981), *Modern Logic: a Survey* (Dordrecht: Reidel).

AHARONI, J. (1972), *Lectures on Mechanics* (Oxford: Clarendon Press).

ALBERT, D., and LOEWER, B. (1988), 'Interpreting the Many-Worlds Interpretation', *Synthese* 77: 195–213.

—— (1990), 'The Measurement Problem: Some "Solutions"', *Synthese*, forthcoming.

AQUINAS, T. (1949), *On Being and Essence*, trans. A. Maurer (Toronto: Pontifical Institute of Medieval Studies).

—— (1964–76), *Summa Theologicae* (London: Blackfriars, in conjunction with Eyre & Spottiswoode).

ARISTOTLE (1984), *Metaphysics*, in Barnes (1984), ii. 1552–728.

ASPECT, A., DALIBARD, J., and ROGER, G. (1982), 'Experimental Test of Bell's Inequalities using Time-Varying Analyzers", *Physical Review Letters* 49: 1804–7.

ASPECT, A., and GRANGIER, P. (1985), 'Tests of Bell's Inequalities with Pairs of Low Energy Correlated Photons: An Experimental Realization of Einstein–Podolsky–Rosen-type correlations", in Lahti and Mittelstaedt (1985), 50–70.

ASPECT, A., GRANGIER, P., and ROGER, G. (1982), 'Experimental Realization of Einstein–Podolsky Rosen–Bohm *Gedankenexperiment*: a New Violation of Bell's Inequalities', *Physical Review Letters* 49: 91–4.

ASQUITH, P., and GIERE, R. (eds.) (1980), *Proceedings of the 1980 Biennial Meeting of the Philosophy of Science Association*, i (East Lansing, Mich.: Philosophy of Science Association).

—— (eds.) (1981), *Proceedings of the 1980 Biennial Meeting of the Philosophy of Science Association*, ii (East Lansing, Mich.: Philosophy of Science Association).

ASQUITH, P., and KITCHER, P. (eds.) (1984), *Proceedings of the 1984 Biennial Meeting of the Philosophy of Science Association*, i (East Lansing, Mich.: Philosophy of Science Association).

ASQUITH, P., and NICKLES, T. (eds.) (1983), *Proceedings of the 1982 Biennial Meeting of the Philosophy of Science Association*, ii (East Lansing, Mich.: Philosophy of Science Association).

BACH, A. (1984), 'On Wave Properties of Identical Particles', *Physics Letters A* 104: 251–4.

—— (1985), 'On the Mixing Measure of the Harmonic Oscillator in Thermal Equilibrium', *Physics Letters A* 111: 356–8.

—— (1987), 'Indistinguishability or Distinguishability of the Particles of Maxwell–Boltzmann Statistics', *Physics Letters A* 125: 447–50.

—— (1988), 'Indistinguishability, Interchangeability and Indeterminism', presented at the International Conference on Statistics in Science, Luino (forthcoming in the *Proceedings*).

—— (1989), *Indistinguishable Classical Particles*, MS circulated December 1989.

BAIRD, D. (1983), 'The Fisher/Pearson Chi-Squared Controversy: A Turning Point for Inductive Inference', *British Journal for the Philosophy of Science* 34: 105–18.

—— (1984), 'Tests of Significance Violate the Rule of Implication', in Asquith and Kitcher (1984), 81–92.

—— (1987), 'Significance Tests: History and Logic', in Johnson and Kotz (1987), viii. 466–71.

BAKER, L. R. (1987), *Saving Belief* (Princeton: Princeton University Press).

BALLENTINE, L. (1970), 'The Statistical Interpretation of Quantum Mechanics', *Reviews of Modern Physics* 42: 358–81.

—— (1989), 'What Do We Learn about Quantum Mechanics from the Theory of Measurement?', forthcoming in *International Journal of Theoretical Physics*.

BALZER, W., PEARCE, D., and SCHMIDT, H.-J. (eds.) (1984), *Reduction in Science* (Dordrecht: Reidel).

BAR-HILLEL, U. (ed.) (1965), *Logic, Methodology, and Philosophy of Science* (Amsterdam: North-Holland).

BARKER, P., and SHUGART, C. G. (eds.) (1981), *After Einstein* (Memphis: Memphis State University Press).

BARNES, J. (ed.) (1984), *The Complete Works of Aristotle* (Princeton: Princeton University Press).

BARNETTE, R. L. (1978), 'Does Quantum Mechanics Disprove the

504 *Bibliography*

Principle of the Identity of Indiscernibles?' *Philosophy of Science* 45: 466–70.

BELINFANTE, F. J. (1973), *A Survey of Hidden Variable Theories* (Oxford: Pergamon Press).

—— (1976), *Measurement and Time Reversal in Objective Quantum Mechanics* (New York: Pergamon Press).

BELL, J. S. (1964), 'On the Einstein–Podolsky–Rosen Paradox', *Physics* 1: 195–200; reprinted in Wheeler and Zurek (1983), 403–8, and in Bell (1987).

—— (1966), 'On the Problem of Hidden Variables in Quantum Mechanics', *Reviews of Modern Physics* 38: 447–75; reprinted in Bell (1987).

—— (1971), 'Introduction to the Hidden Variable Question', in D'Espagnat (1976), 171–81, and in Bell (1987).

—— (1981), 'Bertlmann's Socks and the Nature of Reality', *Journal de Physique*, 42, suppl.: 241–61; also in Bell (1987).

—— (1987), *Speakable and Unspeakable in Quantum Mechanics* (Cambridge: Cambridge University Press).

BELTRAMETTI, E., and CASSINELLI, G. (eds.) (1981a), *The Logic of Quantum Mechanics* (Reading, Mass.: Addison-Wesley).

—— (1981b), 'On the Non-Unique Decomposability of Quantum Mixtures', in Beltrametti and van Fraassen (1981), 455–64.

BELTRAMETTI, E., CASSINELLI, G., and LAHTI, P. (1989), 'Unitary Measurements of Discrete Quantities in Quantum Mechanics', *Journal of Mathematical Physics* 31: 91–8.

BELTRAMETTI, E., and VAN FRAASSEN, B. C. (eds.) (1981), *Current Issues in Quantum Logic* (New York: Plenum).

BERTRAND, J. (1889), *Calcul des probabilités* (Paris: Gauthier-Villars) (3rd edn. New York: Chelsea, 1972).

—— (1907), *Calcul des probabilités*, 2nd edn. (Paris: Gauthier-Villars).

BIRKHOFF, G., and VON NEUMANN, J. (1936), 'The Logic of Quantum Mechanics', *Annals of Mathematics* 37: 823–43; reprinted in Hooker (1975), 1–26.

BLOKHINTSEV, D. I. (1964a), *Principles of Quantum Mechanics* (Russian edn. 1963) (Boston: Allyn and Bacon).

—— (1964b), *Quantum Mechanics* (Dordrecht: Reidel).

BLUMENTHAL, L. M. (1961), *A Modern View of Geometry* (San Francisco: W. H. Freeman; Dover reprint, 1980).

BOREL, E. (1898), *Leçons sur la théorie des fonctions* (Paris: Gauthier-Villars).

—— (1909), *Éléments de la théorie des probabilités* (Paris: Hermann et Fils).

BOSE, D. (1924), 'Plancks Gesetz und Lichtquantenhypothese', *Zeitschrift für Physik* 26: 178–81.

BRAITHWAITE, R. B. (1957), 'On Unknown Probabilities', in Körner (1962), 3–11.

BROWN, H., and HARRE, R. (eds.) (1988), *Philosophical Foundations of Quantum Field Theory* (Oxford: Oxford University Press).

BUB, J. (1968), 'The Danieri–Loinger–Prosperi Quantum Theory of Measurement', *Il Nuovo Cimento* 57B: 503–19.

—— (1973), 'On the Completeness of Quantum Mechanics', in Hooker (1973), 1–65.

—— (1974), *The Interpretation of Quantum Mechanics* (Dordrecht: Reidel).

—— (1976), 'The Statistics of Non-Boolean Event Structures', in Harper and Hooker (1976), iii. 1–16.

—— (1977), 'Von Neumann's Projection Postulate as a Probability Conditionalization in Quantum Mechanics', *Journal of Philosophical Logic* 6: 381–90.

—— (1979), 'The Measurement Problem of Quantum Mechanics', in Toraldo (1979).

—— (1982), 'Quantum Logic, Conditional Probability, and Interference', *Philosophy of Science* 49: 402–21.

—— (1988), 'How to Solve the Measurement Problem of Quantum Mechanics', *Foundations of Physics* 18: 701–22.

—— (1989*a*), 'On Bohr's Response to EPR: A Quantum Logical Analysis', *Foundations of Physics* 19: 793–805.

—— (1989*b*), 'On the Measurement Problem of Quantum Mechanics', in Kafatos (1989).

—— (1989*c*), 'Review Article: The Philosophy of Quantum Mechanics', *British Journal for the Philosophy of Science* 40: 191–211.

—— (1989*d*) 'From Micro to Macro: A Solution to the Measurement Problem of Quantum Mechanics', in Fine and Leplin (1989), 134–44.

—— and DEMOPOULOS, W. (1974), 'The Interpretation of Quantum Mechanics', *Boston Studies in the Philosophy of Science* 13: 92–122.

BURGHARDT, F. J. (1984*a*), 'Modalitäten und Quantenmechanik bei P. Mittelstaedt, B. C. van Fraassen, und M. C. Dalla Chiara', MS, University of Cologne.

—— (1984*b*), 'Modalities and Quantum Mechanics', *International Journal of Theoretical Physics* 23: 171–96.

BUSCH, P., and LAHTI, P. (1984), 'On Various Joint Measurements of Position and Momentum Observables', *Physical Review* D29: 1634–46.

—— (1985), 'A Note on Quantum Theory, Complementarity and Uncertainty', *Philosophy of Science* 52: 64–77.

—— (1989), 'The determination of the Past and Future of a Physical System in Quantum Mechanics', *Foundation of Physics* 19: 633–78.

BUTTERFIELD, J. (1987), 'Substantialism and Determinism', *International Studies in the Philosophy of Science* 2: 10–32.

BUTTERFIELD, J. (1989), 'The Hole Truth', *British Journal for the Philosophy of Science* 40: 1–28.

CARNAP, R. (1950), *Logical Foundations of Probability* (Chicago: University of Chicago Press).

CARTWRIGHT, N. (1974*a*), 'A Dilemma for the Traditional Interpretation of Quantum Mixtures', in Cohen and Schaffner (1974), 251–8.

—— (1974*b*), 'Superposition and Microscopic Observation', *Synthese* 29: 229–42.

—— (1974*c*), 'Van Fraassen's Modal Interpretation of Quantum Mechanics', *Philosophy of Science* 41: 199–202.

—— (1983), *How the Laws of Physics Lie* (New York: Oxford University Press).

—— (1988), 'How to Tell a Common Cause: Generalizations of the Conjunctive Fork Criterion', in Fetzer (1988), 181–8.

CASSINELLI, G., and LAHTI, P. (1989*a*), 'The Measurement Statistics Interpretation of Quantum Mechanics: Possible Values and Possible Measurement Results of Physical Quantities', *Foundations of Physics* 19: 873–90.

—— (1989*b*), 'The Role of Probability in Quantum Theory', lectures at the International Centre for Theoretical Physics, School of Philosophy of Science, Trieste.

CHALMERS, A. F. (1970), 'Curie's Principle', *British Journal for the Philosophy of Science* 19: 133–48.

CHURCHLAND, P., and HOOKER, C. (eds.) (1985), *Images of Science* (Chicago: University of Chicago Press).

COHEN, R., and SCHAFFNER, K. (eds.) (1974), *Proceedings of the 1972 Biennial Meeting of the Philosophy of Science Association* (Dordrecht: Reidel).

COLODNY, R. (ed.) (1965), *Beyond the Edge of Certainty* (Englewood Cliffs, NJ: Prentice-Hall).

—— (ed.) (1972), *Paradigms and Paradoxes: Philosophical Challenges of the Quantum Domain* (Pittsburgh: University Press).

COOKE, R. M., and HILGEVOORD, J. (1980), 'The Algebra of Physical Magnitudes', *Foundations of Physics* 10: 363–73.

COOKE, R. M., KEANE, M., and MORAN, W. (1985), 'An Elementary Proof of Gleason's Theorem', *Mathematical Proceedings of the Cambridge Philosophical Society* 98: 117–28; reprinted as Appendix in Hughes (1989).

CORTES, A. (1976), 'Leibniz's Principle of the Identity of Indiscernibles: A False Principle', *Philosophy of Science* 43: 491–505.

COSTANTINI, D. (1979), 'The Relevance Quotient', *Erkenntnis* 14: 149–57.

——(1987), 'Symmetry and the Indistinguishability of Classical Particles', *Physics Letters A*, 123: 433–6.

——GALAVOTTI, M.-C., and GARIBALDI, U. (1989), 'Indistinguishabilty and Symmetry'. Forthcoming.

——GALAVOTTI, M.-C., and ROSA, R. (1982), 'A Rational Reconstruction of Elementary Particle Statistics', *Scientia* 117: 491–505.

——GALAVOTTI, M.-C., and ROSA, R. (1983), 'A Set of "Ground Hypotheses" for Elementary Particle Statistics', *Il Nuovo Cimento* 74B: 151–8.

——and GARIBALDI, U. (1988), 'Elementary Particle Statistics Reconsidered', *Physics Letters A* 134: 161–4.

——and GARIBALDI, U. (1989), 'Classical and Quantum Statistics as Finite Random Processes', *Foundations of Physics* 19: 743–55.

——GARIBALDI, U., and GALAVOTTI, M. C. (1989), 'Indistinguishability and Symmetry', Genova, MS.

CURIE, P. (1894), 'Symétrie dans les phénomènes physiques', *Journal de Physique* 3, 393–415; reprinted in *Œuvres de Pierre Curie* (Paris: Gauthier-Villars, 1908).

CUSHING, J. T., DELANEY, C. F., and GUTTING, G. M. (eds.) (1984), *Science and Reality: Recent Work in the Philosophy of Science* (Notre Dame: University of Notre Dame Press).

CUSHING, J. T., and McMULLIN, E. (1989), *Philosophical Consequences of Quantum Theory: Reflections on Bell's Theorem* (Notre Dame: University of Notre Dame Press).

DALLA CHIARA, M. L. (1974), 'Quantum Logic and Physical Modalities', *Journal of Philosophical Logic* 6: 391–404.

——(1981), 'Logical Foundations of Quantum Mechanics', in Agazzi (1981).

——(1983), 'Quantum Logic', in D. Gabbay and F. Guenther (eds.), *Handbook of Philosophical Logic* (Dordrecht: Reidel).

——(1985), 'Names and Descriptions in Quantum Logic', in Mittelstaedt and Stachow (1985), 189–202.

——and TORALDO DI FRANCIA, G. (1979), 'Formal Analysis of Physical Theories', in Toraldo (1979).

——and TORALDO DI FRANCIA, G. (1983), 'Individuals, Kinds and Names in Physics', *Logica e Filosofia della Scienza, Oggi*, ii. 49–72; also in *Versus* 40 (1985), 29–50.

DANIERI, A., LOINGER, A., and PROSPERI, G. M. (1962), 'Quantum Theory of Measurement and Ergodicity Conditions', *Nuclear Physics* 33: 297–319; reprinted in Wheeler and Zurek (1983), 657–79.

DAVIES, E. B. (1970), 'On the Repeated Measurement of Continuous Observables in Quantum Mechanics', *Journal of Functional Analysis* 6: 318–46.

508 *Bibliography*

DAVIES, E. B. (1976), *Quantum Theory of Open Systems* (London: Academic Press).

—— and LEWIS, J. T. (1970), 'An Operational Approach to Quantum Probability', *Communications in Mathematical Physics* 17: 239–60.

DAWID, A. P. (1983), 'Invariant Prior Distributions', in Johnson and Kotz (1987), iv. 228–36.

DE BOER, J., DAL, E., and ULFBECK, O. (eds.) (1986), *The Lessons of Quantum Theory* (Amsterdam: Elsevier).

DE FINETTI, B. (1964), 'Foresight: Its Logical Laws, Its Subjective Sources', in Kyburg and Smokler (1964), 95–158.

DEMOPOULOS, W. (1975), 'An Examination of van Fraassen's Modal Interpretation of Quantum Mechanics', MS, University of Western Ontario.

—— (1976), 'The Possibility Structure of Physical Systems', in Harper and Hooker (1976), 55–80.

DE MUYNCK, W. (1975), 'Distinguishable- and Indistinguishable-Particle Descriptions of Systems of Identical Particles', *International Journal of Theoretical Physics* 14: 327–46.

—— (1986a), 'The Bell Inequalities and their Irrelevance to the Problem of Locality in Quantum Mechanics', *Physics Letters A* 114: 65–7.

—— (1986b), 'On the Relation between the Einstein–Podolsky–Rosen Paradox and the Problem of Non-Locality in Quantum Mechanics', *Foundations of Physics* 16: 973–1002.

—— and VAN DEN EIJNDE, J. (1984), 'A Derivation of Local Commutivity from Macrocausality, using a Quantum Mechanical Theory of Measurement', *Foundations of Physics* 14: 111–46.

—— and VAN LIEMPD, G. (1986), 'On the Relation between Indistinguishability of Identical Particles and (Anti)symmetry of the Wave Function in Quantum Mechanics', *Synthese* 67: 477–96.

—— and VAN STEKELENBERG, J. (1986), 'Discussion of a Proof, Given by Sellari and Tarozzi, of the Non-Locality of Quantum Mechanics', *Physics Letters A* 116: 420–2.

D'ESPAGNAT, B. (ed.) (1976), *Conceptual Foundations of Quantum Mechanics*, 2nd edn. (Reading, Mass.: Benjamin).

—— (1983), *In Search of Reality* (New York: Springer-Verlag).

—— (1984), 'Nonseparability and the Tentative Descriptions of Reality', *Physics Reports* 110: 201–64.

DE WITT, B., and GRAHAM, N. (eds.) (1973), *The Many Worlds Interpretation of Quantum Mechanics* (Princeton: Princeton University Press).

DIACONIS, P. (1977), 'Finite Forms of de Finetti's Theorem on Exchangeability', *Synthese* 36: 271–81.

DICKE, R. H. (1989), 'Quantum Measurements, Sequential and Latent',

Foundations of Physics 19: 385–95.

DIEKS, D. (1982), 'Communication by EPR devices', *Physics Letters A* 92: 271–2.

—— (1983), 'Stochastic Locality and Conservation Laws', *Lettere al Nuovo Cimento* 38: 443–7.

—— (1985), 'On the Covariant Description of Wavefunction Collapse', *Physics Letters A* 108: 379–83.

—— (1986), 'On the Relevance of the Bell Inequalities to the Problem of Locality in Quantum Mechanics', *Physics Letters A* 117: 433–5.

—— (1988*a*), 'The Formalism of Quantum Theory: An Objective Description of Reality?', *Annalen der Physik* 7: 174–90.

—— (1988*b*), 'Quantum Mechanics and Realism', *Conceptus* 22: 31–47.

—— (1989*a*), 'Quantum Mechanics without the Projection Postulate and its Realistic Interpretation', *Foundations of Physics* 19: 1397–1423.

—— (1989*b*), 'Resolution of the Measurement Problem through Decoherence of the Quantum State', *Physics Letters A* 142: 439–46.

—— (1990), 'Quantum Statistics, Identical Particles and Correlations', *Synthese* 82: 127–155.

—— and HOEKZEMA, D. (1984), 'The Einstein–Podolsky–Rosen Paradox, the Bell Inequalities, and Locality', *Delft Progress Report* 9: 91–101.

—— and VELTKAMP, P. (1983), 'Distance between Quantum States, Statistical Inference and the Projection Postulate', *Physics Letters A* 97: 24–8.

DIRAC, P. (1926), 'On the Theory of Quantum Mechanics', *Proceedings of the Royal Society of London A* 112: 661–77.

—— (1958), *The Principles of Quantum Mechanics* (Oxford: Oxford University Press).

DORLING, J. (1981), 'How to Rewrite a Stochastic Dynamical Theory so as to Generate a Measurement Paradox', in Beltrametti and van Fraassen (1981), 115–18.

EARMAN, J. (ed.) (1984), *Testing Scientific Theories*, Minnesota Studies in the Philosophy of Science, x (Minneapolis: University of Minnesota Press).

—— (1986), *A Primer on Determinism* (Dordrecht: Reidel).

—— and FRIEDMAN, M. (1973), 'The Meaning and Status of Newton's Law of Inertia and the Nature of Gravitational Forces', *Philosophy of Science* 40: 329–59.

EBERHARD, P. H. (1977), 'Bell's Theorem Without Hidden Variables', *Il Nuovo Cimento* 38B(1): 75–9.

—— (1978), 'Bell's Theorem and the Different Senses of Locality', *Il Nuovo Cimento* 46B(2): 392–419.

EBERHARD, P. H. and ROSS, R. R. (1989), 'Quantum Field Theory Cannot Provide Faster-than-Light Communication', *Foundations of Physics Letters* 2: 127–49.

ECO, U. (1979), *The Role of the Reader: Explorations in the Semiotics of Texts* (Bloomington, Ind.: Indiana University Press).

EDWARDS, C. M. (1970), 'The Algebraic Approach to Quantum Theory I', *Communications in Mathematical Physics* 16: 207–30.

EINSTEIN, A. (1924/5), 'Quantentheorie des einatomigen idealen Gases', *Preussische Akademie der Wissenschaften (Physik–Mathematik Klasse) Sitzungberichte* (= *Berliner Berichte*), 261–7 (1924) and 3–14 (1925).

——PODOLSKY, B., and ROSEN, N. (EPR) (1935), 'Can Quantum-Mechanical Description of Physical Reality Be Considered Complete?', *Physical Review* 47: 777–80; reprinted in Wheeler and Zurek (1983), 138–41.

ELLIS, B. (1965), 'The Origin and Nature of Newton's Laws of Motion', in Colodny (1965), 29–68.

EVERETT, H. (1957), '"Relative State" formulation of Quantum Mechanics', *Reviews of Modern Physics* 29: 454–62.

——(1973), 'The Theory of the Universal Wave Function', in De Witt and Graham (1973), 3–149.

FANO, U. (1957), 'Description of States in Quantum Mechanics by Density Matrix and Operator Techniques', *Reviews of Modern Physics* 29: 74–93.

FARBER, M. (ed.) (1940), *Philosophical Essays in Memory of Edmund Husserl* (Cambridge, Mass.: Harvard University Press).

FEIGL, H., and BRODBECK, M. (eds.) (1953), *Readings in the Philosophy of Science* (New York: Appleton-Crofts).

FERMI, E. (1926), 'Zur Quantelung des idealen einatomigen Gases', *Zeitschrift für Physik* 36: 902–12.

FETZER, J. (1988), *Probability and Causality: Essays in Honour of Wesley Salmon* (Dordrecht: Kluwer).

FEYERABEND, P. (1958), 'Reichenbach's Interpretation of Quantum Mechanics', *Philosophical Studies* 9: 49–59; reprinted in Hooker (1975), 109–22.

FINE, A. (1970), 'Insolubility of the Quantum Measurement Problem', *Physical Review* D2: 2783–7.

——(1981), 'Einstein's Critique of Quantum Theory: The Roots and Significance of EPR', in Barker and Shugart (1981), 147–59; reprinted in Fine (1986), 26–39.

——(1982a), 'Antimonies of Entanglement: The Puzzling Case of the Tangled Statistics', *Journal of Philosophy* 79: 733–47.

——(1982b), 'Hidden Variables, Joint Probability, and the Bell Inequalities', *Physical Review Letters* 48: 291–5.

—— (1982*c*), 'Joint Distributions, Quantum Correlations, and Commuting Observables', *Journal of Mathematical Physics* 23: 1306–10.

—— (1986), *The Shaky Game: Einstein, Realism, and the Quantum Theory* (Chicago: University of Chicago Press).

—— (1989), 'Do Correlations Need To Be Explained?' in Cushing and McMullin (1989).

—— and LEPLIN, J. (eds.) (1989), *Proceedings of the 1988 Biennial Meeting of the Philosophy of Science Association*, i (East Lansing, Mich.: Philosophy of Science Association).

—— and TELLER, P. (1978), 'Algebraic Constraints on Hidden Variables', *Foundations of Physics* 8: 626–9.

FINKELSTEIN, D. (1962/3), 'The Logic of Quantum Physics', *Transactions of the New York Academy of Sciences* 25: 621–37.

FOULIS, D. J., and RANDALL, C. H. (1974), 'Empirical Logic and Quantum Mechanics', *Synthese* 29: 81–111.

FRASER, D. (1961), 'The Fiducial Method and Invariance', *Biometrica* 48: 261–80.

FRENCH, P., UELING, T., JUN., and WETTSTEIN, H. (eds.) (1979), *Midwest Studies in Philosophy*, iv, *Metaphysics* (Minneapolis: University of Minnesota Press).

FRENCH, R. M. (1988), 'The Banach–Tarski Theorem', *Mathematical Intelligencer* 10: 21–8.

FRENCH, S., and REDHEAD, M. (1988), 'Quantum Physics and the Identity of Indiscernibles', *British Journal for the Philosophy of Science* 39: 233–46.

FRIEDMAN, M., and PUTNAM, H. (1978), 'Quantum Logic, Conditional Probability, and Interference', *Dialectica* 32: 305–15.

FURRY, W. H. (1936*a*), 'Note on the Quantum Mechanical Theory of Measurement', *Physical Review* 49: 393–9.

—— (1936*b*), 'Remarks on Measurements in Quantum Theory', Letter to the Editor in *Physical Review* 49: 476.

—— (1966), 'Some Aspects of the Quantum Theory of Measurement', in *Lectures in Theoretical Physics*, iiiA, *Statistical Physics and Solid State Physics* (Boulder: University of Colorado Press).

GAIFMAN, H. (1988), 'A Theory of Higher Order Probabilities', in Skyrms and Harper (1988), 191–220.

GALLISON, P. (1987), *How Experiments End* (Chicago: University of Chicago Press).

GHIRARDI, G. C., RIMINI, A., and WEBER, T. (1980), 'A General Argument against Superluminal Transmission through the Quantum Mechanical Measurement Process', *Lettere al Nuovo Cimento* 27: 293–8.

GIBBS, J. W. (1902), *Elementary Particles in Statistical Mechanics* (New Haven, Conn.: Yale University Press).

GIERE, R. (1985), 'Constructive Realism', in Churchland and Hooker (1985), 75–98.

—— (1988), *Explaining Science* (Chicago: University of Chicago Press).

GINSBERG, A. (1981), 'Quantum Theory and Identity of Indiscernibles Revisited', *Philosophy of Science* 48: 487–91.

—— (1984), 'On a Paradox in Quantum Mechanics', *Synthese* 61: 325–49.

GLEASON, A. M. (1957), 'Measures on the Closed Subspaces of a Hilbert Space', *Journal of Mathematics and Mechanics* 6: 885–93; reprinted in Hooker (1975), 123–34.

GLYMOUR, C. (1980*a*), 'Explanations, Tests, Unity, and Necessity', *Nous* 14: 31–50.

—— (1980*b*), *Theory and Evidence* (Princeton: Princeton University Press).

—— (1985), 'Explanation and Realism', in Churchland and Hooker (1985), 99–117.

—— and FRIEDMAN, M. (1972), 'If Quanta Had Logic', *Journal of Philosophical Logic* 1: 16–28.

GNEDENKO, B. V. (1962), *The Theory of Probability* (New York: Chelsea).

GODAMBE, V. P., and SPROTT, D. A. (eds.) (1971), *Foundations of Statistical Inference* (Toronto: Holt, Rinehart & Winston).

GOLDBLATT, R. (1984), 'Orthomodularity Is Not Elementary', *Journal of Symbolic Logic* 49: 401–4.

GOTTFRIED, K. (1966), *Quantum Mechanics* (New York: W. A. Benjamin).

GREECHIE, R. (1981), 'A Non-Standard Quantum Logic with a Strong Set of States', in Beltrametti and van Fraassen (1981), 375–80.

GREEN, H. J. (1953), 'A Generalized Method of Field Quantization', *Physics Review* 90: 270–3.

GREENBERG, N. F., and RABAY, S. (1982), 'One-Body and Two-Body Operators on Systems of Identical Particles', *American Journal of Physics* 50: 148–55.

GROENEWOLD, H. J. (1952), 'Information in Quantum Measurement', *Koninklijke Nederlandse Akademie van Wetenschappen* B55: 219–27.

—— (1962), 'Objective and Subjective Aspects of Statistics in Quantum Description', in Körner (1962), 197–203.

GROSSMAN, N. (1974), 'The Ignorance Interpretation Defended', *Philosophy of Science* 41: 333–44.

GUDDER, S. (1968*a*) 'Hidden Variables in Quantum Mechanics Reconsidered', *Reviews of Modern Physics* 40: 229–31.

—— (1968*b*), 'Joint Distributions of Observables', *Journal of Mathematics and Mechanics* 18: 325–35.

HACKING, I. (1983), *Representing and Intervening* (Cambridge: Cambridge University Press).

HADJISAVVAS, N. (1981), 'Properties of Mixtures of Non-Orthogonal States', *Letters of Mathematical Physics* 5: 327–32.

HAJEK, A. (1989), 'Probabilities of Conditionals—Revisited', *Journal of Philosophical Logic* 18: 423–8.

HALMOS, P. R. (1956), *Lectures on Ergodic Theory* (New York: Chelsea).

HALPIN, J. (1987), 'Kronz on Quantum Correlations', APA Central (Chicago), May 1987.

HARDEGREE, G. (1976), 'The Modal Interpretation of Quantum Mechanics', in Suppe and Asquith (1976), 82–103.

—— (1979), 'Reichenbach and the Interpretation of Quantum Mechanics', in Salmon (1979), 513–66.

—— (1981), 'Some Problems and Methods of Quantum Logic', in Beltrametti and van Fraassen (1981).

—— and FRASER, P. (1981), 'Charting the Labyrinth of Quantum Logics: A Progress Report', in Beltrametti and van Fraassen (1981).

HARPER, W., and HOOKER, C. A. (eds.) (1976), *Foundations of Probability Theory, Statistical Inference, and Statistical Theories of Science* (Dordrecht: Reidel).

HARPER, W., and SKYRMS, B. (1988), *Causation in Decision, Belief Change, and Statistics* (Dordrecht: Reidel).

HARPER, W., STALNAKER, R., and PEARCE, G. (eds.) (1981), *Ifs: Conditionals, Belief, Decision, Chance, and Time* (Dordrecht: Reidel).

HEALEY, R. (1979), 'Quantum Realism: Naiveté Is No Excuse', *Synthese* 42: 121–44.

HEIDELBERGER, M. (1987), 'Fechner's Indeterminism: From Freedom to Laws of Chance', in Krueger *et al.* (1987), 116–65.

HEISENBERG, W. (1926), 'Schwankungserscheinungen und Quantenmechanik', *Zeitschrift für Physik* 40: 501–6.

HELLMAN, G. (1982), 'Stochastic Einstein-Locality and the Bell Theorems', *Synthese* 53: 461–504.

—— (1987), 'EPR, Bell, and Collapse: A Route around "Stochastic" Variables', *Philosophy of Science* 54: 558–76.

HELLWIG, K. E. (1981), 'Comment on the Contribution by W. Ochs about the Ignorance Interpretation of States', *Erkenntnis* 16: 357–8; see also Ochs (1981).

HEPP, K. (1972), 'Quantum Theory of Measurement and Macroscopic Observables', *Helvetica Physica Acta* 45: 237–48.

HERBERT, N. (1981), 'Flash: A Superluminal Communication Based upon a New Kind of Quantum Measurement', reprint 81-2003,

National Science Foundation, Boulder Creek, Colo.

HERBERT, N. and KARUSH, J. (1978), 'Generalizations of Bell's Theorem', *Foundations of Physics* 8: 313–17.

HERBUT, F. (1969), 'Derivation of the Change of State in Measurement from the Concept of Minimal Measurement', *Annals of Physics* 55: 271–300.

HESTENES, D. (1970), 'Entropy and Indistinguishability', *American Journal of Physics* 38: 840–5.

HOBSON, A. (1971), *Concepts in Statistical Mechanics* (New York: Gordon and Breach).

HOOKER, C. (1972), 'The Nature of Quantum Mechanical Reality: Einstein vs. Bohr', in Colodny (1972), 67–302.

—— (ed.) (1973), *Contemporary Research in the Foundations and Philosophy of Quantum Theory* (Dordrecht: Reidel).

—— (ed.) (1975), *The Logico-Algebraic Approach to Quantum Mechanics* (Dordrecht: Reidel).

—— (ed.) (1979), *Physical Theory as Logico-Operational Structure* (Dordrecht: Reidel).

HOUTAPPEL, R. M. F., VAN DAM, H., and WIGNER, E. P. (1965), 'The Conceptual Basis and Use of the Geometrical Invariance Principle', *Reviews of Modern Physics* 37: 595–632.

HUDSON, R. L. (1981), 'Analogues of de Finetti's Theorem and Interpretative Problems in Quantum Mechanics', *Foundations of Physics* 11: 805–8.

HUGHES, R. I. G. (1989), *The Structure and Interpretation of Quantum Mechanics* (Cambridge, Mass.: Harvard University Press).

JAMMER, M. (1961), *Concepts of Mass* (New York: Harper & Row).

JARRETT, J. P. (1984), 'On the Physical Significance of the Locality Conditions in the Bell Arguments', *Nous* 18: 569–89.

—— (1989), 'Bell's Theorem: A Guide to the Implications', in Cushing and McMullin (1989), 60–79.

JAUCH, J. M. (1968), *Foundations of Quantum Mechanics* (New York: Addison-Wesley).

JAYNES, E. T. (1957), 'Information Theory and Statistical Mechanics', *Physical Review* 108: 171–90.

—— (1968), 'Prior Probabilities', *IEEE Transactions on Systems Science and Cybernetics* SSC4: 227–41; reprinted in Jaynes (1983), 114–30.

—— (1973), 'The Well-Posed Problem', *Foundations of Physics* 3, 477–92; reprinted in Jaynes (1983), 131–48.

—— (1983), *E. T. Jaynes: Papers on Probability, Statistics and Statistical Physics*, ed. R. Rosencrantz (Dordrecht: Reidel).

JEFFREY, R. (1983), *The Logic of Decision*, 2nd edn. (Chicago: University Press) (1st edn. 1965).

JEFFREYS, H. (1939), *Theory of Probability* (Oxford: Clarendon Press).

JOHNSON, N. L., and KOTZ, S. (eds.) (1987), *Encyclopedia of Statistical Sciences* (New York: John Wiley).

JOHNSON, R. W. (1979), 'Axiomatic Characterization of the Directed Divergences and their Linear Combinations', *IEEE Transactions on Information Theory* IT-25: 709–16.

JORDAN, T. F. (1969), *Linear Operators for Quantum Mechanics* (New York: John Wiley).

—— (1983), 'Quantum Correlations Do Not Transmit Signals', *Physics Letters A* 94: 264.

JOSEPHSON, B. D. (1962), 'Possible New Effects in Superconductive Tunnelling', *Physics Letters A* 1: 251–3.

KAEMPFER, F. A. (1965), *Concepts in Quantum Mechanics* (New York: Academic Press).

KAFATOS, M. (ed.) (1989), *Bell's Theorem, Quantum Theory, and Conceptions of the Universe* (Boston: Kluwer).

KALMBACH, G. (1984), *Orthomodular Lattices* (London: Academic Press).

KAMLAH, A. (1987), 'The Decline of the Laplacean Theory of Probability: A Study of Stumpf, von Kries, and Meinong', in Krueger *et al.* (1987), 91–116.

KAPLAN, I. G. (1976), 'The Exclusion Principle and Indistinguishability of Identical Particles in Quantum Mechanics', *Soviet Physics Uspekhi* 18: 988–94 (translation of *Uspekhi Fizicheskikh Nauk* 117 (1975), 691–704).

KENDALL, M. G., and MORAL, P. A. P. (1963), *Geometrical Probability* (New York: Hafner).

KLEIN, A. G., and OPAT, G. I. (1975), 'Observability of 2π Rotations: A Proposed Experiment', *Physical Review* D11: 523–8.

—— (1976), 'Observation of 2π Rotations by Fresnel Diffraction of Neutrons', *Physical Review Letters* 37, 238–40.

KNORR-CETTINA, K. (1981), *The Manufacture of Knowledge: An Essay on the Constructivist and Contextual Nature of Science* (New York: Pergamon Press).

KOCHEN, S. (1979), 'The Interpretation of Quantum Mechanics', Princeton University MS.

—— (1985*a*), 'Identical Particles', Princeton University MS.

—— (1985*b*), 'A New Interpretation of Quantum Mechanics', in Lahti and Mittelstaedt (1985), 151–70.

—— and SPECKER, E. (1965*a*), 'Logical Structures Arising in Quantum Theory', in Addison *et al.* (1965); reprinted in Hooker (1975).

—— and SPECKER, E. (1965*b*), 'The Calculus of Partial Propositional Functions', in Bar-Hillel (1965); reprinted in Hooker (1975).

Kochen, S. and Specker, E. (1967), 'The Problem of Hidden Variables in Quantum Mechanics', *Journal of Mathematics and Mechanics* 17: 59–87; reprinted in Hooker (1975), 293–328.

Körner, S. (ed.) (1962), *Observation and Interpretation in the Philosophy of Physics* (New York: Dover).

Kraus, K. (1985*a*), 'Quantum Theory, Causality and EPR Experiments', in Lahti and Mittelstaedt (1985), 461–80.

—— (1989), 'Quantum Theory Does Not Require Action at a Distance', *Foundations of Physics Letters* 2: 1–6.

Kronz, F. M. (1988), 'EPR: The Correlations Are Still a Mystery', *Philosophy of Science* 55: 631–9.

—— (1989), 'Hidden Locality, Conspiracy and Superluminal Signals', forthcoming in *Philosophy of Science*.

Krueger, L., Daston, L. J., and Heidelberger, M. (1987), *The Probabilistic Revolution* (Cambridge, Mass.: MIT Press).

Kyburg, H., Jun., and Smokler, H. (eds.) (1964), *Studies in Subjective Probability* (New York: John Wiley).

Lahti, P. (1988), 'Quantum Mechanical Limitations on the Measurability of Physical Quantities', *Acta Polytechnica Scandinavica* El 63: 33–64.

—— and Mittelstaedt, P. (eds.) (1985), *Symposium on the Foundations of Modern Physics* (Singapore: World Scientific Publishing Co.).

—— and Ylinen, K. (1987), 'On Total Non-Commutativity in Quantum Mechanics', Report Series Turku-FTL-R119.

Landau, L. D., and Lifshitz, E. M. (1959), *Quantum Mechanics* (London: Pergamon Press).

Latour, B. (1986), *Laboratory Life: The Construction of Scientific Facts* (Princeton: Princeton University Press).

Lebesgue, H. (1902), . . . *Intégrale, longueur, aire . . .* (Milan: Rebeschini).

Lefschetz, S. (1977), *Differential Equations: Geometric Theory* (New York: Dover).

Leggett, A. J. (1980), 'Macroscopic Quantum Systems and the Quantum Theory of Measurement', *Progress of Theoretical Physics Supplement* 69: 80–100.

—— (1986), 'Quantum Mechanics at the Macroscopic Level', in de Boer *et al.* (1986), 35–56.

—— and Garg, A. (1985), 'Quantum Mechanics versus Macroscopic Realism: Is the Flux There When Nobody Looks?' *Physical Review Letters* 54: 857–60.

Leibniz, G. (1973), *Discourse on Metaphysics*, 2nd edn., trans. G. R. Montgomery (Peru, Ill.: Open Court).

—— (1981), *New Essays Concerning Human Understanding*, trans. R.

Remnant and J. Bennett (Cambridge: University Press); reprinted with corrections 1982.

LEWIS, D. (1979), 'Attitudes *de dicto* and *de se*', *Philosophical Review* 88; 513–43; reprinted with postscript in Lewis (1983).

—— (1983), *Collected Papers*, i (New York: Oxford University Press).

—— (1986), *Collected Papers*, ii (New York: Oxford University Press).

LLOYD, E. (1988), *The Structure and Confirmation of Evolutionary Theory* (New York: Greenwood).

LOÈVE, M. (1955), *Probability Theory: Foundations, Random Sequences* (New York: van Nostrand).

LUCKENBACH, S. (ed.) (1972), *Probabilities, Problems, and Paradoxes: Readings in Inductive Logic* (Encino, Cal.: Dickenson).

LUCRETIUS (1985), *On the Nature of the Universe*, trans. R. E. Latham (New York: Viking Penguin).

LUEDERS, G. (1951), 'Über die Zustandsenderung durch den Messprozess', *Annalen der Physik* 8: 322–8.

LUDWIG, G. (1967), 'Attempt of an Axiomatic Foundation of Quantum Mechanics and More General Theories II', *Communications in Mathematical Physics* 4: 331–48.

—— (1983/5), *Foundation of Quantum Mechanics*, 2 vols. (New York and Berlin: Springer-Verlag).

—— (1985/6), *An Axiomatic Basis for Quantum Mechanics*, 2 vols. (New York and Berlin: Springer-Verlag).

LYUBOSHITZ, V. L., and PODGORETSKII, M. I. (1969), 'Interference of Non-Identical Particles', *Soviet Physics JETP* 28: 469–75.

—— (1971), 'The Question of the Identity of Elementary Particles', *Soviet Physics JETP* 33: 5–10.

MACH, E. (1974), *The Science of Mechanics* (LaSalle, Ill.: Open Court).

MARCUS, R. B., DORN, G., and WEINGARTNER, P. (eds.) (1986), *International Congress of Logic, Methodology, and Philosophy of Science*, vii (Amsterdam: North-Holland).

MARGENAU, H. (1936), 'Quantum-Mechanical Description', *Physical Review* 49: 240–2.

—— (1944), 'The Exclusion Principle and its Philosophical Importance', *Philosophy of Science* 11: 187–208.

—— (1950), *The Nature of Physical Reality* (New York: McGraw-Hill).

—— (1963), 'Measurement and Quantum States', *Philosophy of Science* 30: 1–16 and 138–57.

MARLOW, A. R. (ed.) (1978*a*), *Mathematical Foundations of Quantum Theory* (New York: Academic Press).

—— (1978*b*), 'Quantum Theory and Hilbert Space', *Journal of Mathematical Physics* 19: 1841–6.

MARLOW, A. R. (1978c), 'Orthomodular Structures and Physical Theory', in Marlow (1978a).

MARTIN, G. E. (1982), *Transformation Geometry: An Introduction to Symmetry* (New York: Springer-Verlag).

MARTINEZ, S. (1989), 'Minimal Disturbance in Quantum Logic', in Fine and Leplin (1989), 83–8.

MATTHEWS, W. N., JUN., and ESRICK, M. A. (1980), 'Locality of the Field Operator of Many-Body Theory', *American Journal of Physics* 48: 782–3.

McGRATH, J. (1977), 'A Formal Statement of the Einstein–Podolsky –Rosen Argument', *International Journal of Theoretical Physics* 17: 557–71.

—— (1980), 'A Formal Statement of Schrödinger's Cat Paradox', in Asquith and Giere (1980), 251–63.

McKINNON, E. (1979), 'Scientific Realism: The New Debates', *Philosophy of Science* 46: 501–32.

—— (1981), 'The Interpretation of Quantum Mechanics: A Critical Review', *Philosophia* 10: 89–124.

McKINSEY, J. C. C., and SUPPES, P. (1955), 'On the Notion of Invariance in Classical Mechanics', *British Journal for the Philosophy of Science* 5: 290–302.

McMULLIN, E. (1989), 'The Explanation of Distant Action: Historical Notes', in Cushing and McMullin (1989), 272–302.

MERZBACHER, E. (1970), *Quantum Mechanics*, 2nd edn. (New York: John Wiley).

MESSIAH, A. (1958), *Quantum Mechanics* (Amsterdam: North-Holland).

—— and GREENBURG, O. W. (1964), 'Symmetrization Postulate and its Experimental Foundation', *Physical Review* B136: 248–67.

MIELNIK, B. (1968), 'Geometry of Quantum States', *Communications in Mathematical Physics* 9: 55–80.

MILLIKAN, R. (1917), *The Electron*, ed. J. W. M. Durmond (Chicago: University of Chicago Press).

MILNE, P. (1983), 'A Note on Scale Invariance', *British Journal for the Philosophy of Science* 34: 49–55.

MIRMAN, R. (1973), 'Experimental Meaning of the Concept of Identical Particles', *Il Nuovo Cimento* 18B: 110–22.

MITTELSTAEDT, P. (1976), *Philosophical Problems of Modern Physics* (Dordrecht: Reidel).

—— (1978), *Quantum Logic* (Dordrecht: Reidel).

—— (1981), 'Classification of Different Areas of Work Afferent to Quantum Logic', in Beltrametti and van Fraassen (1981), 3–16.

—— (1985), 'Constitution Naming and Identity in Quantum Logic', in Mittelstaedt and Stachow (1985), 215–34.

—— and STACHOW, E. W. (1983), 'Analysis of the Einstein–Podolsky–Rosen Experiment by Relativistic Quantum Logic', *International Journal of Theoretical Physics* 22: 517–40.

—— and STACHOW, E. W. (eds.) (1985), *Recent Developments in Quantum Logic* (Mannheim: Bibliographisches Institut).

MOORE, G. (1982), *Zermelo's Axiom of Choice* (New York: Springer-Verlag).

MORGENBESSER, S. (ed.) (1967), *Philosophy of Science Today* (New York: Basic Books).

MOSTELLER, F. (1965), *Fifty Challenging Problems in Probability* (Reading, Mass.: Addison-Wesley).

MOULINES, C. U. (1984), 'Ontological Reduction in Natural Science', in Balzer *et al.* (1984), 51–70.

MUNITZ, M. (ed.) (1973), *Logic and Ontology* (New York: New York University Press).

NERLICH, G. (1976), *The Shape of Space* (Cambridge: Cambridge University Press).

NORTHRUP, E. P. (1944), *Riddles in Mathematics* (New York: von Nostrand).

OCHS, W. (1981), 'Some Comments on the Concept of State in Quantum Mechanics', *Erkenntnis* 16: 339–56; see also Hellwig (1981).

OZAWA, M. (1984), 'Quantum Measuring Processes of Continuous Observables', *Journal of Mathematical Physics* 25: 79–87.

PARRY, W. (1981), *Topics in Ergodic Theory* (Cambridge: Cambridge University Press).

PAULI, W. (1925), 'Über den Zusammenhang des Abschlusses der Elektronengruppen im Atom mit der Komplexstruktur der Spektren', *Zeitschrift für Physik* 31: 765–83.

—— (1940), 'The Connection between Spin and Statistics', *Physical Review* 58: 716–22.

PEACOCKE, C. (1981), 'Not Real but Observable', *The Times Literary Supplement* 30: 121.

PENROSE, R., and ISHAM, C. J. (eds.) (1986), *Quantum Concepts in Space and Time* (Oxford: Oxford University Press).

PERES, A. (1978), 'Pure States, Mixtures, and Compounds', in Marlow (1978*a*), 357–64.

PEREY, F. G. (1982), 'Application of Group Theory to Data Reduction', Oak Ridge National Laboratory Report ORNL5908 (September 1982).

PIRON, C. (1972), 'Survey of General Quantum Physics', *Foundations of Physics* 2: 287–314; reprinted in Hooker (1975), 513–44.

—— (1976), *Foundations of Quantum Physics* (New York: W. A. Benjamin).

PITOWSKY, I. (1989), *Quantum Probability–Quantum Logic* (New York: Springer-Verlag).

POINCARÉ, H. (1905), *Science and Hypothesis* (New York: Dover, 1952 reprint).

—— (1912), *Calcul des probabilités* (Paris: Gauthier).

PULMANNOVA, S. (1980), 'Relative Compatibility and Joint Distributions of Observables', *Foundations of Physics* 10: 641–55.

PUTNAM, H. (1969), 'Is Logic Empirical?', *Boston Studies in the Philosophy of Science* 5: 199–215.

—— (1978), *Meaning and the Moral Sciences* (London: Routledge & Kegan Paul).

—— (1988), *Representation and Reality* (Cambridge, Mass.: MIT Press).

QUINE, W. V. (1963), *From a Logical Point of View* (New York: Harper & Row).

RASTAL, P. (1983), 'The Bell Inequalities', *Foundations of Physics* 16: 555–70.

—— (1985), 'Locality, Bell's Theorem, and Quantum Mechanics', *Foundations of Physics* 15: 963–73; see also Stapp (1985).

REDHEAD, M. (1975), 'Symmetry in Intertheory Relations', *Synthese* 32: 77–112.

—— (1983), 'Quantum Field Theory for Philosophers', in Asquith and Nickles (1983), 57–99.

—— (1987), *Incompleteness, Non-Locality, and Realism* (Oxford: Oxford University Press).

—— (1988), 'A Philosopher Looks at Quantum Field Theory', in Brown and Harré (1988), 9–24.

—— and TELLER, P. (1989), 'Particles, Particle Labels, and Quanta: The Toll of Unacknowledged Metaphysics', MS.

REICHENBACH, H. (1944), *The Philosophical Foundations of Quantum Mechanics* (Berkeley, Cal.: University of California Press).

—— (1948), 'The Principle of Anomaly in Quantum Mechanics', *Dialectica* 2: 337–50.

—— (1956), *The Direction of Time* (Berkeley, Cal.: University of California Press).

—— (1957), *The Philosophy of Space and Time* (German edn., 1928) (New York: Dover).

—— (1959), *Modern Philosophy of Science* (New York: Humanities Press).

RESCHER, N. (ed.) (1968), *Studies in Logical Theory* (Oxford: Basil Blackwell).

ROBERTSON, B. (1973), 'Introduction to Field Operators in Quantum Mechanics', *American Journal of Physics* 41: 678–90.

ROSENKRANZ, R. D. (1977), *Inference, Method and Decision* (Dordrecht: Reidel).

—— (1981), *Foundations and Applications of Inductive Probability* (Atascadero, Cal.: Ridgeview).

ROTH, L. M., and INOMATA, N. (eds.) (1986), *Fundamental Questions in Quantum Mechanics* (New York: Gordon & Breach).

RUSSELL, B. (1953), 'On the Notion of Cause with Applications to the Free Will Problem', in Feigl and Brodbeck (1953), 387–407.

SALMON, W. (1969), Comments on van Fraassen (1969), presented at the Annual Meeting of the American Philosophical Association (East), December 1969; summarized (by van Fraassen) in Luckenbach (1972), 135–8.

—— (ed.) (1979), *Hans Reichenbach: Logical Empiricist* (Dordrecht: Reidel).

—— (1984), *Scientific Explanation and the Causal Structure of the World* (Princeton: Princeton University Press).

SARRY, M. F. (1979), 'Permutation Symmetry of Wave Functions of a System of Identical Particles', *Soviet Physics JETP* 50: 678–80.

SCHEIBE, E. (1981), 'Quantentheorie und verborgene Parameter', *Der Physikerunterricht* 15: 56–74.

—— (1986), 'What Kind of Hidden Variables Are Excluded by Bell's Inequality?', in Weingartner and Dorn (1986), 251–71.

—— (1991), 'EPR Situation and Bell's Inequality', in W. Spohn *et al.* (eds.) *Existence and Explanation* (Dordrecht: Kluwer).

SCHIFF, I. L. (1955), *Quantum Mechanics* (New York: McGraw-Hill).

SCHROEDINGER, E. (1935*a*), 'Discussion of Probability Relations between Separated Systems', *Proceedings of the Cambridge Philosophical Society* 31: 555–63.

—— (1935*b*), 'Die gegenwärtige Situation in der Quantenmechanik I–III', *Die Naturwissenschaften* 23: 807–12, 823–8, 844–9; translated as 'The Present Situation in Quantum Mechanics' by J. D. Trimmer, in Wheeler and Zurek (1983), 152–67.

—— (1936), 'Probability Relations between Separated Systems', *Proceedings of the Cambridge Philosophical Society* 32: 446–52.

SCHROECK, F. (1989), 'Coexistence of Observables', *International Journal of Theoretical Physics* 28: 247–62.

SHADMI, Y. (1978), 'Teaching the Exclusion Principle with Philosophical Flavor', *American Journal of Physics* 46: 844–8.

SHIMONY, A. (1983), 'Controllable and Uncontrollable Locality', *Proceedings of the International Symposium on the Foundations of Quantum Mechanics* (Tokyo: ISFQM), 225–30.

—— (1984), 'Contextual Hidden Variable Theories and Bell's Inequalities', *British Journal for the Philosophy of Science* 35: 25–45.

SHIMONY, A. (1986), 'Events and Processes in the Quantum World', in Penrose and Isham (1986), 182–203.

—— (1988), 'The Reality of the Quantum World', *Scientific American* 258: 46–53.

—— (1989), 'Search for a Worldview which can Accommodate our Knowledge of Microphysics', in Cushing and McMullin (1989), 25–37.

SIGMAN, J. (1989), 'Science and its Relation to Art: An Exploration of Problems in the Philosophy of Science', Senior thesis, Princeton University.

SKYRMS, B. (1980), *Causal Necessity* (New Haven, Conn.: Yale University Press).

—— (1982), 'Counterfactual Definiteness and Local Causation', *Philosophy of Science* 49: 43–50.

—— and HARPER, W. (1988), *Causation, Chance and Credence* (Dordrecht: Kluwer).

SMITH, G. J., and WEINGARD, R. (1987), 'A Relativistic Formulation of the Einstein–Podolsky–Rosen Paradox', *Foundations of Physics* 17: 149–71.

SNEED, J. (1966), 'Von Neumann's Argument for the Projection Postulate', *Philosophy of Science* 33: 22–39.

—— (1971), *The Logical Structure of Mathematical Physics* (Dordrecht: Reidel).

STAIRS, A. (1983*a*), 'On the Logic of Pairs of Quantum Systems', *Synthese* 56: 437–60.

—— (1983b) 'Quantum logic, realism, and value definiteness', *Philosophy of Science* 50: 578–602.

—— (1984), 'Sailing into Charybdis: van Fraassen on Bell's Theorem', *Synthese* 61: 351–59.

STALNAKER, R. (1968), 'A Theory of Conditionals', in Rescher (1968); reprinted in Harper *et al*. (1980), 41–56.

—— (1979), 'Anti–Essentialism', in French *et al*. (1979), 343–55.

STAPP, H. P. (1971), 'S–matrix Interpretation of Quantum Theory', *Physical Review* D3: 1303–20.

—— (1985), 'Comments on "Locality, Bell's Theorem, and quantum mechanics"', *Foundations of Physics* 15: 973–76; see also Rastal (1985).

STEGMÜLLER, W. (1976), *The Structure and Dynamics of Theories* (New York: Springer–Verlag).

SUDARSHAN, E., and MEHRA, J. (1970), 'Classical Statistical Mechanics of Identical Particles and Quantum Effects', *International Journal of Theoretical Physics* 3: 245–51.

SUPPE, F. (1972), 'Theories, the Formulations and the Operational Imperative', *Synthese* 25: 129–59.

SUPPE, F. (1974), *The Structure of Scientific Theories* (Urbana, Ill.: University of Illinois Press).

—— (1989), *The Semantic Conception of Theories and Scientific Realism* (Urbana, Ill.: University of Illinois Press).

—— and ASQUITH, P. (eds.) (1976), *Proceedings of the 1976 Biennial Meeting of the Philosophy of Science Association* (East Lansing, Mich.: Philosophy of Science Association).

SUPPES, P. (1967), 'What Is a Scientific Theory?', in Morgenbesser (1967), 55–67.

—— (1974), 'The Structure of Theories and the Analysis of Data', in Suppe (1974), 266–83.

—— (ed.) (1976), *Logic and Probability in Quantum Mechanics* (Dordrecht, Reidel).

—— (ed.) (1980), *Studies in the Foundations of Quantum Mechanics* (East Lansing, Mich.: Philosophy of Science Association).

—— and ZANOTTI, M. (1976), 'On the Determinism of Hidden Variable Theories with Strict Correlation and Conditional Statistical Independence of Observables', *Synthese* 29: 311–30; reprinted in Suppes (1976), 445–55.

—— and ZANOTTI, M. (1980), 'A New Proof of the Impossibility of Hidden Variables Using the Principles of Exchangeability and Identity of Conditional Distributions', in Suppes (1980).

SVENONIUS, L. (1960), *Some Problems in Logical Model Theory*, Library of *Theoria*, iv (Lund: CWK Gleerup).

TELLER, P. (1979), 'Quantum Mechanics and the Nature of Continuous Physical Quantities'. *Journal of Philosophy* 76: 345–61.

—— (1981), 'The Projection Postulate and Bohr's Interpretation of Quantum Mechanics', in Asquith and Giere (1981), 201–23.

—— (1983a), 'The Projection Postulate as a Fortuitous Approximation', *Philosophy of Science* 50: 413–31.

—— (1983b), 'Quantum Physics, the Identity of Indiscernibles, and Some Unanswered Questions', *Philosophy of Science* 50: 309–19.

—— (1984), 'The Projection Postulate of Quantum Mechanics: A New Perspective', *Philosophy of Science* 51: 369–95.

—— (1989), 'Relativity, Relational Holism, and the Bell Inequalities', in Cushing and McMullin (1989).

TERSOFF, J., and BAYER, D. (1983), 'Quantum Statistics for Distinguishable Particles', *Physical Review Letters* 50: 553–4.

THOMPSON, P. (1989), *The Structure of Biological Theories* (Albany, NY: State University of New York Press).

TODHUNTER, I. (1865), *A History of the Mathematical Theory of Probability* (London: Macmillan).

TOOLEY, M. (1987), *Causation: A Realist Approach* (Oxford: Oxford University Press).

TORALDO DI FRANCIA, G. (ed.) (1979), *Problems in the Foundations of Physics* (Amsterdam: North-Holland).

USPENSKY, J. V. (1937), *Introduction to Mathematical Probability* (New York: McGraw-Hill).

VAN AKEN, J. (1985), 'Analysis of Quantum Probability Theory I', *Journal of Philosophical Logic* 14: 267–96.

—— (1986), 'Analysis of Quantum Probability Theory II', *Journal of Philosophical Logic* 15: 333–67.

VAN FRAASSEN, B. C. (1969), 'Probabilities and the Problem of Individuation', presented at the Annual Meeting of the American Philosophical Association (East), December 1969; in Luckenbach (1972), 121–38; see also Salmon (1969).

—— (1970), 'On the Extension of Beth's Semantics of Theories', *Philosophy of Science* 37, 325–34.

—— (1972*a*), 'A Formal Approach to the Philosophy of Science', in Colodny (1972), 303–66.

—— (1972*b*), 'The Labyrinth of Quantum Logics', *Boston Studies in the Philosophy of Science* 13: 224–54; reprinted in Hooker (1975).

—— (1973*a*), 'Extension, Intension, and Comprehension', in Munitz, (1973), 101–32.

—— (1973*b*), 'A Semantic Analysis of Quantum Logic', in Hooker (1973), 80–113.

—— (1974*a*), 'The Einstein–Podolsky–Rosen Paradox', *Synthese* 29: 291–309.

—— (1974*b*), 'Hidden Variables in Conditional Logic', *Theoria* 40: 176–90.

—— (1977), 'The Only Necessity is Verbal Necessity', *Journal of Philosophy* 74: 71–85.

—— (1978), 'Essence and Existence', *American Philosophical Quarterly Monograph*, series no. 12.

—— (1978), 'Time, Physical and Experiential', *Epistemologia* 1: 323–38.

—— (1979*a*), 'Foundations of Probability: Modal Frequency Interpretation', in Toraldo (1979), 344–87.

—— (1979*b*), 'Hidden Variables and the Modal Interpretation of Quantum Statistics', *Synthese* 42: 155–65.

—— (1980*a*), 'A Re-examination of Aristotle's Philosophy of Science', *Dialogue* 19: 20–45.

—— (1980*b*), *The Scientific Image* (Oxford: Oxford University Press).

—— (1980*c*), 'A Temporal Framework for Conditionals and Chance', *Philosophical Review* 89: 91–108; reprinted in Harper *et al.* (1980).

—— (1981*a*), 'Assumptions and Interpretations of Quantum Logic', in Beltrametti and van Fraassen (1981), 17–34.

—— (1981*b*), 'The End of the Stalnaker Conditional?' Unpublished typescript, Princeton University.

—— (1981*c*), 'A Modal Interpretation of Quantum Mechanics', in Beltrametti and van Fraassen (1981), 229–58.

—— (1981*d*), 'Theory Construction and Experiment: an Empiricist View', in Asquith and Giere (1981), 663–78.

—— (1982), 'The Charybdis of Realism: Epistemological Implications of Bell's Inequality', *Synthese* 5: 25–38; reprinted with postscript in Cushing and McMullin (1989).

—— (1984*a*), 'Glymour on Evidence and Explanation', in Earman (1984), 165–76.

—— (1984*b*), 'The Problem of Indistinguishable Particles', in Cushing *et al.* (1984), 153–72.

—— (1984*c*), 'Theory Comparison and Relevant Evidence', in Earman (1984), 27–42.

—— (1985*a*), 'Empiricism in the Philosophy of Science', in Churchland and Hooker (1985), 245–308.

—— (1985*b*), *An Introduction to the Philosophy of Space and Time*, 2nd edn. (New York: Columbia University Press).

—— (1985*c*), 'Salmon on Explanation', *Journal of Philosophy* 82: 639–51.

—— (1985*d*), 'EPR: When Is a Correlation Not a Mystery?' in Lahti and Mittelstaedt (1985).

—— (1985*e*), 'Statistical Behaviour of Indistinguishable Particles: Problems of Interpretation', in Mittelstaedt and Stachow (1985), 161–87.

—— (1986), 'Aim and Structure of Scientific Theories', in Marcus *et al.* (1986), 307–18.

—— (1989), *Laws and Symmetry* (Oxford: Oxford University Press).

VARADARAJAN, V. S. (1968), *Geometry of Quantum Theory*, i (New York: van Nostrand).

VILLEGAS, C. (1971), 'On Haar Priors', in Godambe and Sprott (1971), 409–16.

—— (1977), 'Inner Statistical Inference', *Journal of the American Statistical Association* 72: 453–8.

—— (1981), 'Inner Statistical Inference ii', *Annals of Statistics* 9: 768–76.

VOLKOV, D. V. (1959), 'On the Quantization of Half–Integer Spin Fields', *Soviet Physics JETP* 9: 1107–11.

—— (1960), 'S-matrix in the Generalized Quantization Method', *Soviet Physics JETP* 11: 375–8.

VON MISES, R. (1957), *Probability, Statistics and Truth* (New York: Macmillan).

—— (1964), *Mathematical Theory of Probability and Statistics*, ed. H. Geiringer (New York: Academic Press).

VON NEUMANN, J. (1955), *Mathematical Foundations of Quantum Mechanics* (German edn. 1932) (Princeton: Princeton University Press).

VON PLATO, J. (1988), 'Ergodic Theory and the Foundations of Probability', in Skyrms and Harper (1988), 257–78.

WAN, K. (1980), 'Superselection Rules, Quantum Measurement, and Schrödinger's Cat', *Canadian Journal of Physics* 58: 976–82.

WEAVER, W. (1963), *Lady Luck: The Theory of Probability* (Garden City, NY: Doubleday-Anchor).

WEINGARTNER, P., and DORN, G. (1986), *Foundations of Physics: Papers contributed to the VIIth International Congress for Logic, Methodology, and Philosophy of Science* (Vienna).

WESSELS, L. (1981), 'The "EPR" Argument: A Post-Mortem', *Philosophy of Science* 40: 3–30.

—— (1989), 'The Way the World Isn't: What the Bell Theorems Force Us to Give Up', in Cushing and McMullin (1989).

WEYL, H. (1929), 'The Spherical Symmetry of Atoms', Rice Institute Pamphlet XVI, no. 4; reprinted in Weyl (1968), 268–81.

—— (1931), *Theory of Groups and Quantum Mechanics* (New York: Dover).

—— (1940), 'The Ghost of Modality', in Farber (1940), 278–303.

—— (1946), *The Classical Groups* (Princeton: Princeton University Press).

—— (1952), *Symmetry* (Princeton: Princeton University Press).

—— (1968), *Gesammelte Abhandlungen*, iii (New York: Springer-Verlag).

WHEELER, J. A. (1957), 'Assessment of Everett's "Relative State" Formulation of Quantum Theory', *Reviews of Modern Physics* 29: 463–5.

—— and ZUREK, W. H. (eds.) (1983), *Quantum Theory and Measurement* (Princeton: Princeton University Press).

WHEWELL, W. (1840), *The Philosophy of the Inductive Sciences Founded on Their History* (London: Parker).

WICK, G. C., WIGHTMAN, A. S., and WIGNER, E. P. (1952), 'The Intrinsic Parity of Elementary Particles', *Physical Review* 88: 101–5.

WIGNER, E. P. (1926), 'Über nicht kombinierende Terme in der neueren Quantentheorie', *Zeitschrift für Physik* 40: 492–500 (1926); 883–92 (1927).

—— (1931), *Gruppentheorie und ihre Anwendung auf die Quantenmechanik der Atomspectren* (Braunschweig: Fr. Vieweg).

—— (1970), 'On Hidden Variables and Quantum Mechanical Probabilities', *American Journal of Physics* 39: 1005–9.

WILSON, M. (1989), 'Critical Notice: John Earman's *A Primer on Determinism*', *Philosophy of Science* 56: 502–32.

YALE, P. (1968), *Geometry and Symmetry* (San Francisco: Holden-Day) (Dover reprint, 1988).

YLINEN, K. (1985), 'On a Theorem of Gudder on Joint Distributions of Observables', in Lahti and Mittelstaedt (1985), 691–4.

ZABELL, S. L. (1988), 'Symmetry and its Discontents', in Skyrms and Harper (1988), 155–90.

ZECCA, A. (1981), 'Products of logics', in Beltrametti and van Fraassen (1981), 405–12.

ZUREK, W. H. (1981), 'Pointer Basis of Quantum Apparatus: Into What Mixture Does the Wave Packet Collapse?', *Physical Review* D24: 1516–25.

INDEX